Paula Arcari

Making Sense of 'Food' Animals

A Critical Exploration of the Persistence of 'Meat'

palgrave
macmillan

Paula Arcari
Department of Social Sciences
Centre for Human Animal Studies (CfHAS)
Edge Hill University
Ormskirk, UK

ISBN 978-981-13-9587-1   ISBN 978-981-13-9585-7   (eBook)
https://doi.org/10.1007/978-981-13-9585-7

This Palgrave Macmillan imprint is published by the registered company Springer Nature Singapore Pte Ltd.
The registered company address is: 152 Beach Road, #21-01/04 Gateway East, Singapore 189721, Singapore

Making Sense of 'Food' Animals

"This is a fascinating and important work that stretches the field of critical animal studies in new ways, takes existing scholarship forward in its synthesis of ideas within Foucauldian approaches to studying humans and animals. Original interviews with consumers and producers are handled with exemplary ethical standards by Dr Arcari, making this a timely book for students and scholars to support their own explorations of the ways in which humans currently treat animals within food production systems."

—Alex Lockwood, *The University of Sunderland, UK*

"This is an excellent book which is beautifully written and genuinely a pleasure to read. Paula Arcari presents us with a highly original and significant contribution to a number of fields: sociological understandings of food and eating practices, thinking about 'food animals' in sociological and cultural animal studies, and ultimately, the persistence of "meaty practices" despite public concern and welfarist moves. It is an important contribution to Foucauldian scholarship and stretches Foucauldian insights in innovative ways for critical posthumanist theory."

—Erika Calvo, *University of East London, UK*

"*Making Sense of 'Food' Animals* is the first systematic investigation of its kind … a rigorous, substantial, original and significant contribution to knowledge, with implications for how we understand discourses surrounding animal-based food consumption, and the role of narratives of self in relation to social change. This wonderfully clear and accessible book is a valuable tool for understanding the kinds of knowledge constructions which support continued animal product consumption, despite the mounting ethical, health and environmental evidence which suggests that this consumption is pernicious and not sustainable."

—Dinesh Wadiwel, *University of Sydney, Australia*

Preface

I want to begin by openly owning the assumptions and values I bring to this research and declaring it to be an attempt at a "moral science, a critical enquiry into the everyday conditions of domination for the purposes of altering them" (Seidman 1998: 329). With this declaration, I am acknowledging that my theoretical and methodological choices are themselves "rhetoric"—"encoding certain assumptions and values about the social world" (Agger 1991: 114). Accounting for myself in this way, making explicit my a priori knowledge, beliefs, and biases, is an attempt at reflexive research as described by Cutcliffe (2003). This reflexivity is not a post-analysis addition to the book, but was part of the entire research process, as will become apparent. Hence, while included here as a point of reference for the reader, a constant awareness of my own personal and contextual orientation has informed every phase of this research, allowing me to "ponder the ways in which who [I am] may both assist and hinder the process of co-constructing meanings" (Berger 2014: 221).

Meat does not occupy a prominent position in my personal foodscape. I am vegan for all the reasons associated with veganism as an intersectional and therefore necessarily political movement and not simply a diet. These reasons include the materialisation of animal bodies, the aggregation and de-personalisation of living beings, the 'othering' and oppression of animals based on species, their use as a human resource, the intersection of issues of race and gender in our (mis-)use of animals, and many others (Adams 2010 [1990]; Cudworth 2011; Fiddes 1991; Harper 2011; Nibert 2013). My position is best summed up by the main character in Emma Geen's 2016 novel, *The Many Selves of Katherine North*, who observes, "other subjectivities aren't a consumer item. Their habitats aren't playgrounds" (190).

I do not consider meat to be an essential or even necessary part of the human diet. Whole cultures have existed for generations without it. However, many, and indeed the majority of people, do consider it both essential and necessary. Putting aside arguments relating to the (un)availability of alternatives, which pertain to broader mechanisms of distribution and access rather than necessity, my personal view is that a plant-based diet is a viable proposition for the global population. Given the breadth of scholarly work covering animal rights, critical animal studies, veganism, and notions of the next (nonhuman) social contract (Gabardi 2017), literary explorations that challenge the edibility of nonhuman animals (e.g. Vint 2010; Atwood 1998, 2004), and filmic endeavours that draw attention to our current treatment of animals (e.g. Noah 2014; Okja 2017; Carnage 2017), I am certainly not isolated in this view.

A critical approach to animal studies is gaining traction across multiple social science and humanities disciplines including history, geography, anthropology, literature, the creative arts, cultural studies, sociology, development studies, political science, law, criminology, environmental sciences, and many more. These approaches, and the thinking they reflect, antecede the non-, post- and more-than-human turns across the social sciences, which broadly (cl)aim to decentre humans and also, in some cases, to fundamentally reshape human relations with living nonhuman others. They also go further in their de-centring by extending it across all domains of social life, challenging the most spurious conceptions of human-animal 'entanglements', including, but not limited to, those where animals are used for food, entertainment, education, sport, service, research, and fashion. Such human-animal 'relations' are characterised, above all, by commodity and exchange values that are fundamentally human-centric.

Reflecting the unparalleled numbers of 'livestock' bred and killed every year, and the scope and scale of the impacts of their production and consumption, the use of animals for food is a consistent focus across this critical body of work. Efforts to understand human's attachment to meat cut across a range of disciplines, as listed above, and fall into three broad categories. First are those that seek to identify the attitudes, motivations, and belief systems (e.g. Melanie Joy's 'carnism') that make individuals 'choose' to eat meat, and from there encourage people to change, or at least question, their consumption practices (Joy 2011; King 2017). Others take a philosophical approach that aims to clarify whether or not eating animals can be morally and ethically justified (Bramble and Fischer 2015; Visak and Garner 2015). These often include recommendations for how individual consumers can make 'better' choices (Singer and Mason 2007). Finally, there are those that take a more systemic view of 'meat culture' as part of broader social, cultural, economic, and political complexes and explore how it is variously constituted and mobilised (Adams 1991; Fiddes 1991; Nibert 2013; Wadiwel 2015; Potts 2016). Addressing the alternate side of the equation, a growing body of scholarship is exploring vegan practices and 'transitions', encompassing their recent mainstreaming and what that means for different cultures (e.g. Wright 2015; Castricano and Simonson 2016; Quinn and Westwood 2018; and work by Richard Twine, among others). This work is critical for understanding the various ways in which veganism is being articulated and practised, and for working through the range of questions and contradictions it can present.

This book follows the same broad intent as the work listed above. It contributes to the growing mountain of efforts seeking to find more effective ways to disrupt human's persistent adherence to these ultimately violent practices. However, unlike these volumes, this book tries to zoom in on the most resilient defences of the Western 'meat culture' by examining what has favoured the emergence of so-called ethical and sustainable meat, as opposed to no meat, as a response to environmental and ethical challenges. Apart from the focus on emerging *practices* (rather than individual choices) involving 'better' meat, what this volume also does differently from most other analyses is specifically examine how 'food' animals, as well as 'meat', continue to be re-constituted as edible materials, even when those constitutions are radically tested. This understanding is vital to associated efforts to effect a more extensive vegan transition, and maintains a focus on dismantling normalised systems of animal use as fundamental to ensuring that any steps towards the hoped-for vegan transition are not easily reversed. It therefore aligns with, and hopefully complements, recent work that similarly problematises human-animal relations and the ways animals' usability as well as edibility is socially constituted (e.g. Potts 2016; Lopez and Gillespie 2015; Wadiwel 2015; Chrulew and Wadiwel 2016; Cudworth and Hobden 2018). In this way, my central concern lies with the animal and realising their right to a non-humanised life.

Ormskirk, UK   Paula Arcari

References

Adams, C. J. (1991). Ecofeminism and the Eating of Animals. *Hypatia, 6*(1), 125–145.

Adams, C. J. (2010). *The Sexual Politics of Meat: A Feminist-vegetarian Critical Theory*. London; New York: Continuum International Publishing Group.

Agger, B. (1991). Critical Theory, Poststructuralism, Postmodernism. *Annual Review of Sociology, 17*, 105–131.

Atwood, M. (1998). *The Edible Woman*. New York: Anchor Books.

Atwood, M. (2004). *Oryx and Crake*. New York: Anchor Books.

Berger, R. (2014). Now I See It, Now I Don't: Researcher's Position and Reflexivity in Qualitative Research. *Qualitative Research, 15*(2), 219–234.

Bramble, B., & Fischer, B. (2015). *The Moral Complexities of Eating Meat*. Oxford: Oxford University Press.

Castricano, J., & Simonsen, R. R. (2016). *Critical Perspectives on Veganism*. Cham, Switzerland: Palgrave Macmillan.

Chrulew, M., & Wadiwel, D. (2016). Editors' Introduction: Foucault and Animals. In *Foucault and Animals* (pp. 1–18). Brill: Leden.

Cudworth, E. (2011). *Social Lives with Other Animals*. Basingstoke; New York: Palgrave Macmillan.

Cudworth, E., & Hobden, S. (2018). *The Emancipatory Project of Posthumanism*. Routledge.

Cutcliffe, J. R. (2003). Reconsidering Reflexivity: Introducing the Case for Intellectual Entrepreneurship. *Qualitative Health Research, 13*(1), 136–148.

Fiddes, N. (1991). *Meat: A Natural Symbol*. London; New York: Routledge.

Gabardi, W. (2017). *The Next Social Contract: Animals, the Anthropocene, and Biopolitics*. Philadelphia: Temple University Press.

Harper, A. B. (2011). Vegans of Color, Racialized Embodiment, and Problematics of the 'Exotic'. In A. H. Alkon & J. Agyeman (Eds.), *Cultivating Food Justice: Race, Class, and Sustainability* (pp. 221–238). Cambridge, MA; London: The MIT Press.

Lopez, P. J., & Gillespie, K. (2015). *Economies of Death: Economic Logics of Killable Life and Grievable Death*. London: Routledge.

Nibert, D. A. (2013). *Animal Oppression and Human Violence: Domesecration, Capitalism, and Global Conflict*. New York; Chichester: Columbia University Press.

Seidman, S. (1998). *Contested Knowledge: Social Theory in the Postmodern Era*. New York: John Wiley and Sons.

Visak, T., & Garner, R. (2015). *The Ethics of Killing Animals*. New York; Oxford: Oxford University Press.

Wadiwel, D. J. (2015). *The War Against Animals*. Leiden; Boston: Brill Rodopi.

Wright, L. (2015). *The Vegan Studies Project: Food, Animals, and Gender in the Age of Terror*. Athens, Georgia: University of Georgia Press.

Acknowledgements

It is only through the encouragement of Dinesh Wadiwel and Erika Cudworth that this book came about. Their positive assessments of the doctoral thesis on which the book is based, and joint assurance of its value in published format, provided the inspiration I needed to embark on the transformation process. In that endeavour, I thank the editorial and production team at Palgrave Macmillan, especially Joshua Pitt, for their wonderful support and guidance from proposal through to production. I also thank Alex Lockwood, the (at the time) anonymous reviewer who provided many insightful and constructive comments that helped refine the final manuscript, and Jane Daly and Cecily Maller for casting a final critical eye over associated additions to the text.

In terms of the doctoral research, my primary supervisor Cecily Maller, plus Yolande Strengers and Tania Lewis, loomed large as guides, mentors, and overall support throughout my candidature. I learned so much as they steered and honed my academic efforts and they continue to provide the best exemplars of the kind of researcher I would like to become, academically, professionally, and personally. I also thank everyone who was part of the Social Practice Theory Reading Group between 2012 and 2018. Under the leadership of Yolande Strengers and Cecily Maller, this group poked, prodded, and wrestled my early thinking out from comfy corners to embrace a much more complex and curious world. My research benefitted immeasurably from this constant unsettling of my existing orders.

Above all, the 41 people who voluntarily shared their time, selves, homes, and food with me deserve the utmost thanks as I could not have cleared the first bend without them. I am deeply grateful for their generosity and their passion, which breathed life and energy through the entire research process, with plenty left over to fuel the revisions for this publication. Finally, I am grateful to both the Australian Government, under the Australian Post-Graduate Award (APA) scheme, and RMIT University, for providing the funding and institutional support that enabled me to undertake this research.

On a more personal front, I am fortunate to have such dear friends and intellectual touchstones as Shae Hunter and Jane Daly who frequently volunteered their time and brainpower in ways that sharpened the writing, thinking, and speaking associated with my research. They also added the personal support and well-timed comic relief that make everything so much more do-able. Special mentions go to Bhavna Middha, Larissa Nicholls, and Tamzin Rollason for their constant support, insights, and provocations, and for creating a stimulating and collegiate academic environment.

I thank my parents who didn't get to see this point, but helped me arrive, and taught me to question and challenge, especially when power was involved. Also my lovely siblings, who I feel fortunate to have in my life, and who excel, in strikingly different ways, at being the best older brother and sister I could wish for. Last but never least, to my canine friends present and past—the pleasingly nonhuman presences who keep me motivated, distracted, and provide company. Without them, and our aimless meanderings through nature-filled spaces, I don't know that I would have thought half the thoughts that ended up in this book.

This book is dedicated to my eternal friend and guide Viva Sali (Jillian Gibb), who created a space for me that can never be replaced but I will cherish always. Her artistry, knowledge, wisdom, and fearless compassion for nonhuman animals are my constant inspiration.

I acknowledge that the work associated with this book was carried out on Country that belongs to the people of the Kulin Nation. I pay respects to all Elders past, present, and future, and recognise that the lands and water they care for were never ceded.

Contents

List of Figures

List of Tables

1

Introduction

Vector image of cow with cuts of meat delineated (Image by FoxysGraphic on Vectorstock)

© The Author(s) 2020   **1**
P. Arcari, *Making Sense of 'Food' Animals*,
https://doi.org/10.1007/978-981-13-9585-7_1

I use this introduction to explain the question that underpins and guides this book, which asks why cows, pigs, chickens, sheep, and goats[1] are not regarded as anything other than 'food' animals? Why do cartographies of meat, such as illustrated in the above figure, dominate humans' understandings of these animals' lives and bodies, thereby normalising and supporting the intensification of environmentally harmful practices that demand the termination of millions of lives every hour? My point of observation is an entirely different and hypothetical territory, one that is perhaps quixotic—and that is a vegan pantopia. This pantopia is conceived in contrast to utopia, described by Foucault (1967: 3) as "fundamentally unreal spaces". Literally denoting "a place that does not exist … the 'forever nowhere'" (Bauman 2002: 238), utopian imaginings conjure idealistic but also seemingly unattainable dreams (Bauman 2005). Pantopia is instead "the place of everywhere" (Jacobson 2013: 233), evocative of creative possibilities rather than impossibilities.

Though still a largely imagined, hoped-for future, the seeds of a vegan pantopia are present in very real but diffuse sites of resistance, or "effectively enacted utopias" (Foucault 1967: 3), which Foucault terms 'heterotopia'. Heterotopia are sites that 'disturb' or 'detonate' the normalised social order, logic, and language (Foucault 1989; Dehaene and De Cauter 2008). Vegan heterotopia are therefore construed as sites where normalised constitutions of meat consumption and 'food' animals are at once "represented, contested and inverted" (Foucault 1967: 3; see also Meininger 2013). Crucially for my purposes, both Foucault and Meininger highlight the inherent power of heterotopia, simply by occupying the anomalous zones that normality negates, and from which the "self-propelling power of the othering and excluding normality" can be critiqued, disturbed, and challenged (Meininger 2013: 28).

Being myself an occupant of vegan heterotopia, I can exercise the power this offers to observe how 'normal' space is constituted and maintained, and subsequently challenge the associated mechanisms of inclusion and exclusion. As Hook (2007: 184) states, heterotopia "represent a point of destabilization for current socio-political or discursive orders of power". In doing this, I intend to "trouble habitual ways of thinking and

---

[1] As the animals that are overwhelmingly used for meat, above all others.

acting" (Castree and Nash 2004: 1342) in alignment with an openly emancipatory agenda with regard to humans' current use of 'food' animals. As part of this, I imagine what it would take for vegan heterotopia to become a unified vegan pantopia. Therefore, building on my first question, more specifically, the central question that guides this book is, what is it about meat consumption and the use of animals as 'food' that keeps these animals persistently edible? This is what I endeavoured to understand through my research. For answering this question has revealed mechanisms of inclusion that are simultaneously mechanisms of exclusion that prevent the *expan*sion of heterotopic sites of veganism. Chief amongst these is the ontological mapping of nonhuman bodies.

1 Animal Mapping, Practices, and Power

In *Science and Sanity*, Alfred Korzybski made the statement "a map is not the territory" (1958 [1933]: 58). He also said, "words are not the things we speak about" (222). Yet, for all the ways that animals used as food are spoken about, and their bodies literally mapped, their territory or undesignated 'thingness' remains enduringly unknown. In order to apprehend a life for them not predicated on the value of their bodies to humans—an unmapped territory of other possibilities—the current map would be discarded and the words used to speak about them would change.

In his investigation of Australian explorers' texts, Simon Ryan notes that space is socially produced and describes maps as an imperial technology used to establish colonial space (1996: 5). Mapping, then, is centrally about control and colonisation—a distinctly non-neutral process which Ryan represents through his conception of the 'cartographic eye'. Through the course of this book, this notion of the cartographic eye and its entitled, territorialising mission will resonate with references to the 'arrogant eye' in cinematic studies, and to theorisations of the gaze—patriarchal, colonial, and, ultimately, human. Indeed, since Ryan's articulation, the 'cartographic gaze'[2] has been defined by John Pickles as "a controlling

---

[2] The cartographic gaze is a recognised topic of scholarly attention (Ellis and Waterton 2005; Jacob 2006; Wilson 2011).

gaze rendering the broad swathes of worldly complexity and enormity in miniature form for a discrete purpose" (2004: 80). Pickles' reference here to visual rendering and reduction to miniature form becomes doubly meaningful when applied to animals used as food. Indeed, the lead character in Italo Calvino's 1983 novel *Mr Palomar* makes the same connection:

> On the wall a chart shows an outline of a bull, like a map covered with frontier lines that mark off the areas of consuming interest, involving the entire anatomy of the animal excepting only horns and hooves. The map of the human habitat is this, no less than the planisphere of the planet; both are protocols that should sanction the rights man has attributed to himself, of possession, division, and consumption without residue of the terrestrial continents and of the loins of the animal body. (69)

Drawing on Ryan's terminology to consider cartographies of meat, the process of designating and charting the various locations, features, and landmarks in the life of a cow, sheep, pig, or chicken, or the journey of a steak, cutlet, loin, or breast, is the outcome of a long history of interconnected systems of human value and use that are not easily disentangled or dismantled. Eating meat is universally embroiled in social, cultural, and religious practices and recruits sensory and biological processes to create even more complex entanglements. Equally, its production is part of a tightly meshed, globalised network of economic, political, and scientific arrangements. Together they comprise the animal-industrial complex[3] (Noske 1989; Twine 2012). If this extensive complex is to be dismantled, there first needs to be some understanding of what, besides a certain path dependency, makes eating meat such an enduring, and persistently alluring, part of social practices.

Practice theories comprise a vast and heterogeneous body of literature (Warde 2005) arguably characterised by four different 'types' (Schatzki 2001) and two generations of theorists (Postill 2010). In this book, I rely

---

[3] Noske (1997) describes rather than defines the animal-industrial complex. Based on Noske's account, Twine (2012: 23) offers the following definition: "a partly opaque and multiple set of networks and relationships between the corporate (agricultural) sector, governments, and public and private science. With economic, cultural, social and affective dimensions it encompasses an extensive range of practices, technologies, images, identities and markets". Regarding the human-animal relationship, Noske describes it as "embedded in a web of exploitative practices, in which one type of exploitation is carried over onto another" (1997: 38).

on the theoretical lineage traced by Warde (2005)—from Giddens and Bourdieu, through Schatzki, Reckwitz, and Shove—up to and including recent theorists who are applying social practice theories in diverse and distinctive ways to gain insight into the workings of the social world and how its less desirable trajectories might be altered (Hui et al. 2016; Strengers and Maller 2014; Shove and Spurling 2013). Social practice theories are united by a shared understanding of practices as the unit of focus rather than individual behaviour. Across the body of work I am drawing on, practices are commonly understood as a diverse nexus of bodily activities or routinised "doings and sayings" (Schatzki 1996: 89) such as driving, keeping warm, cooking, or eating breakfast (Maller 2015). Individuals are conceived as being recruited into, reproducing, or defecting from practices (Shove et al. 2012). As this account suggests, social practices are conceived as both entities and performances. This enables the constituent elements necessary to the existence and performance of practices to be determined, while also providing the opportunity to address questions of change—for example, exploring how or why certain practices persist, evolve, or die out, and how to approach more purposive intervention.

Practices as entities generally comprise material objects and infrastructures (living and non-living), common understandings and meanings, and competencies and skills (Shove et al. 2012; Reckwitz 2002). Rules and 'teleoaffective' structures, the latter understood as "orientations toward ends and how things matter" (Schatzki 1997: 302) or "emotion and motivational knowledge" (Reckwitz 2002: 249), are also commonly included. Practices emerge and are recognised as entities through their repeated performance in generally cohesive ways, or, to invert Warde's phrasing, 'a practice presupposes a performance' (2005: 134). Many practices may share one or more elements, or be temporally and/or spatially shaped to greater and lesser degrees by other practices. For example, roads are an element of driving, cycling, walking, and other forms of transport, with many shared meanings, competencies, and teleoaffectivities. Roads are also part of planning practices, shaped by population changes, development, and increased car ownership. 'Meat' can be part of practices of shopping, cooking (e.g. roasting, stewing, BBQ-ing), eating, religious and historical celebration (e.g. Christmas, Easter, Eid al-Adah,[4]

---

[4] The annual Islamic 'Festival of sacrifice'.

Gadhimai,[5] Thanksgiving,[6] Australia Day),[7] commensality, and overseas travel, again with many overlapping meanings, competencies, and teleo-affectivities. The availability of 'meat' relies on practices associated with animal agriculture, trading, and markets. Such groups of practices are commonly referred to as a bundle or complex of practices with varying dimensions of co-dependency (Shove et al. 2012). Any one practice, *and its constituent elements*, is therefore constantly being shaped by bodily per-formances across a wide array of practices.

The understandings, meanings, competencies, rules, and teleoaffective structures that variably characterise a practice are understood as belong-ing to, or associated with, that practice rather than being qualities of the individual that performs it (Reckwitz 2002: 250). This includes dis-course, which is understood as an implicit part of social practices (Jorgensen and Phillips 2002; Halkier et al. 2011). Indeed, Reckwitz explains that practice and discourse cannot be conceived separately, more especially "if we want to trace the ramifications of affects as formed in specific discourses at specific times" (2016: 121–122). However, as I emphasised in the previous paragraph, what I have described—the mean-ings, discourses, rules, affects, and so on—are not fixed, pre-existing ele-ments that are then enlisted in practices, but are themselves constituted through the ongoing performance not only of said practice, but of all other practices that are in any way related.[8] It can therefore be seen how the normalisation and assumed stability of certain material objects as practice elements can be critiqued and challenged, especially when those same objects are constituted and understood differently across alternate, or counter-practices, for instance in heterotopia.

---

[5] A Hindu sacrificial ceremony held every five years.

[6] An annual holiday marking the landing of the Mayflower in Cape Cod in 1620. The beginning of colonisation is celebrated by many, and is observed by others, including indigenous peoples, as 'a national day of mourning'.

[7] Australia Day marks the declaration of British sovereignty on 26th January 1788. As with Thanksgiving in the US, it is also known amongst indigenous peoples and their allies as 'Invasion Day'. It was declared a national public holiday in 1994 and was quickly co-opted by Meat and Livestock Australia (MLA) in a strategic advertising campaign associating eating 'lamb' with patrio-tism. The meat-centric Aussie BBQ has become synonymous with Australia Day, while not eating meat on this day is construed (by MLA) as Un-Australian.

[8] Practices are not closed systems, and so all practices can be said to be related to greater and lesser degrees.

Although I use practices as my unit of focus or foundation for understanding how the social world is (re)produced, the research presented in this book does not undertake an analysis of practices of the same order as previously cited works. Rather it is concerned with this question of how it is that certain objects, in this case living 'food' animals and their non-living bodies, are persistently (re)constituted (or made sense of) as material elements across a vast nexus of interconnected bundles of practices. The persistence and maintenance of these constitutions contribute to the stability of *all* associated practices, and vice versa. In other words, I consider these constitutions central to understanding how to destabilise practices involving 'meat' and 'food' animals, and set them on an alternate (vegan-pantopian) trajectory. To my knowledge, the a priori constitution of material objects as practice elements has not been a focus of practice analyses to date. By highlighting opportunities to problematise the constitution of practice elements, especially normalised 'objects', my analysis therefore introduces a further way in which social practice theories can approach questions of change and intervention.

Opening up the constitution of material 'objects' as part of social practices for scrutiny in this way potentially leads to an engagement with power, especially when those 'objects' are living organisms—human or nonhuman. Becoming part of practices involves the materialisation of some aspect of an organism's being, whether their physicality, biology, body, disaggregated parts, and/or their 'other-ness' in general (Arcari 2018). From the perspective of the 'practitioners' around whom the practice is conceived, these (once) living objects, or parts thereof, are a resource that is more or less essential to the practice. They are effectively subaltern, understood as "a condition of subordination brought about by colonization or other forms of economic, social, racial, linguistic, and/or cultural dominance" (Beverley 1999: Frontmatter)—to which I would add species as another basis for subordination. Directing critical attention at constitutions of living or once living 'objects' of practices foregrounds these systemic forms of dominance, or mechanisms of power. In the absence of this attention, these constitutions—or how these objects make sense—continue to be unquestioningly reproduced and further normalised, and the power relations that shape them are effectively being maintained through their concealment in everyday 'doings and sayings' (Fig. 1.1).

Fig. 1.1 A meat map observed hanging on a restaurant wall (Author's own image)

Supporting this conception, Foucault asserts that in order to be successful, mechanisms of power cannot be recognised as such, and that part of their concealment depends on foregrounding power's positive aspects: "[Power] needs to be considered as a productive network which runs through the whole social body, much more than as a negative instance whose function is repression" (1980: 119). In the same way, meat consumption and the use of animals as food need to be understood as part of a productive network of social practices that persistently maintains the edibility of animals. It is no good attempting to discredit the production and consumption of meat on economic, environmental, health, social, or even ethical grounds if this productive basis of its endurance has not been recognised—this is like targeting the many heads of the Greek hydra while leaving the one immortal head intact.[9] Moreover, the use of ani-

---

[9] A multi-headed, serpentine water monster from Greek mythology. For every head that was severed, the monster was said to grow two or three more.

mals for food cannot be conceived as a deliberate act of domination, whereby individuals recognise and wield their power over these animals' lives solely for their own gain, whether economic, sensory, or otherwise, and without any regard for the animals' wellbeing. In other words, power does not result from "the choice or decision of an individual subject" (Foucault 1978: 95). It is rather a dispersed, indeterminate phenomenon of social relations that is "employed and exercised through a net-like organisation" and "never localised here or there, never in anybody's hands, never appropriated as a commodity or piece of wealth" (Foucault 1980: 98). In Foucault's conception then, power comes from everywhere and individuals are conceived as "vehicles of power, not its points of application" (1980: 93, 98).

However, Foucault does not discount that there are 'points of application', as exemplified, for example, by meat maps. He describes two primary forms of power, the first being a "mode of action which does not act directly and immediately on others. Instead, it acts upon *their actions*; an action upon an action, on existing action or on those which may arise in the present or the future" (1982: 789, emphasis added). This describes power in its more covert aspect, where it is variously persuasive, coercive, disciplinary, and limiting. The other is "a relationship of violence [which] acts upon *a body* or upon things; it forces, it bends, it breaks on the wheel, it destroys, or it closes the door on all possibilities" (ibid., emphasis added). This describes a relation of overt power where the possibility of an individual's resistance, which Foucault identifies as one of the necessary conditions for the exercise of power, has been removed. Foucault calls this a state of domination (1994: 283). Animals used for food, or 'food' animals, can be seen as being acted upon by both modes of power as they traverse the various landmarks on the map that chronicles their lives and deaths. Ultimately though, a state of domination always presides in their relations with humans.

In relation to systemic power, Foucault (1982: 787) goes on to describe certain 'blocks' in which relationships of power become so regulated and formulaic that they form a concerted system. From a practices perspective, these relationships could almost be considered another element, or a common effect of several elements, that is shared, or faithfully reproduced, across a bundle of practices.[10] He identifies religious orders, prison systems, apprenticeship, medical care, and military service as illustrative of such blocks, emphasising that they are characterised by an invigilated process of rational and economic adjustment between "productive activities, resources of communication, and the play of power relations" (1982: 788). I suggest that the network of relations, institutions, and practices that support animals' state of domination, consolidated in the animal-industrial complex, constitute another block where the domination of 'food' animals has been similarly codified, ordered, professionalised, regulated, structurally realised, and systemically normalised so that associated relations of power are similarly "rooted in the system of social networks" (Foucault 1982: 791). However, like Foucault (1980: 92) and also Noske (1997: 39), it is not power as such that I want to focus on, as if it were a reified force, but rather its mechanisms, or the *how* of power: how it gains expression, how it comes to be exercised, how it is sustained, how it is legitimated, and thus how it endures.

In adopting this Foucauldian approach to the embodied, productive mechanisms of power, this book adds to existing theorisations of how power *and* affect can be understood as part of practices by focusing on how they shape the constitution of individual elements, particularly living, human and nonhuman, material 'objects'.

This book is therefore primarily concerned with understanding what nourishes the roots of our domination of 'food' animals[11]—a domination that is fully embodied and not simply an external phenomenon—making it thrive and survive, especially through current challenges to its legitimacy. More specifically, returning to Korzybski (1958 [1933]), what makes this the only useable map we have with which to view and navigate the wider territory from which 'food' animals are derived? Is it possible to eliminate the map altogether, rather than simply rearrange its contents (which I will demonstrate is the outcome of efforts to produce meat that

---

[10] Although it is of course how these relationships are constituted and maintained that is being systemically reproduced across practices.

[11] All animals, including microbes, in their ontological relation to humans, exist in a state of domination, in that their bodies are always justifiably available and their lives legitimately extinguishable. In this book, however, I am focusing only on animals commonly used for food, or 'food' animals for short.

is considered in some way 'better')? Is it possible that 'food' animals might be permitted to exist in this wider territory, shrugging off the shackles of their various designations to become more than the human words used to describe them? These are the more philosophical questions I hold in mind throughout the ensuing chapters, rather than ones I intend to answer directly. The specific aims and objectives I have set out to address in order to shed light on these questions are described in the following section.

2   The Aim of This Book

The focal point of my analysis is the persistence of meat consumption and the use of animals as food in spite of significant challenges to their environmental and ethical legitimacy (Steinfeld et al. 2006; Adams 2010 [1990]; Nibert 2013; Eisnitz 2006). I intend to identify what contributes to this persistent edibility of 'food' animals even, and indeed particularly, as this edibility is being increasingly critiqued. I therefore focus only on meat that is labelled, promoted, and understood as being ethical and sustainable, and purposefully sought by consumers as a 'better' alternative to factory farmed meat, better even than organic and free range. In so doing, I make an a priori assumption that it is particularly where there is awareness and appreciation of the issues associated with meat consumption, and yet consumption persists, that there is the opportunity to access fundamental, legitimising criteria supporting persistent animal use.

My primary research question is: **How do 'food' animals remain persistently and ethically edible?**

Focusing on an area of consumption where this edibility is most likely to be questioned, the aim of my research is, then: **to explore how 'food' animals remain persistently edible specifically when the environmental and ethical legitimacy of their consumption is challenged.** I approach this exploration via the analysis of empirical data generated through interviews with self-identified producers and consumers of ethical and sustainable meat located in the greater Melbourne region of the state of Victoria, Australia. Interviews focused on how 'food' animals and meat were constituted as part of participants' everyday activities or social practices. A focus on social practices allows these constitutions to be seen as part of broader mechanisms and arrangements of social life, rather than solely the result of individual choice and behaviour. My research objectives, listed below, reflect the iterative and reflexive process by which I set about exploring how 'food' animals' persistent, ethical edibility is constituted by my participants, and how this exploration was in turn shaped and re-directed as the research proceeded.

1. How does certain 'knowledge' of animals contribute to their edibility and non-edibility?
2. How do sensory and emotional associations with animals and meat shape their edibility or non-edibility?
3. Where and how is this embodied knowledge of 'food' animals challenged and how is their edibility maintained?
4. What effect have the increased visibility of 'food' animals and increased transparency of meat production processes had on the edibility of animals, and what does this say about how animals are 'made sense' of?
5. What mechanisms of power can be discerned in relation to the persistence of meat consumption and the use of animals as food?
6. How, then, do consumers of ethical and sustainable meat 'make sense' of animals and meat?
7. Is it possible that 'food' animals could be 'made sense of' in other ways, or permitted to make no (human) sense? What sort of dis-ordering of Foucault's nexus of power/knowledge/pleasure would this require?

In terms of the book's academic orientation, there is now a large body of literature on ethical consumption, particularly in relation to food. Reflecting three broadly distinct approaches, this literature encompasses philosophical perspectives that draw on (utilitarian) 'theories of the good', (deontological) 'theories of the right', and ideas of altruism and morality (Wilk 2001; Barnett et al. 2005a; Pellandini-Simanyi 2014); psychology-based attempts to characterise and motivate the 'ethical consumer' (Barnett et al. 2005c; Eckhardt et al. 2010; Memery et al. 2012); and sociological and cultural explorations of the everyday or 'ordinary ethics' of ethical, moral, or 'good' consumption (encompassing politics, materialities and practices) (Lewis and Potter 2011; Guthman 2003; Clarke et al. 2008; Carrier 2008; Goodman et al. 2010; Hall 2011).

Addressing each of these approaches in turn: I am not concerned with philosophic theorising about the type of ethics or morality that might be said to characterise understandings of animals and their flesh. While certainly not denigrating the value and importance of abstract theories, I am keen to remain focused on, and foreground, the living and experiencing being-ness of animals rather than 'philosophising them away', to paraphrase Cudworth (2003). I also do not hold that there is such a thing as a psychologically identifiable 'ethical consumer' but rather that variable 'truths' or 'knowledges' of what is ethical (following Foucault's use of these terms) become attached to certain material objects, human and nonhuman figures, and practices. As 'truths' change, attachments might weaken or break altogether and new ones form with other objects, figures, and practices. However, the 'truth' of consuming meat and using animals for food, as it characterises traditionally meat-eating societies (which constitute the majority; see Chemnitz and Becheva 2014), is one whose attachments have not altered significantly for centuries. Hence it is these persistent 'truths' or 'knowledges' regarding animals and meat that sustain their associations through understandings of what is ethical and/or sustainable that I want to investigate. I therefore follow the third method of enquiry by taking a broadly sociological/cultural approach to critically explore, with reference to empirical data, a specific rendering of ethical consumption in terms of what it reveals about humans' enduring relation with edible animals.

The research described in this book contributes to an embodied understanding of the persistence of meat consumption *and* the use of animals as food by focusing on how meat and animals are made sense of within 'ethical' and 'sustainable' practices. This understanding can be used firstly to reflect on the minimal impact of past and current strategies intended to intervene in meat consumption and the use of animals as food, and secondly to indicate what else needs to be acknowledged and accounted for to constitute a more effective and radical challenge to these practices. Addressing the wider importance of this research, I situate human's use of 'food' animals within a broader field of study concerned with multiple forms of oppression and the connections between them.

3   Book Outline

The book is organised into five parts. Over two brief chapters, Part I provides further background on the issues associated with meat production and consumption, followed by an account of the theory and methodology that informed my analysis. The next three parts of the book each correspond to one of three key aspects of Foucault's work that I have chosen as a framework for my analysis. Hence, Part II focuses on power/knowledge; Part III addresses the pleasure of knowing; and Part IV explores the power of transparency. Each part addresses one or more of my first four objectives, and together they are directed at my fifth objective which is **to identify mechanisms of power in relation to the persistent of meat consumption and the use of animals as food**. These three parts comprise the book's major components, with the addition of this introduction, two background chapters (Part I), and a conclusion (Part V, comprising Chap. 10) that draws together my overall findings with reference to the aim and objectives before exploring the possibility of a future that has no need for meat maps.

Chapter 2 begins with an overview of the key issues and agendas that are increasingly shaping the production and consumption of meat, and at the same time foregrounding the persistence of associated practices. Briefly, these include environmental degradation, pollution, habitat loss, contribution to greenhouse gases and climate change, health issues relating to meat consumption, ethical and social justice issues relating to industry practices and the mistreatment and abuse of animals, and broader intersectional issues relating to the ongoing objectification and use of 'others'. I discuss how these issues and agendas are variously defined and approached in the literature, with particular reference to conceptualisations of the 'Anthropocene' and certain extensions of non-dualistic (non- and more-than-human) and particularly posthuman approaches.[12]

---

[12] Posthumanism is an increasingly diverse field of study, encompassing approaches based in techno-science, biology, philosophy, literature, art, the social sciences, the humanities, and many others. Within the social sciences, further distinctions are shaped by environmental, feminist, eco-feminist, cultural, and other perspectives. The strand of posthumanism I am referring to here is characterised by a largely eco-feminist, but also techno-science and environmental approach to posthumanism. It is characterised by notions of 'becoming with', 'vibrant matter', 'hybridity', and 'mortal/vital/material entanglement' (Haraway 2008; Bennett 2010; Whatmore 2002; Barad 2007; Braidotti 2013).

With respect to recent shifts in approaches to producing and consuming meat, I explain why it is important that associated practices are challenged in ways that confront not only the consumption of 'meat' but also the use of animals as food. From here, in Chap. 3 I draw directly on Cudworth (2005, 2011) and Cudworth and Hobden's (2014a, b) accounts of anthroparchy and critical posthumanism, in combination with Foucault's theorisations of power, knowledge, and pleasure, to describe my theoretical architecture. The chapter concludes with a brief account of my research design and methodology.

In Part II, I address the first of my objectives and explore **how certain 'knowledge' of animals contributes to their edibility and non-edibility**. I approach this over two chapters. The first, Chap. 4, is concerned with animal categories and the maintenance of order. Here I explore designations of animal kinds, or "order and its modes of being" (Foucault 1989: xxiii), and specifically how different animals are more broadly constructed as edible or not. Extending this enquiry, I focus on the language and discourses my participants use in relation to animals, and the knowledge they enlist to support orders of edibility, which together contribute to how they make sense of 'food' animals and 'meat'. From this taxonomical analysis of my participants' sense-making, I identify what I term 'validating discourses'. These are common discourses used by producers and consumers that support and reinforce overarching orders of animal edibility. Validating discourses are understood as the first **mechanism of power**, and the three identified in this chapter—invoking nature, the benevolence of the natural contract, and the value of contingent life—are therefore considered key to 'food' animals' persistent edibility. An awareness of these discourses is carried through to the next chapter, where their contribution to shaping further designations of (ethical) edibility becomes apparent.

Chapter 5 further examines these orders, or rather degrees, of edibility—that is, the orders within the a priori category of 'edible', to show how 'food' animals are additionally constituted by my participants as variously 'good' or 'bad', ethical or unethical to eat. The validating discourses discussed in the previous chapter suffuse these negotiations of edibility, contributing to understandings of what is 'natural', 'proper', 'right', 'genuine', 'authentic', and therefore by default also ethical and 'good'. After identifying exactly what my participants understand to be 'good', 'bad', and 'ethical', I undertake an in-depth interrogation of notions around the 'kill-ability' and 'better' killing of animals. A group of interview questions was specifically oriented to this topic with the intention of targeting what are perhaps the most unavoidably problematic practices associated with eating meat, and thereby getting to the heart of how 'meat' and 'food' animals maintain their ethical edibility. This section demonstrates how effectively the validating discourses work to maintain the normative order, and also serves as a point of entry to my discussion, in Part III of emotional discomfort and particularly transgressions.

Part III turns to the pleasure of knowing to explore my second objective, which is **how sensory and emotional associations with animals and meat shape their edibility or non-edibility.** I tackle this across two chapters, beginning, in Chap. 6, by situating this part of the book in the broader literature on senses, emotions, and affect. I then demonstrate how my participants' senses inform their determinations of 'good' and 'bad' meat, conceiving the senses as the link between Foucault's knowing and pleasure. This sets up the main focus of the first half of this chapter, which is emotion—specifically, how the emotions expressed and identified by my participants, in addition to their senses, contribute to an embodied mapping, or 'making sense' of 'food' animals and 'meat'. In the second half, I explore how *particular* emotions, associated with comfort and discomfort, become associated with different animals and meat. Here, I start to more specifically address my third objective, to explore **where and how an embodied knowledge of 'food' animals is challenged and how their edibility is maintained.**

Extending this line of enquiry, Chap. 7 explores circumstances in which more intense challenges lead to transgression. First, I examine when 'good' meat becomes emotionally inedible—when 'meat' or 'food' animals become something else—and also how such transgressions are corrected, policed, and can be seen as normalised and even necessary features of the existing map. As an outcome of this, I introduce the first of two emotions of 'distinction' that I identify in my data. As with the validating discourses, these emerge as common emotional associations with meat, 'food' animals and/or associated practices, which support and reinforce meat consumption and the use of 'food' animals. Emotions of distinction, the first being requisite bravery, are therefore similarly recognised as **mechanisms of power**.

The second part of this chapter examines the reverse transgression—when 'bad' or unethical meat become edible—and particularly the circumstances, or broader practices, which these transgressions are part of. The further insight this provides on how my participants make sense of meat and animals leads to the identification of a second emotion of distinction, identified as cultural omnivorousness. The final part of this chapter identifies a third 'type' of mechanism of power. In addition to the validating discourses and emotions of distinction, there are also ethico-aesthetic mechanisms that support meat consumption and the use of animals as food. Defined as an embodied ethic that encompasses a consideration for the life and pleasure of the 'other', I identify two ethico-aesthetics. The first, described as moral approval, is addressed in this chapter, while the second is covered in Part IV.

In sum, Part III demonstrates how knowledge, senses, and emotions operate rhizomatically to constitute and maintain a Foucauldian nexus of power/knowledge/pleasure that casts 'food' animals as persistently edible. So far, over Parts II and III, my analysis has revealed six **mechanisms of power**, across three 'types' of mechanism, that contribute to the persistence of meat consumption and the use animals as food, thereby cumulatively speaking to my fifth objective: to identify **mechanisms of power in relation to the persistence of meat consumption and the use animals as food**.

In Part IV, the final data analysis component of the book, I turn to Foucault's insights regarding the power of transparency and Laura Mulvey's seminal work on the pleasure of looking before bringing my analysis of the power/knowledge/pleasure of meat and 'food' animals to an integrated conclusion in the notion of the 'entitled gaze'. This section is equally concerned with my fourth objective to identify **mechanisms of power**. Divided again into two chapters, Chap. 8 focuses on the deeply normalised sense of entitlement to animals, their lives, and their bodies that permeates producers' and consumers' constitutions of good/bad, ethical/unethical, right/wrong meat—as these are described in the preceding two sections. Entitlement, as it manifests in my participants' accounts of their practices, is discussed with reference to the wider literature on privilege, oppression, and intersectionality. This leads to a critique of 'respect' as a concept widely employed by producers and consumers in relation to the use of 'natural resources'. Respect is identified as the second ethico-aesthetic, and subsequently discussed in more detail in relation to the constituted notion of a 'natural' contract.

Next, in Chap. 9, my focus is the increased material as well as discursive visibility of 'food' animals, especially in the 'ethical' foodscape. Here, I attend to my fourth objective by examining **what effect increased visibility of 'food' animals and increased transparency of meat production processes have on the edibility of animals, and what this says about how animals are 'made sense' of**. I demonstrate how visibility has become a marketable commodity that is promoted and used as brand leverage. Contrary to popularly held assumptions, I show how visibility does not challenge my participants' constructions of animals as food. Power, knowledge, and pleasure are brought to bear by producers and consumers alike on the animal subject of their gaze, and consequently, under the weight of what I describe as this 'entitled gaze', they remain firmly mapped as edible resources.

To conclude, Chap. 10 brings together the discussion and arguments from the book's three main parts to show how orders of knowledge, socialised senses and emotions, and the entitled gaze work rhizomatically, in a nexus of power/knowledge/pleasure, to trap 'food' animals in their 'rightful' place—that being a state of domination. I provide a summary of the key findings of my research which aimed to elucidate **mechanisms of power** and, in so doing, understand **how consumers of ethical and sustainable meat 'make sense' of animals and meat**—my sixth objective. Together, my findings in response to these six objectives address the book's overall aim, which is to explore how 'food' animals remain persistently edible despite significant challenges to their environment and ethical legitimacy. Following this, and attending to my final objective, I consider whether it is **possible that 'food' animals could be 'made sense of' in other ways, or permitted to make no (human) sense, and what** sort of dis-ordering of power/knowledge/pleasure this would require. I end with a critical reflection on the book's overall contribution and the sort of further research it might prompt into questions of animal use using a critical posthumanist lens.

A brief note is necessary here on my use of single quotation marks around the words 'food', in reference to 'food' animals, and also 'meat'. My intention is to disturb normalised associations of animals as both food and meat, and emphasise that these are human constructions that are reinforced through discourse and practice. However, recognising that they can be distracting for the reader, I persist in using these quotation marks only in the introduction. Through the remainder of the book, the reader can assume that these words are still viewed critically. I occasionally return to their use when referring to 'dairy', 'broiler', or other kinds of 'food' animals as a reminder of this point.

4   Exclusions and Limitations

This research contains a number of exclusions and limitations. Perhaps the most obvious exclusion is my focus on terrestrial animals and the production and consumption of their 'meat' to the exclusion of marine animals, their 'meat', and other terrestrial animal products such as eggs and dairy. Although the production and consumption of 'seafood', eggs, and dairy products raise significant issues worthy of investigation, it is not possible within the scope of this study to give sufficient consideration to all kinds of animal products. My decision to focus on terrestrial 'meat' is supported by a number of observations about 'food' animals, their products, and the broader utility of this research.

Terrestrial 'meat', eggs, and dairy products, more than 'seafood', are implicated in long, interconnected genealogies of domestication (of animals and nature), colonialism, and environmental change. Practices associated with the production and consumption of these 'foodstuffs' are equally 'sticky' in terms of their persistence, geographical spread, and per capita increase (OECD-FAO 2017; Conway 2015). Indeed, the production and consumption of eggs and dairy products cannot be isolated from the production and consumption of meat. Associated practices are intertwined and often inextricably interlocked, with 'spent' animals, and wastage (e.g. male chicks and calves) constituting a significant proportion of the trade in slaughtered animal products, including meat. Exploring how 'egg' and 'dairy' animals and their products continue to be 'made sense of' in the midst of concerns about increasing demand, the environmental impacts of production, associated health issues, and ethical issues surrounding the treatment of animals, would therefore provide a rich topic for analysis.

However, terrestrial 'meat' has been on the receiving end of more direct and sustained challenges on all these fronts for several decades such that it leads eggs and dairy in the variety and extent of responses. Rejecting 'mainstream' options, consumers now have access to ethical, sustainable, environmentally friendly, organic, grass-fed, biodynamic, humane, free range, happy, and many other iterations of 'better' meat. Egg producers have adopted some of these production and branding practices, but not nearly to the same extent, while dairy products, besides organic, A2-protein and lactose-free products, remain comparatively untouched by the broader environmental, health, and ethical concerns associated with terrestrial 'meat'.[13] Therefore, compared with both eggs and dairy, so-called ethical and/or sustainable 'meat', in all its various guises, is a widespread and recognisably distinct trend that is shaping significant changes in mainstream as well as alternative markets.

For these reasons, and for the fact that I am interested in exploring how understandings of 'food' animals and their edibility respond to direct and sustained challenges (relating to the environment, health, and ethics), this study focuses solely on terrestrial 'meat'. However, the same approach may be extended to all animals used to produce food for humans, including 'layer' chickens, 'dairy' cows, and marine animals, to similarly explore their ongoing (re)constitution as edible. Indeed, concerns about the environmental and ethical implications of increasing production and consumption of 'seafood' (Walsh 2011; Silver and Hawkins 2014; Larsen and Roney 2013) have tended to overshadow those associated with the treatment of animals in the egg and dairy industries. Yet at

---

[13] Notwithstanding the rise of plant-based 'milks' that are gaining increasing traction. It is not clear whether this is being shaped primarily by environmental, health, ethical, and/or other concerns.

the same time, 'seafood' is often promoted as the healthy and more sustainable alternative to 'meat'. Combined with the lower regard afforded to marine life, and ongoing questions concerning their sentience (Brown 2015; Bergqvist and Gunnarsson 2013), the increasing production and per capita consumption of seafood (FAO 2016) and also fish oil supplements present a particularly interesting case for exploring the persistent edibility of nonhuman animals.

A further two exclusions relating to the scope of my research, commonly understood as the system boundary (Braschel and Posch 2013), are that I do not explore how 'food' animals and 'meat' are 'made sense of' across practices that are *not* associated with ethical, sustainable, or otherwise 'better' meat, nor how the broader nexus of social practices that constitute the animal-industrial complex shape this sense-making. A comparison of the discourses and practices associated with 'mainstream' meat, plant-based diets, *and* ethical/sustainable meat would allow for a more comprehensive analysis, and understanding, of how the edibility of 'meat' and animals is differently constituted between practices, and perhaps reconstituted over time as practices change. Likewise, locating these discourses and practices, and those explored in this book, in the broader nexus of economic, political, religious, social, and cultural practices that support the meat apparatus is an essential component of a more thorough analysis and understanding. Once again, within the parameters of this research, a study of this scale is not possible.

Further limitations of this study relate to the type of data collected and the orientation of my analysis rather than its scope. In order to allocate sufficient resources to exploring understandings of 'meat' and 'food' animals, I was limited in the extent to which I could explore how constitutions of gender, and other social 'locations' such as socio-economic status and 'race', might also shape these understandings.

Within the broad field of research that addresses consumption, which stretches across the social, cultural, and political sciences, a burgeoning body of literature explores lifestyle politics and patterns of consumption relating to race, class, and gender (Lewis 2011; Barnett et al. 2005b; Wilk 2001; Thompson 2011). Relatedly, the agency of consumption as a positive expression and active means of social change through, for example, conscientious or political consumption, green consumption, and ethical consumption, is another rich area of research (Micheletti et al. 2012; Lewis 2008; Littler 2008; Humphery 2011; Spaargaren and Mol 2008). Studies of 'Alternative Food Networks' (AFNs) such as Slow Food, local, organic, fair trade, vegetarianism, and others, feature strongly among this research (Gray 2013; Micheletti and Stolle 2010; Clarke et al. 2008; Goodman 2004), and many of these studies highlight the different ways in which these forms of consumption intersect with holistic conceptions of race, class, and gender. Of particular relevance for my research are notions of ethical consumption as appealing to, and creating, privileged niche groups of predominantly white, middle class consumers (Lury 2011; Guthman 2008; Slocum 2007). This raises questions regarding how these, and perhaps other, social locations might shape practices associated with ethical and sustainable meat.

This body of work only indirectly informs my analysis, and, as with the exclusions above, the parameters of the research process and the priorities of my research precluded me from including aspects of social location in my data collection activities. Nevertheless, while acknowledging there may be a range of factors relating to my recruitment process that contributed to my sample, the fact that only two of the total 41 participants could be described as non-white (both at least second-generation) suggests there is more to explore and understand about the way these practices are constituted, and also even conceived.

Similarly, it is interesting to note that 19 of the 26 consumers I interviewed were female. In fact, of the 65 consumers who originally expressed interest in being part of the study, 54 were female, and only 11 were male. The producer sample tells a slightly different story, though perhaps more similar than expected based on the long-standing tradition of the male-dominated 'livestock' industry. Of the 15 producers interviewed, six were female, of which three appeared as more equal partners in the business, while the remaining three were the sole or primary operators.

Though I cannot make any robust comment in this regard, Parry (2010) highlights the role of gender norms in shaping and even perhaps enforcing the rules around a new type of ethical meat eating. For instance, contrary to the association of meat eating and butchery with masculinity, across the 'New Carnivore' literature, which bears many of the characteristics associated with ethical and sustainable meat, he observes "a bold revision of traditional feminine gender norms, as well as performance of female empowerment" (74). Simultaneously, across the body of literature dealing with the ethics of care (human and nonhuman), care is theorised as an alternately feminine and feminised trait (Gilligan 1982; Noddings 1984; Donovan and Adams 2007). Shifts in understandings of 'meat', 'food' animals and their edibility therefore warrant further investigation from the perspective of binary gender norms, particularly in light of Comninou's somewhat prophetic enquiry over 20 years ago as to whether "animal exploitation [will] become the ultimate symbol of equality with the white male?" (1995: 14).

References

Adams, C. J. (2010). *The Sexual Politics of Meat: A Feminist-vegetarian Critical Theory*. London; New York: Continuum International Publishing Group.

Arcari, P. (2018). 'Dynamic' Nonhuman Animals in Theories of Practice: Views from the Subaltern. In C. Maller & Y. Strengers (Eds.), *Social Practices and Dynamic More-than-humans: Living Things, Unbounded Materials, and Automation*. Basingstoke; New York: Palgrave Macmillan.

Barad, K. (2007). *Meeting the Universe Halfway: Quantum Physics and the Entanglement of Matter and Meaning*. Durham; London: Duke University Press.

Barnett, C., Cafaro, P., & Newholm, T. (2005a). Philosophy and Ethical Consumption. In R. Harrison, T. Newholm, & D. Shaw (Eds.), *The Ethical Consumer* (pp. 11–24). London: SAGE Publications Ltd.

Barnett, C., Cloke, P., Clarke, N., & Malpass, A. (2005b). Consuming Ethics: Articulating the Subjects and Spaces of Ethical Consumption. *Antipode, 37*(1), 23–45.

Barnett, C., Clarke, N., Cloke, P., & Malpass, A. (2005c). The Political Ethics of Consumerism. *Consumer Policy Review, 15*(2), 45–51.

Bauman, Z. (2002). *Society under Siege*. Oxford; Malden: Polity Press.

Bauman, Z. (2005). *Living in Utopia*. Presentation: London School of Economics. 27 October.

Bennett, J. (2010). *Vibrant Matter: A Political Ecology of Things*. Durham; London: Duke University Press Books.

Bergqvist, J., & Gunnarsson, S. (2013). Finfish Aquaculture: Animal Welfare, the Environment, and Ethical Implications. *Journal of Agricultural and Environmental Ethics, 26*(1), 75–99.

Beverley, J. (1999). *Subalternity and Representation: Arguments in Cultural Theory*. Durham; London: Duke University Press.

Braidotti, R. (2013). *The Posthuman*. Cambridge; Malden: Polity Press.

Braschel, N., & Posch, A. (2013). A Review of System Boundaries of GHG Emission Inventories in Waste Management. *Journal of Cleaner Production, 44*(April), 30–38.

Brown, C. (2015). Fish Intelligence, Sentience and Ethics. *Animal Cognition, 18*(1), 1–17.

Calvino, I. (1983). *Mr Palomar*. San Diego; New York; London: A Harvest Book.

Carrier, J. G. (2008). Think Locally, Act Globally: The Political Economy of Ethical Consumption. *Research in Economic Anthropology, 28*, 31–51.

Castree, N., & Nash, C. (2004). Mapping Posthumanism: An Exchange. *Environment and Planning A, 36*(8), 1341–1363.

Chemnitz, C., & Becheva, S. (Eds.). (2014). *Meat Atlas: Facts and Figures about the Animals We Eat*. Berlin; Brussels: Heinrich Boll Stiftung and Friends of the Earth Europe.

Clarke, N., Cloke, P., Barnett, C., & Malpass, A. (2008). The Spaces and Ethics of Organic Food. *Journal of Rural Studies, 24*(3), 219–230.

Comninou, M. (1995). Speech, Pornography, and Hunting. In C. J. Adams & J. Donovan (Eds.), *Animals and Women: Feminist Theoretical Explorations* (pp. 126–148). Durham; London: Duke University Press Books.

Conway, A. (2015, November 23). Global Egg Consumption to Rise Worldwide through 2024. *WATTAgNet*. Online. October 2017.

Cudworth, E. (2003). *Environment and Society*. London: Routledge.

Cudworth, E. (2005). *Developing Ecofeminist Theory: The Complexity of Difference*. Basingstoke; New York: Palgrave Macmillan.

Cudworth, E. (2011). *Social Lives with Other Animals*. Basingstoke; New York: Palgrave Macmillan.

Cudworth, E., & Hobden, S. (2014a). Civilisation and the Domination of the Animal. *Millennium - Journal of International Studies, 42*(3), 746–766.

Cudworth, E., & Hobden, S. (2014b). Liberation for Straw Dogs? Old Materialism, New Materialism, and the Challenge of an Emancipatory Posthumanism. *Globalizations, 12*(1), 134–148.

Dehaene, M., & De Cauter, L. (2008). *Heterotopia and the City*. London; New York: Routledge.

Donovan, J., & Adams, C. J. (Eds.). (2007). *The Feminist Care Tradition in Animal Ethics: A Reader.* New York; Chichester: Columbia University Press.

Eckhardt, G. M., Belk, R., & Devinney, T. M. (2010). Why Don't Consumers Consume Ethically? *Journal of Consumer Behaviour, 9*(6), 426–436.

Eisnitz, G. A. (2006). *Slaughterhouse.* New York: Prometheus Books.

Ellis, R., & Waterton, C. (2005). Caught between the Cartographic and the Ethnographic Imagination: The Whereabouts of Amateurs, Professionals, and Nature in Knowing Biodiversity. *Environment and Planning D: Society and Space, 23*(5), 673–693.

FAO. (2016). *The State of World Fisheries and Aquaculture: Contributing to Food Security and Nutrition for All.* Food and Agricultural Organisation of the United Nations, Rome, 200 pp.

Foucault, M. (1967). Of Other Spaces: Utopias and Heterotopias. *Architecture/Mouvement/Continuite* (October), 1–9.

Foucault, M. (1978). *The History of Sexuality.* New York: Pantheon Books.

Foucault, M. (1980). *Power/Knowledge: Selected Interviews and Other Writings, 1972–1977.* New York: Pantheon Books.

Foucault, M. (1982). The Subject and Power. *Critical Inquiry, 8*(4), 777–795.

Foucault, M. (1989). *The Order of Things.* London; New York: Routledge.

Foucault, M. (1994). *Ethics, Subjectivity and Truth* (P. Rabinow, Ed.). New York: The New Press.

Gilligan, C. (1982). *In a Different Voice.* Cambridge, MA; London: Harvard University Press.

Goodman, M. K. (2004). Reading Fair Trade: Political Ecological Imaginary and the Moral Economy of Fair Trade Foods. *Political Geography, 23*(7), 891–915.

Goodman, M. K., Maye, D., & Holloway, L. (2010). Ethical Foodscapes?: Premises, Promises, and Possibilities. *Environment and Planning A, 42*(8), 1782–1796.

Gray, M. (2013). *Labor and the Locavore - The Making of a Comprehensive Food Ethics.* Berkeley; Los Angeles; London: University of California Press.

Guthman, J. (2003). Fast Food/Organic Food: Reflexive Tastes and the Making of 'Yuppie Chow'. *Social & Cultural Geography, 4*(1), 45–58.

Guthman, J. (2008). Bringing Good Food to Others: Investigating the Subjects of Alternative Food Practice. *Cultural Geographies, 15*(4), 431–447.

Halkier, B., Katz-Gerro, T., & Martens, L. (2011). Applying Practice Theory to the Study of Consumption: Theoretical and Methodological Considerations. *Journal of Consumer Culture, 11*(1), 3–13.

Hall, S. M. (2011). Exploring the 'Ethical Everyday': An Ethnography of the Ethics of Family Consumption. *Geoforum, 42*(6), 627–637.

Haraway, D. (2008). *When Species Meet.* Minneapolis; London: University of Minnesota Press.

Hook, D. (2007). *Foucault, Psychology and the Analytics of Power.* Basingstoke; New York: Palgrave Macmillan.

Hui, A., Schatzki, T., & Shove, E. (2016). *The Nexus of Practices: Connections, Constellations, Practitioners.* London; New York: Routledge.

Humphery, K. (2011). The Simple and the Good: Ethical Consumption as Anti-consumerism. In T. Lewis & E. Potter (Eds.), *Ethical Consumption: A Critical Introduction* (pp. 40–53). London; New York: Routledge.

Jacob, C. (2006). *The Sovereign Map: Theoretical Approaches in Cartography Throughout History.* Chicago; London: University of Chicago Press.

Jacobson, M. H. (2013). Solid Modernity, Liquid Utopia – Liquid Modernity, Solid Utopia. In A. Elliott (Ed.), *The Contemporary Bauman* (pp. 217–240). London; New York: Taylor & Francis.

Jorgensen, M., & Phillips, L. J. (2002). *Discourse Analysis as Theory and Method.* SAGE Publications Ltd.

Korzybski, A. (1958). *Science and Sanity: An Introduction to Non-Aristotelian Systems and General Semantics.* Englewood, NJ: Institute of General Semantics.

Larsen, J., & Roney, J. M. (2013, June 12). Farmed Fish Production Overtakes Beef. *Earth Policy Institutei.* Online. February 2017.

Lewis, T. (2008). Transforming Citizens? Green Politics and Ethical Consumption on Lifestyle Television. *Continuum, 22*(2), 227–240.

Lewis, T. (2011). The Ethical Turn in Commodity Culture: Consumption, Care and the Other. *sic: Journal of Literature, Culture and Literary Translation*, Vol. 2. Online. March 2013.

Lewis, T., & Potter, E. (Eds.). (2011). *Ethical Consumption: A Critical Introduction.* London; New York: Routledge.

Littler, J. (2008). *Radical Consumption: Shopping for Change in Contemporary Culture.* Maidenhead, UK: Open University Press.

Lury, C. (2011). *Consumer Culture.* Cambridge; Malden: Policy Press.

Maller, C. J. (2015). Understanding Health Through Social Practices: Performance and Materiality in Everyday Life. *Sociology of Health & Illness, 37*(1), 52–66.

Meininger, H. P. (2013). Inclusion as Heterotopia: Spaces of Encounter between People with and without Intellectual Disability. *Journal of Social Inclusion, 4*(1), 24–44.

Memery, J., Megicks, P., Angell, R., & Williams, J. (2012). Understanding Ethical Grocery Shoppers. *Journal of Business Research, 65*(9), 1283–1289.

Micheletti, M., & Stolle, D. (2010). Vegetarianism - A Lifestyle Politics? In M. Micheletti & A. S. McFarland (Eds.), *Creative Participation: Responsibility-Taking in the Political World* (pp. 125–145). Boulder, CO; London: Paradigm Publishers.

Micheletti, M., Cheng, S.-L., Stolle, D., Olsen, W., et al. (2012). Habits of Sustainable Citizenship: The Example of Political Consumerism. In A. Warde & D. Southerton (Eds.), *The Habits of Consumption* (Collegium - Studies across Disciplines in the Humanities and Social Sciences) (pp. 141–163). Helsinki: Helsinki Collegium for Advanced Studies.

Nibert, D. A. (2013). *Animal Oppression and Human Violence: Domesecration, Capitalism, and Global Conflict.* New York; Chichester: Columbia University Press.

Noddings, N. (1984). *Caring, a Feminine Approach to Ethics & Moral Education.* Berkeley; Los Angeles; London: University of California Press.

Noske, B. (1989). *Humans and Other Animals: Beyond the Boundaries of Anthropology.* London: Pluto Press.

Noske, B. (1997). *Beyond Boundaries: Humans and Animals.* Montreal: Black Rose Books.

OECD-FAO. (2017). *OECD-FAO Agricultural Outlook 2017–2026.* Paris: OECD Publishing. 142 pp. https://doi.org/10.1787/agr_outlook-2017-en.

Parry, J. (2010). *The New Visibility of Slaughter in Popular Gastronomy.* Masters Thesis, Cultural Studies: University of Canterbury.

Pellandini-Simanyi, L. (2014). *Consumption Norms and Everyday Ethics.* Basingstoke; New York: Palgrave Macmillan.

Pickles, J. (2004). *A History of Spaces: Cartographic Reason, Mapping and the Geo-Coded World.* London; New York: Routledge.

Postill, J. (2010). Introduction: Theorising Media and Practice. In B. Brauchler & J. Postill (Eds.), *Theorising Media and Practice* (pp. 1–33). New York; Oxford: Berghahn Books.

Reckwitz, A. (2002). Toward a Theory of Social Practices: A Development in Culturalist Theorizing. *European Journal of Social Theory, 5*(2), 243–263.

Reckwitz, A. (2016). Practices and Their Affects. In A. Hui, T. Schatzki, & E. Shove (Eds.), *The Nexus of Practices: Connections, Constellations, Practitioners* (pp. 114–125). London; New York: Taylor & Francis.

Ryan, S. (1996). *The Cartographic Eye: How Explorers Saw Australia.* Cambridge, UK; New York: Cambridge University Press.

Schatzki, T. R. (1996). *Social Practices: A Wittgensteinian Approach to Human Activity and the Social.* Cambridge, UK; New York: Cambridge University Press.

Schatzki, T. R. (1997). Practices and Actions: A Wittgensteinian Critique of Bourdieu and Giddens. *Philosophy of the Social Sciences, 27*(3), 283–308.

Schatzki, T. R. (2001). *Practice Theory.* London; New York: Routledge.

Shove, E., & Spurling, N. (Eds.). (2013). *Sustainable Practice: Social Theory and Climate Change.* London; New York: Routledge.

Shove, E., Pantzar, M., & Watson, M. (2012). *The Dynamics of Social Practice: Everyday Life and How It Changes.* London; Thousand Oaks; New Delhi: SAGE Publications Ltd.

Silver, J. J., & Hawkins, R. (2014). I'm Not Trying to Save Fish, I'm Trying to Save Dinner': Media, Celebrity and Sustainable Seafood as a Solution to Environmental Limits. *Geoforum, 84*(August), 218–227.

Slocum, R. (2007). Whiteness, Space and Alternative Food Practice. *Geoforum, 38*(3), 520–533.

Spaargaren, G., & Mol, A. (2008). Greening Global Consumption: Redefining Politics and Authority. *Global Environmental Change, 18*(3), 350–359.

Steinfeld, H., Gerber, P., Wassenaar, T., Castel, V., et al. (2006). *Livestock's Long Shadow: Environmental Issues and Options.* Rome: Food and Agricultural Organisation of the United Nations. 416 pp.

Strengers, Y., & Maller, C. (Eds.). (2014). *Social Practices, Intervention and Sustainability: Beyond Behaviour Change.* New York; London: Routledge.

Thompson, C. J. (2011). Understanding Consumption as Political and Moral Practice: Introduction to the Special Issue. *Journal of Consumer Culture, 11*(2), 139–144.

Twine, R. (2012). Revealing the 'Animal-Industrial Complex' - A Concept & Method for Critical Animal Studies. *Journal for Critical Animal Studies, 10*(1), 12–39.

Walsh, B. (2011, July 7). The End of the Line. *Time Magazine - Science and Space.* Online. March 2013.

Warde, A. (2005). Consumption and Theories of Practice. *Journal of Consumer Culture, 5*(2), 131–153.

Whatmore, S. (2002). *Hybrid Geographies: Natures Cultures Spaces.* London; Thousand Oaks; New Delhi: SAGE Publications Ltd.

Wilk, R. (2001). Consuming Morality. *Journal of Consumer Culture, 1*(2), 245–260.

Wilson, M. W. (2011). 'Training the Eye': Formation of the Geocoding Subject. *Social & Cultural Geography, 12*(4), 357–376.

Part I

Background

2

The Problem with 'Food' Animals

1   Meat, Animals, Environment, and Ethics

Agricultural production of our food is recognised as the source of a significant proportion of the total global greenhouse gases that are contributing to climate change—between 14% and 22% according to various sources (Barker et al. 2007; McMichael et al. 2007; Schwarzer et al. 2012; Smith et al. 2007).[1] Disaggregating the impacts of the sector as whole, it is revealed that 80% of global agricultural emissions of greenhouse gases are attributable to livestock (Schwarzer et al. 2012)—that is 11% to 18% of total emissions. Other studies have suggested that meat[2] production on its own contributes 15% to 24% of total global emissions (Fiala 2008; Steinfeld et al. 2006), higher than many of the estimates for the total agricultural sector, and comparable to, or higher than, the percentage of

---

[1] For their fifth assessment report (2014), the Intergovernmental Panel on Climate Change (IPCC) changed the scope of the agriculture sector to include agriculture, forestry, and other land use (AFOLU). The proportion of overall emissions from AFOLU is now estimated to be 24%, although the authors say that the main contributions are from "deforestation and agricultural emissions from livestock, soil and nutrient management" (25). The sixth assessment report is expected 2022.

[2] 'Meat' as referred to in these studies and reports, and in this book, refers to the flesh of animals most commonly used as food, which are cows, pigs, chickens, sheep, and goats. It excludes fish.

© The Author(s) 2020
P. Arcari, *Making Sense of 'Food' Animals*,
https://doi.org/10.1007/978-981-13-9585-7_2

estimated emissions from industry, forestry and transport (19%, 17% and 13% respectively, IPCC 2007). Goodland and Anhang (2009) claim that due to routine underestimation, omission and mis-assignment of various emissions sources, the figure is closer to 51%.[3]

Although emissions estimates vary depending on the system boundaries used, it is widely agreed that 30% of the earth's surface, or 70–75% of all agricultural land, is allocated to livestock either directly through grazing or for growing the 35–40% of the world's total grain used as feed (Bailey et al. 2014; IAASTD 2009; Pimentel and Pimentel 2007; Ripple et al. 2014; Steinfeld et al. 2006; Worldwatch Institute 2013). To support these animals and produce their feedstock requires almost a third of the total water footprint attributed to agriculture globally (Mekonnen and Hoekstra 2012). Considered another way, Mekonnen and Hoekstra estimate 12% of the global consumption of groundwater and surface water for irrigation is allocated to growing feed for livestock, as opposed to food, fibres, or other crop products (408). In the prominent Food and Agriculture Organization (FAO) publication *Livestock's Long Shadow*, the authors declare that the contribution of livestock to environmental problems, including deforestation, desertification, water pollution, eutrophication of freshwater and marine ecosystems, oceanic dead zones, acidification, soil and nutrient loss, loss of habitats, species extinction[4]/ endangerment, and other land use changes, is "on a massive scale" (Steinfeld et al. 2006: xx; see also Koneswaran and Nierenberg 2008; UNEP 2012; Ripple et al. 2014; Leip et al. 2015). This all serves as evidence that meat production is, directly and indirectly, the leading cause of environmental degradation associated with agriculture. Demanding the lion's share of our available arable land and water, and producing the most greenhouse gases, it is the most inefficient way of converting food into nutritional energy (Pimentel et al. 2003; Clune et al. 2017).

---

[3] Goodland and Anhang (2009) identify several sources of emissions that they claim have been underestimated, overlooked, or misallocated in past studies. While subsequent authors have accepted some of their claims, there remains controversy around others, especially livestock respiration, which comprises 13.7% of their total figure of 51%.

[4] Fiona Probyn-Rapsey (2017) highlights how the word 'extinction' obscures the agency (human and nonhuman) behind it, arguing instead that a discourse and cultural politics of 'eradication' foregrounds how species are rendered eradicable.

In addition, an estimated 50–84% of the antibiotics produced globally are allocated to the meat industry as growth promoters and to prevent and treat infection (Laxminarayan 2002; Worldwatch Institute 2015; Robinson et al. 2016). This recognised overuse of antibiotics has led to ongoing 'calls to action' from health experts in Europe, the US, the UK and elsewhere, who share concerns about a "return to the preantibiotic era for many types of infections" and cite misguided regulation of antibiotic use in food animals as one of the primary causes (Spellberg et al. 2008: 155; also, Carlet et al. 2012; Laxminarayan et al. 2013; Martin et al. 2015; UK Veterinary Medicines Directorate 2019). Antimicrobial resistance (AMR) is now identified by the Food and Agriculture Organization of the United Nations (FAO) as posing a "major global threat of increasing concern" with "implications for both food safety and food security".[5] With reference to the sheer (and increasing) numbers of animals bred for food, 'livestock' (mostly cattle and pigs) now comprise approximately 60% of the earth's total mammalian biomass, humans 36%, and just 4% are wild mammals—a loss of 83% since human civilisation first appeared (Bar-On et al. 2018). Agricultural production and expansion (and implicitly livestock production) has been described as the main driver of current land extinctions threatening 26% of remaining mammals (McKie 2014; Monastersky 2014) and causing leading scientists to predict the cow could become the earth's largest mammal in just a few hundred years (Smith et al. 2018).

Aside from environmental impacts, meat production and consumption has been associated with a range of implications for human health that over time are becoming less contested. Mounting evidence demonstrates direct links to cancer (Walker et al. 2007; Joshi et al. 2009; zur Hausen 2012; Pan et al. 2012), coronary heart disease, and stroke (Walker et al. 2007; Pan et al. 2012; Feskens et al. 2013), and indirect links to both diabetes (Pan et al. 2012; Feskens et al. 2013) and obesity (Wang and Beydoun 2009; Vergnaud et al. 2010). There are, in addition, health impacts to workers and wider communities from exposure to pesticides, fertilisers, and waste from the animal industry (Walker et al. 2007), as well as the threat of 'super-bugs' associated with the overuse of antibiotics

---

[5] www.fao.org/antimicrobial-resistance/background/what-is-it/en/.

(Silbergeld et al. 2008). Most health and medical bodies now urge people to reduce the proportion of these foods in their diets while also providing assurances of the adequacy of a well-balanced vegetarian or vegan diet for all life stages (Craig and Mangels 2009; NHMRC 2013; www.nhs.uk). Less well known or acknowledged are the psychological effects of acts of normalised and other violence carried out by those involved in killing these food animals. This violence has been linked to increased crime rates, interhuman violence, and a high propensity for aggression (Beirne 2004; Fitzgerald et al. 2009; Stull and Broadway 2012; Richards et al. 2013).

From a certain perspective, the United Nations Environment Programme's (UNEP) concern that meat (and dairy) production "undermines the ecological foundation of food security" (2012: 31) seems to be widely shared, and is based on increasingly unavoidable scientific evidence. A number of organisations and prominent researchers have expressed concern regarding the impacts associated with the production and consumption of meat, and its future sustainability, particularly in the context of steadily increasing rates of per capita consumption and the FAO's predicted doubling of worldwide consumption by 2050 (FAO 2014). Encompassing affiliations with the UN, the FAO, Oxford University's Food Climate Research Network and Programme on the Future of Food, Chatham House, the World Resources Institute, the World Wildlife Fund, and the 2006 Stern Review, there is growing consensus that moving to a meat- and dairy-free diet is necessary to avoid the worst impacts of climate change (Springmann et al. 2018; UNEP 2018; Waite and Vennard 2018; WWF 2017; Wellesley et al. 2015; Steinfeld et al. 2006). On its current trajectory, the livestock industry represents "one of the greatest challenges to global food security and to the environment" (Giovannucci et al. 2012: 12), while a growing number of studies demonstrate that vegetarian and vegan diets are associated with the lowest environmental impacts (Scarborough et al. 2014; Baroni et al. 2006; Clune et al. 2017; Harwatt and Hayek 2019).

Environmental and social/health challenges are not the only ones confronting the global meat industry[6] and an increasingly meat-consuming, and growing, world population. Meat production has increased almost

---

[6] The global meat industry includes all types and scales of meat production.

fivefold since 1961 in order to keep pace with demand (Ritchie and Roser 2017), and the treatment of animals within these expanding systems of progressively more industrialised and mechanised use is also coming under increased scrutiny. This has dovetailed with technological advances in surveillance equipment, meaning that cameras can now be easily carried into most premises undetected. Consequently, visual footage of food animals in industrialised systems of production and slaughter has become much more commonplace and also more widely disseminated through both mainstream and social media.

Much of this footage has been obtained undercover and shows the crowded, concrete and steel, and primarily indoor conditions in which the majority of these animals are raised, without natural light or freedom to move; and conditions on denuded feedlots[7] where mostly cattle but also sheep spend between two and four months being intensively 'fattened' in order to reach slaughter weight more efficiently than they would on pasture. The physical and mental conditions of the animals themselves are also a focus, with oversized chickens commonly seen struggling to stand and pigs exhibiting stress-related behaviours such as tail biting or bar chewing. What such footage may also reveal is the treatment of these animals at the hands of industry workers, especially at the slaughterhouse. An extended news report, broadcast in 2011 by the ABC—one of Australia's leading national public television networks—on their investigative journalism/current affairs programme, Four Corners, graphically documented the treatment of Australian cattle in Indonesian slaughterhouses and sparked a period of vigorous public debate about the ethics of live animal export that is frequently reignited (Ferguson 2011; Burke 2011; Vidot and Conifer 2016; Wahlquist 2017; Evershed and Wahlquist 2018). Similarly graphic footage of animal cruelty in Australian slaughterhouses has provoked action, with many businesses being investigated, some facing formal charges, and two closed as a direct result (Cannane 2012; March 2011;

---

[7] More often associated with US meat production processes, there are at least 450 feedlots in Australia, mostly in New South Wales and Queensland, supplying around 80% of the beef sold in Australian supermarkets (ALFA website—About the Australian Feedlot Industry. Accessed April 2018).

see also Noone 2014; Thomas 2016; Walls 2016; Carlyon 2016; Aird 2017; Sullivan 2018). Outside Australia, abuse and violence directed at food animals seem equally commonplace.

Improved surveillance techniques have thus provided a window into the formerly hidden world of industrial meat production. New forms of social media have allowed this knowledge to disseminate more widely and rapidly than ever before so that increasing numbers of consumers are becoming aware of ethical (and not just environmental and health) issues associated with the production and consumption of meat, even if not the principle of using animals as food. Animal advocacy organisations (both welfare- and rights-oriented)[8] make frequent use of visual images showing the rearing, transport, holding, and slaughter of food animals in their campaigns, including associated acts of negligence, abuse, and violence that animals suffer at the hands of industry owners and workers.

A growing concern with the environmental, health, and ethical implications of 'food choices' is reflected in, and further fuelled by, popular media. Over just the last ten years, there has been a swathe of new literature and visual media focusing in some way on questions involving meat. Some of the more prominent of these are listed in Table 2.1. Over the same time, a slew of dietary prescriptions have emerged representing a variety of positions on the nexus of concerns between environment, health, and ethics. What they all share is a focus on eating less and/or 'better' meat. These include: flexitarian, reducetarian, climatarian, ethitarian, palaeo, semi-vegetarian, part-time carnivore, and part-time vegan. Kangatarians, vegeroos, and cameltarians even make occasional appearances in Australia. The question now is whether and how all of these evidence, discussions, and dietary manoeuvrings have changed the situation for food animals.

---

[8] An animal welfare perspective focuses on the treatment of (food) animals within systems of rearing, transport, holding, and slaughter, seeking to reduce suffering and eliminate 'abuse'. The use of animals for food is therefore not problematised—only their treatment. An animal rights perspective regards any use of animals as unethical, however well regulated and monitored.

Table 2.1 Prominent meat-related non-fiction literature and visual media from 2006 to present

Title	Year	Author/Director/Producer
Books		
The Omnivore's Dilemma	2006	Michael Pollan
Slaughterhouse	2007	Gail Eisnitz
The Ethics of What We Eat	2007	Peter Singer & Jim Mason
Eating Animals	2009	Jonathan Safran Foer
The Compassionate Carnivore	2009	Catherine Friend
The Meat Crisis	2010	John Webster
Meat: A Benign Extravagance	2011	Simon Fairlie
Why We Love Dogs, Eat Pigs and Wear Cows	2011	Melanie Joy
The Ethical Butcher	2013	Berlin Reed
Meatonomics	2013	David son Simon
Farmageddon	2014	Phillip Lymbery & Isabel Oakeshott
The Carnivore's Manifesto	2014	Patrick Martins
Defending Beef	2014	Nicolette Hahn Niman
The Meat Racket	2014	Christopher Leonard
The Ethical Meat Handbook	2015	Meredith Leigh
Meathooked	2016	Marta Zaraska
Eat This Book: A Carnivore's Manifesto	2016	Dominique Lestel
The Ethical Carnivore: My Year Killing to Eat	2016	Louise Gray
What's the Matter with Meat	2017	Katy Kieffer
Big Chicken	2017	Maryn McKenna
TV and Documentary		
Earthlings	2005	Shaun Monson
Meat the Truth	2007	Karen Soeters & Gertjan Zwanikken
Kill It, Cook It, Eat It	2008	Firefly Productions; BBC (UK)
Hugh's Chicken Run	2008	KEO Films; Channel 4
Jamie's Fowl Dinners	2008	Firefly Productions; BBC (UK)
Food Inc.	2009	Robert Kenner
Planeat	2010	Shelley Lee Davies & Or Shlomi
LoveMEATender	2011	Manu Coeman
Farmageddon	2011	Kristin Canty
Forks over Knives	2011	Lee Fulkerson
Speciesism	2013	Mark Devries
Cowspiracy	2014	Kip Andersen & Keegan Kuhn
For the Love of Meat	2016	Matthew Evans; SBS (Australia)
Meat	2017	David White
Meat the Future	2019	Liz Marshall

2   The Persistence of Meat

Given that the production and consumption of meat is under the spotlight more than ever before, both scientifically and publicly, it might reasonably be expected that its status on the perceived hierarchy of foods might be seriously questioned. However, these same ten years have seen global production and consumption of meat increase steadily, on both a net and per capita basis. This means that meat consumption is continuing to increase faster than population growth.[9] The conservative estimation of over 65 billion food animals slaughtered every year for human consumption (almost 7.5 million per hour), more realistically estimated at 100–150 billion by other sources,[10] does not include 'acceptable' industry losses due to transport, contamination, disease, and other factors (Chambers and Grandin 2001; Greger 2007)—at least another 1.3 billion—nor the 33% of an estimated 90 billion marine animals killed annually for food that are used as feedstock (Alder et al. 2008). The FAO's (2014) predicted doubling translates to just under two billion more animals slaughtered per year, for every year until 2050. Other figures suggest an average 25 million more domestic ruminants have been added to the planet every year for the past 50 years (Ripple et al. 2014). Either way, considering the environmental resources that 65 billion food animals currently require, and the environmental impacts of raising them for meat, not to mention the ethical concerns, more effective ways to challenge the enduring (and increasing) use of animals as food are urgently needed.

Despite the growth in meat consumption, at the same time, recent media reports suggest that the numbers of vegans and vegetarians are increasing specifically in the UK, Australia, the US, and China (Watters 2015; Dean 2014; Quinn 2016; Anon 2016). In the UK, almost half of the increase is being observed in the younger demographic, aged between 15 and 34. Similarly, it is mainly China's teenagers and 'youth' that are

---

[9] There are shifts in the global meat market with growth slowing slightly in some nations and increasing in others, and changes in the types of meat being consumed. See Christine Chemnitz and Stanka Becheva, *Meat Atlas: Facts and Figures About the Animals We Eat*; also Rousseau 2016.

[10] See the Animal Kill Counter which is based on FAO statistics: http://www.adaptt.org/killcounter.html.

reportedly turning to a vegan diet, citing both health and ethical concerns (Anon 2014). It is impossible to predict what the future will look like for food animals—whether the number annually slaughtered for meat will continue to increase as statistically based trends indicate, or whether the numbers of people reportedly turning vegetarian or vegan will continue to grow and start to have an impact on these numbers. More important, then, is whether these self-reported vegetarians and vegans will actually remain so.

Contrary evidence has suggested that 84% of vegetarians and vegans return to eating animals (Herzog 2014). In addition, a number of books and a huge number of media articles document their author's or other peoples' rejection of vegetarianism or veganism, sometimes after several decades (Keith 2009; Nicholson 2012; Reed 2013; Younger 2015; Lennon 2007; Wheal 2013; Roberts 2016; English 2014; Woginrich 2011; Scott 2017; Blair 2017; Lein 2017; Moody 2018; Hunt 2018; Donaldson 2017). Admittedly, there are vegan structures and markets now in place, in 2019, that did not exist several years ago, so it is perhaps unreasonable to extrapolate from this earlier evidence. Ongoing research in this rapidly changing space is much needed. Nevertheless, as well as health issues, the emergence and availability of 'better' meat that is produced 'locally', 'ethically', and 'sustainably', and where animals are treated 'humanely' is cited as one of the main factors supporting a shift *away* from plant-based diets (Wheal 2013; Lennon 2007; Woginrich 2011; Tone 2014). These authors indicate, sometimes explicitly, that their original rejection of meat was because they found the *ways* in which animals are used in industrial systems problematic, and not their use in itself.

Indeed, notwithstanding the growing market for alternative protein in Western diets,[11] the 'ethicalisation' of the broader, mainstream meat industry seems to have been the primary outcome of increasing concerns for the environmental, health and ethical impacts of meat production

---

[11] During the course of conducting this research, the range of alternative meat products has become more extensive and attracted significant investment. This is, of course, another outcome of these concerns. However, as this research will highlight, there are persistent associations with 'real' meat, from living animals, and an associated resistance to 'unnatural' ways of producing 'meat' that I suspect will persist, regardless of how well new products mimic the 'real' thing (whether in look, taste, texture, nutrition, or other qualities). This also relates to the persistence of normalised understandings of protein, which are also beginning to be explored (Sexton 2018; Wilson 2019).

and consumption, circumventing, for many, any question of eschewing animals as food altogether. In 2014, Mcdonalds announced their intention to source "a portion" of their beef from accredited sustainable sources by 2016 and is now setting specific targets for ten countries including the US, Australia, New Zealand, and Canada (Katsnelson 2015; Beef Central 2018). Australia's two leading supermarkets, Coles and Woolworths, have also responded to an identified trend towards "more humane and ethical shopping habits" (Coles 2011) by introducing ethical, sustainable, and/ or responsible sourcing policies. These supermarkets do not explicitly label their products as being ethical or sustainable; it is rather implied through association with animal welfare organisations, such as the RSPCA, and reference to responsible sourcing policies and farming practices.[12] Fast-food chains Subway, Burger King, and KFC in Australia all variously state their commitment to sustainable and ethical sourcing, responsible farming, animal welfare, and the environment.

These moves by leading fast-food chains and supermarkets indicate that 'ethics' has been recognised as a commodifiable and economically valuable asset by big retailers keen to capitalise on this, as yet, additional market. By additional I mean that it has not so far been associated with a relative decline in the production and consumption of factory farmed meat. Any suspicion that maintaining profit margins is considered more important here than actual change could be well founded.

Explicitly prioritising perceptions over actual practices, in their *2010–2015 Strategic Plan*, Meat and Livestock Australia Limited (MLA), the leading industry service body that delivers marketing and research programmes for 47,500 of Australia's cattle, sheep, and goat producers, declared the development of a set of environment, welfare, and ethical standards to be an "imperative" for the industry so that "consumer *perceptions* of animal welfare and environmental issues do not become a major barrier to red meat consumption" (MLA 2010, emphasis added). Similarly, the Cattle Council of Australia, which represents the interests of Australia's beef cattle producers through their member organisations,

---

[12] However, in their account of supermarket wars and ethical consumption, Lewis and Huber note that in the case of Coles, it is the improved taste outcomes of these measures for consumers, rather than any alleged benefits for animals or the environment, that are emphasised (2015: 14).

states in its 2013 strategy framework document *Beef: 2015 and Beyond* that the "industry's environmental and ethical credentials *as perceived* by government, the community and its customers are of immediate and future importance" (CCA 2013: 12, emphasis added). That the director of McDonalds worldwide supply chain sustainability similarly stated, "The long term goal is to drive continuous improvement that *enhances and maintains the social license* to operate and sell more beef" (Katsnelson 2015, emphasis added), suggests that this is a strategic approach to *perception* with broad corporate appeal.

These various mainstream measures, effectively 'ethical light', are widely portrayed as offering a 'balanced' and positive way to address the environmental and ethical problems associated with large-scale meat production and consumption (Lennon 2007; Meryment 2011; Woginrich 2011)—or at least a step in the right direction (Healy 2014; Matsumoto 2016). It is assumed that by raising consumer awareness about the environmental and ethical consequences of our food, people will make better choices and the broader market will respond accordingly. However, returning to the question of whether and how this changes anything for cows, pigs, chickens, sheep, and goats, it seems that the cartography of their bodies—the food animal map—remains the same; it has merely gained some new locations, features, and landmarks, while some existing ones have been rearranged and rebranded. However, another response to concerns about the environmental and ethical impacts of industrial production processes, and also to concerns regarding the health impacts of the associated meat, is the shift to a different model of production altogether.

2.1 'Genuinely' Ethical and Sustainable Meat

Meat specifically labelled and promoted as being ethical and sustainable is a fairly recent but, as yet, largely unexamined emergence in the broader market for meat. Google Trends data indicate that searches for the terms 'ethical' or 'sustainable meat' started to register sporadically in Australia, the US, Canada, and the UK sometime before 2004 and have increased steadily since then. Currently in Australia, it is primarily a niche product with small, local producers supplying directly to consumers through farmers markets, online purchasing and delivery, farm-gate, and specialist retailers. However, it is a growing trend evidenced by reports of increasing demand for mobile butchers and slaughtermen in Australia, France, and the US (Mackenzie 2015; son 2016; Layton 2010).

Australian producers of 'ethical' and 'sustainable' meat claim that their products offer a more ethical and environmentally friendly alternative to mainstream, factory farmed meat, and consumer demand for their products is growing (Akerman 2010; Cansdale 2018; McCosker 2018). Consumers in the global north and Australasia appear enthusiastic about supporting this new trend,[13] again reflecting and contributing to the dominant view evident in popular and academic discourse that it represents, at the very least, a 'step in the right direction' (Cole 2011; Freeman 2010; LaVeck 2006; Faruqi 2015; Cuthbert 2016; Monaco 2017). However, this particular step—ethical and sustainable meat—is just one of several, besides genetic and technological innovations, that have emerged in opposition primarily to factory farming and not the production and consumption of meat per se. Essentially, challenges to the meat industry and to meat consumption based on environmental, health, and ethical concerns are being countered with efforts to produce healthier meat from happy, well-treated animals, using sustainable and even environmentally friendly farming practices. Meat promoted and labelled as ethical and sustainable promises the complete package, variously claiming that animals are reared in non-intensive, more humane conditions, that farmers practise 'regenerative' farming with less or no use of growth hormones and non-therapeutic antibiotics, using improved farm and waste management techniques, and encouraging lower consumption of 'better' meat.

---

[13] There are farms across the US, the UK, Canada, and New Zealand whose websites promote their variously ethical, sustainable, humane, high-welfare, 'mindful', and 'conscientious' products and overall approach to raising animals for meat. The trend is also recognised by big business: US food chains such as Chipotle now supply their customers with meat from 'happy' animals, and new standards and labelling schemes, which assure consumers that an animal led "a good life", received "higher welfare", and was "humanely treated", are becoming more widespread (e.g. the Global Animal Partnership's 5-step Animal Welfare Rating used by Whole Foods Market in the US, and the UK RSPCA's Freedom Food label).

However, the global demand for high quantities of cheaply produced meat persists while only a small proportion of discerning and privileged consumers participate in practices that prioritise alternative and 'better' meat products such as organic and free range, and even fewer in practices that involve specifically ethical and sustainable meat. This raises a number of questions that I cannot address in this book. For instance, in relation to the alleged mainstreaming of 'better' meat practices, how do such efforts interact with global market dynamics and the economic objectives of the animal-industrial complex? What is it that actually ends up being mainstreamed? In what ways is this meat then environmentally, socially, nutritionally, and/or ethically 'better'? And in relation to niche practices involving alternative and 'better' meat, what do these practices reveal about broader food politics? Do their associated systems of production and provision merely reflect and reproduce structural inequalities along lines of class, gender, and race (as well as species)? In what ways and for whom are they 'better'? Such questions highlight opportunities for further research to explore how different agendas are brought together and prioritised, and by what terms one 'system' is considered better than another. In other words, they emphasise the need to examine *the politics* involved in the politicisation of food.

3 Respecting 'Our' Resources

What the foregoing discussion indicates is that however much they are 'respected' through improved production methods, animals remain persistently edible despite damning and undeniable evidence of the environmental impacts of using them as food, and of the systems of domination that are used to control and ultimately kill them. It might be argued that this is largely on account of this evidence being filtered through various human-centric discourses[14] (scientific, economic, political, public), through the media, and then through everyday normalised practices, so that the majority of consumers remain unaware of the extent of these impacts, and the actual terms of our 'relationship' with food animals. The

---

[14] See Arcari (2017).

findings of a UK-based study (Bailey et al. 2014) suggest this is likely to be the case. Based on at least 12,000 survey responses from 12 countries, the researchers found "a significant lack of understanding about the links between livestock and climate change among publics—an awareness gap" (22).[15]

According to this line of thought, if consumers were to become more aware, that is, were provided with more information, then a transformation in practices would inevitably follow. However, a large, and arguably growing, proportion of consumers are gaining considerable knowledge and awareness—often first-hand—of exactly how meat is produced and of the associated issues. This trend is being shaped by, and also further shaping, the emergence of 'ethical' and 'sustainable' meat, of ethical or 'conscious' consumption in general, and the associated de- and re-fetishisation of production processes in favour of authenticity, tradition, transparency, and artisanal 'skills' (Pratt 2008; Parry 2010; Ocejo 2014; Matchar 2015). And yet, this knowledge and awareness has failed to fundamentally alter practices. Animals have remained enduringly edible in the face of all of these challenges. This is why I have chosen to focus on this specific model of meat consumption, being one that implicitly and explicitly foregrounds the problems associated with the mainstream meat industry and offers a point at which these are somehow satisfactorily resolved. How meat and food animals are reconstituted as part of this resolution will, I contend, reveal those persistently 'productive' aspects of humans' relationship with food animals that Foucault indicates are key to the successful exercise of power, and to its substantial concealment.

Navigating towards my theoretical approach, which I describe in the next chapter, the continuing treatment of food animals as a resource— just one of many 'natural' resources that the earth (or God) 'provides' for humans—can be viewed as part of a package of symptoms accredited to the 'human age' or the Anthropocene. This marks an age where

---

[15] The study was based on the results of an online survey conducted in "Brazil, China, France, Germany, India, Italy, Japan, Poland, Russia, South Africa, the UK and the US, with a minimum of 1000 participants in each country" (Bailey et al. 2014: 17).

humans stand apart from all other animals and planetary processes in the ways we are "remaking the planet" (Crutzen 2002; Monastersky 2015; Jamieson and Nadzam 2015: 10). Not yet formalised, the onset of the Anthropocene is generally placed around the time of the Industrial Revolution and is associated with a barrage of environmental changes attributed to human activities, including the hole in the ozone layer, a 30% increase in CO_2 (to levels not seen in 400,000 years), accelerating rates of erosion, the progressive loss of Arctic Ice, loss of habitats and biodiversity, and associated extinctions (Monastersky 2015: 145; WWF 2017). Most, if not all, of these changes have been caused by humans' increasing effectiveness at mining the planet's perceived 'resources'. The cartographic eye used in this endeavour is decidedly anthropocentric. Water, forests, land, air space, plant-life, and (nonhuman) animals—all are designated terra (and bestia) nullias until recognised under this eye whereupon the 'resources' emerge, such as gas, coal, oil, minerals, wood, and meat. Human value makes 'things' infinitely accessible, 'own-able', and mappable.

Rather than simply describing and demonstrating how meat consumption and the use of animals for food fit the criteria of the Anthropocene, I am interested in what underpins this age of unbridled extraction and exploitation. For the sensibility which maintains animals as persistently edible bears similarities to that which supports humans in their sovereign right over the entire planet. If the basis of this fundamental entitlement, which is manifest through mechanisms of power and domination, can be understood, then perhaps the 'map' and the 'word' can be seen for what they are, that is, neither the territory nor the thing. Thus freed from the constraints of normalisation and naturalisation, the perception of broader territories and 'things' might then be possible. In prioritising this endeavour, what is at stake, as John Sanbonmatsu argues, "is not simply a set of eating guidelines, but a total critique of society—of a way of life that has become inimical to life" (2014a). In the next chapter, I suggest that a critical posthumanism offers a way to imagine, and enact, this emancipatory vision.

References

Aird, H. (2017, July 5). Delay in Abattoir Animal Cruelty Investigation Angers Welfare Groups. *ABC News.*

Akerman, P. (2010, November 16). Ethical Beef Eaters Move against the Herd. *The Australian.* Online. January 2014.

Alder, J., Campbell, B., Karpouzi, V., Kaschner, K., et al. (2008). Forage Fish: From Ecosystems to Markets. *Annual Review of Environment and Resources, 33*(1), 153–166.

Anon. (2014, February 1). China's Vegetarian Population Touches 50 Million: Report. *Times of India.* Online. February 2017.

Anon. (2016, August 17). Study: The Lamb Ads Aren't Working as Aussies Turn Vego in Droves. *B&T Magazine.*

Arcari, P. (2017). Normalised, Human-Centric Discourses of Meat and Animals in Climate Change, Sustainability and Food Security Literature. *Agriculture and Human Values, 34*(1), 69–86.

Bailey, R., Froggatt, A., & Wellesley, L. (2014). *Livestock – Climate Change's Forgotten Sector* (Chatham House: Research Paper). London: The Royal Institute of International Affairs.

Barker, T., Bashmakov, I., Bernstein, L., Bogner, J. E., et al. (2007). Technical Summary. In B. Metz, O. R. Davidson, P. R. Bosch, R. Dave, et al. (Eds.), *Climate Change 2007: Mitigation. Contribution of Working Group III to the Fourth Assessment Report of the Intergovernmental Panel on Climate Change.* Cambridge; New York: Cambridge University Press.

Bar-On, Y. M., Phillips, R., & Milo, R. (2018). The Biomass Distribution on Earth. *Proceedings of the National Academy of Sciences, 115*(25), 6506–6511.

Baroni, L., Cenci, L., Tettamanti, M., & Berati, M. (2006). Evaluating the Environmental Impact of Various Dietary Patterns Combined with Different Food Production Systems. *European Journal of Clinical Nutrition, 61*(2), 279–286.

Beef Central. (2018, April 2). *Strong Early Results Claimed in Canada's Certified Sustainable Beef Model.* Beef Central.

Beirne, P. (2004). From Animal Abuse to Interhuman Violence? A Critical Review of the Progression Thesis. *Society and Animals, 12*(1), 39–65.

Blair, O. (2017, October 18). 6 Former Vegetarians Explain Why They Started Eating Meat Again. *Cosmopolitan.*

Burke, K. (2011, May 31). Shocking Slaughterhouse Abuse Sparks Investigation - Comments. *The Sydney Morning Herald.* Online. January 2014.

Cannane, S. (2012, February 9). NSW Abattoir Closed over Slaughter Practices. *ABC News Lateline*. Online. January 2014.

Cansdale, D. (2018, June 17). Ethically-Sourced Meat: Kill Your Own Pig and Reap the Tasty Reward. *ABC News*.

Carlet, J., Jarlier, V., Harbarth, S., Voss, A., et al. (2012). Ready for a World without Antibiotics? The Pensières Antibiotic Resistance Call to Action. *Antimicrobial Resistance and Infection Control, 1*(1), 11.

Carlyon, P. (2016, October 14). Tasmanian Abattoir Investigated over Animal Cruelty Claims. *ABC News*. Online. February 2017.

CCA. (2013). *Beef 2015 & Beyond*. Cattle Council of Australia. Report.

Chambers, P. G., & Grandin, T. (2001). *Guidelines for Humane Handling, Transport and Slaughter of Livestock*. Food and Agricultural Organisation of the United Nations and the Humane Society International. Regional Office for the Asia Pacific.

Clune, S., Crossin, E., & Verghese, K. (2017). Systematic Review of Greenhouse Gas Emissions for Different Fresh Food Categories. *Journal of Cleaner Production, 140*(Part 2), 766–783.

Cole, M. (2011). From "Animal Machines" to "Happy Meat"? Foucault's Ideas of Disciplinary and Pastoral Power Applied to 'Animal-Centred' Welfare Discourse. *Animals, 1*(4), 83–101.

Craig, W. J., & Mangels, A. R. (2009). Position of the American Dietetic Association: Vegetarian Diets. *Journal of the American Dietetic Association, 109*(7), 1266–1282.

Crutzen, P. J. (2002). Geology of Mankind. *Nature, 415*(3 January), 23.

Cuthbert, D. (2016, September 21). Mobile Abattoir Improves Animal Welfare and Meat Quality in France. *Food & Drink International*.

Dean, T. (2014, January 13). China's Vegan Population Is Largest in the World. *VegNews Daily*. Online. February 2017.

Donaldson, S. (2017, December 31). Five Years a Vegetarian - and Now I'm Back Eating Meat. *The Independent*.

English, N. (2014, January 17). I Was a Vegetarian for Five Years and Today I Slaughtered a Chicken. *Greatist*. Online. June 2015.

Evershed, N., & Wahlquist, C. (2018, April 10). Live Exports: Mass Animal Deaths Going Unpunished as Holes in System Revealed. *The Guardian*.

FAO. (2014, November). *Meat & Meat Products*. Food and Agricultural Organisation of the United Nations: Animal Production and Health. Online. June 2015.

Faruqi, S. (2015). *Project Animal Farm: An Accidental Journey into the Secret World of Farming and the Truth About Our Food*. Pegasus Books.

Ferguson, S. (2011, 30 May). A Bloody Business. *abc.net.au*. Online. January 2014.

Feskens, E. J. M., Sluik, D., & van Woudenbergh, G. J. (2013). Meat Consumption, Diabetes, and Its Complications. *Current Diabetes Reports, 13*(2), 298–306.

Fiala, N. (2008). Meeting the Demand: An Estimation of Potential Future Greenhouse Gas Emissions from Meat Production. *Ecological Economics, 67*(3), 412–419.

Fitzgerald, A. J., Kalof, L., & Dietz, T. (2009). Slaughterhouses and Increased Crime Rates: An Empirical Analysis of the Spillover from 'The Jungle' into the Surrounding Community. *Organization and Environment, 22*(2), 158–184.

Freeman, C. P. (2010). Meat's Place on the Campaign Menu: How US Environmental Discourse Negotiates Vegetarianism. *Environmental Communication: A Journal of Nature and Culture, 4*(3), 255–276.

Giovannucci, D., Scherr, S., Nierenberg, D., Hebebrand, C., et al. (2012). *Food and Agriculture: The Future of Sustainability*. A Strategic Input to the Sustainable Development in the 21st Century (SD21) Project. New York: United Nations Department of Economic and Social Affairs, Division for Sustainable Development. 80 pp.

Goodland, R., & Anhang, J. (2009, November/December). Livestock and Climate Change. *Worldwatch*.

Greger, M. (2007). The Long Haul: Risks Associated With Livestock Transport. *Biosecurity and Bioterrorism: Biodefense Strategy, Practice, and Science, 5*(4), 301–312.

Harwatt, H., & Hayek, M. (2019). *Eating Away at Climate Change with Negative Emissions: Repurposing UK Agricultural Land to Meet Climate Goals*. Cambridge, MA: Harvard Law School, Animal Law and Policy Program.

zur Hausen, H. (2012). Red Meat Consumption And Cancer: Reasons to Suspect Involvement of Bovine Infectious Factors in Colorectal Cancer. *International Journal of Cancer (Journal international du cancer), 130*(11), 2475–2483.

Healy, S. (2014). Animal Farming in Australia: Consumer Awareness, Concern and Action. In G. L. Burns & M. Paterson (Eds.), *Engaging with Animals: Interpretations of a Shared Existence* (pp. 185–204). Sydney: Sydney University Press.

Herzog, H. (2014, December 2). 84% of Vegetarians and Vegans Return to Meat. Why? *Psychology Today*. Online. February 2017.

Hunt, B. (2018, February 1). Stories from People Who Have Given Up Vegan Diets. *benhunt.com*.

IAASTD. (2009). Agriculture at a Crossroads: Global Report. In B. D. McIntyre, H. R. Herren, & J. Wakhungu (Eds.), *International Assessment of Agricultural Knowledge, Science and Technology for Development*. Washington: Island Press. 590 pp.

IPCC. (2007). *Climate Change 2007: Synthesis Report* (Core Writing Team, Bernstein et al., Eds.). Geneva: Intergovernmental Panel on Climate Change. 73 pp.

Jamieson, D., & Nadzam, B. (2015). *Love in the Anthropocene*. New York; London: OR Books.

Joshi, A. D., Corral, R., Siegmund, K. D., Haile, R. W., et al. (2009). Red Meat and Poultry Intake, Polymorphisms in the Nucleotide Excision Repair and Mismatch Repair Pathways and Colorectal Cancer Risk. *Carcinogenesis, 30*(3), 472–479.

Katsnelson, A. (2015, January 13). Will McDonald's "Sustainable Beef" Burgers Really Be any Better? *The Guardian*. Online. June 2015.

Keith, L. (2009). *Vegetarian Myth: Food, Justice, and Sustainability*. Oakland: PM Press.

Koneswaran, G., & Nierenberg, D. (2008). Global Farm Animal Production and Global Warming: Impacting and Mitigating Climate Change. *Environmental Health Perspectives, 116*(5), 578–582.

LaVeck, J. (2006, September 8–11). Compassion for Sale? Doublethink Meets Doublefeel as 'Happy Meat' Comes of Age. *Satya*.

Laxminarayan, R. (2002). How Broad Should the Scope of Antibiotics Patents Be? *American Journal of Agricultural Economics, 84*(5), 1287–1292.

Laxminarayan, R., Duse, A., Wattal, C., Zaidi, A. K. M., et al. (2013). The Lancet Infectious Diseases Commission: Antibiotic Resistance—the Need for Global Solutions. *The Lancet, 13*(12), 1057–1098.

Layton, L. (2010, June 20). As Demand Grows for Locally Raised Meat, Farmers Turn to Mobile Slaughterhouses. *The Washington Post*.

Lein, A. (2017). After Being Vegan for 3 Years, I Went Back to Meat. And This Happened… Healthline. Online, December 19.

Leip, A., Billen, G., Garnier, J., Grizzetti, B., et al. (2015). Impacts of European Livestock Production: Nitrogen, Sulphur, Phosphorus and Greenhouse Gas Emissions, Land-Use, Water Eutrophication and Biodiversity. *Environmental Research Letters, 10*(11), 115004.

Lennon, C. (2007, August). Why Vegetarians Are Eating Meat. *Food and Wine*. Online. July 2015.

Lewis, T., & Huber, A. (2015). A Revolution in an Eggcup? Supermarket Wars, Celebrity Chefs, and Ethical Consumption. *Food, Culture and Society: An International Journal of Multidisciplinary Research, 18*(2), 289–307.

Mackenzie, M. (2015, January 17). Mobile Butchers in Demand as Consumers Seek Fresh, Ethically Slaughtered Meat. *ABC Radio National*. Online. June 2015.

March, S. (2011, November 30). Abattoir Closed Over Animal Cruelty Concerns. *ABC News*. Online. February 2017.

Martin, M. J., Thottathil, S. E., & Newman, T. B. (2015). Antibiotics Overuse in Animal Agriculture: A Call to Action for Health Care Providers. *American Journal of Public Health, 105*(12), 2409–2410.

Matchar, E. (2015). *Homeward Bound: Why Women Are Embracing the New Domesticity*. New York: Simon & Schuster.

Matsumoto, N. (2016, June 14). The Grassfed Burger Gap. *Civil Eats*. Online. August 2016.

McCosker, A. (2018, May 8). Beef Australia 2018: What Lies Ahead for the Industry as 'Locavores' and Digital Disruption Loom Closer. *ABC News: Rural*.

McKie, R. (2014, December 14). Earth Faces Sixth 'Great Extinction' with 41% of Amphibians Set to Go the Way of the Dodo. *The Guardian*. Online. December 2014.

McMichael, A. J., Powles, J. W., Butler, C. D., & Uauy, R. (2007). Food, Livestock Production, Energy, Climate Change, and Health. *The Lancet, 370*(9594), 1253–1263.

Mekonnen, M. M., & Hoekstra, A. Y. (2012). A Global Assessment of the Water Footprint of Farm Animal Products. *Ecosystems, 15*(3), 401–415.

Meryment, E. (2011, June 5). Sickened Meat-Lovers Turn to 'Ethical' Beef. *The Sunday Telegraph*. Online. April 2013.

MLA. (2010). Strategic Plan 2010–2015. *Meat and Livestock Australia*. 48 pp.

Monaco, E. (2017, April 28). Why It's More Important to Be an Ethical Omnivore than a Vegetarian. *Good Housekeeping*.

Monastersky, R. (2014, December 10). Biodiversity: Life – a Status Report. *Nature: News Feature*. Online. December 2014.

Monastersky, R. (2015). The Human Age. *Nature, 519*(7542), 144–147.

Moody, L. (2018, May). 8 Real People Share Why They Stopped Being Vegan. *Mindbodygreen*.

NHMRC. (2013). *Australian Dietary Guidelines.* Australian Government, National Health and Medical Research Council. Department of Health and Aging. Canberra: Commonwealth of Australia. 226 pp.

Nicholson, J. (2012). *The Meat Fix: How a Lifetime of Healthy Eating Nearly Killed Me!* Biteback Publishing.

Noone, R. (2014, April 2). Video Shows Abattoir Staff Abusing Pigs. *The Daily Telegraph.* Online. February 2017.

Ocejo, R. E. (2014). Show the Animal: Constructing and Communicating New Elite Food Tastes at Upscale Butcher Shops. *Poetics, 47,* 106–121.

Pan, A., Sun, Q., Bernstein, A., Schultze, M. B., et al. (2012). Red Meat Consumption and Mortality. *Archives of Internal Medicine, 172*(7), 555–563.

Parry, J. (2010). *The New Visibility of Slaughter in Popular Gastronomy.* Masters Thesis, Cultural Studies: University of Canterbury.

Pimentel, D., & Pimentel, M. H. (2007). *Food, Energy, and Society.* London; New York: Taylor & Francis.

Pimentel, D., Pimentel, D., Pimentel, M., & Pimentel, M. (2003). Sustainability of Meat-Based and Plant-Based Diets and the Environment. *The American Journal of Clinical Nutrition, 78*(3), 660S–663S.

Pratt, J. (2008). Food Values: The Local and the Authentic. In G. De Neve, P. Luetchford, J. Pratt, & D. C. Wood (Eds.), *Research in Economic Anthropology* (pp. 53–70). Bingley: Emerald Group Publishing Limited.

Probyn-Rapsey, F. (2017). *The Cultural Politics of Eradication.* Conference Keynote Lecture. Seventh Australian Animal Studies Association Conference: Animal Intersection. 3–5 July, Adelaide, Australia.

Quinn, S. (2016, April 18). Number of Vegans in Britain Rises by 360% in 10 Years. *The Telegraph.* Online. February 2017.

Reed, B. (2013). *The Ethical Butcher: How Thoughtful Eating Can Change Your World.* Berkeley, CA: Soft Skull Press.

Richards, E., Signal, T., & Taylor, N. (2013). A Different Cut? Comparing Attitudes toward Animals and Propensity for Aggression within Two Primary Industry Cohorts—Farmers and Meatworkers. *Society and Animals, 21*(4), 395–413.

Ripple, W. J., Smith, P. T., Haberl, H., Montzka, S. A., et al. (2014). Ruminants, Climate Change and Climate Policy. *Nature Climate Change, 4,* 2–5.

Ritchie, H., & Roser, M. (2017). Meat and Seafood Production & Consumption. *OurWorldInData.org.* Online. April 2018.

Roberts, A. M. (2016, May 26). The True Confessions of an Ex-Vegan. *POPSUGAR.* Online. February 2017.

Robinson, A. (2016, October 22). 'Ethical' Meat Production Concept Launched in France. *Farmers Guardian*.

Robinson, T. P., Bu, D. P., Carrique-Mas, J., Fèvre, E. M., et al. (2016). Antibiotic Resistance Is the Quintessential One Health Issue. *Transactions of The Royal Society of Tropical Medicine and Hygiene, 110*(7), 377–380.

Rousseau, O. (2016, April 13). Food Trends: Meat Consumption up, Beef Declines. *GlobalMeatnews.com*. Online. February 2017.

Sanbonmatsu, J. (2014a). Interview with John Sanbonmatsu (S. Rodriguez, Ed., pp. 1–12). Online. Direct Action Everywhere.

Scarborough, P., Appleby, P. N., Mizdrak, A., Briggs, A. D. M., et al. (2014). Dietary Greenhouse Gas Emissions of Meat-Eaters, Fish-Eaters, Vegetarians and Vegans in the UK. *Climatic Change, 125*(2), 179–192.

Schwarzer, S., Witt, R., & Zommers, Z. (2012, October 1–10). Growing Greenhouse Gas Emissions Due to Meat Production. *UNEP Global Environmental Alert Services (GEAS)*. Online. June 2013.

Scott, J. (2017, September 5). Why I Gave Up Being Vegan. *BBC News*.

Sexton, A. E. (2018). Eating for the Post-Anthropocene: Alternative Proteins and the Biopolitics of Edibility. *Transactions of the Institute of British Geographers, 43*(4), 586–600.

Silbergeld, E. K., Graham, J., & Price, L. B. (2008). Industrial Food Animal Production, Antimicrobial Resistance, and Human Health. *Annual Review of Public Health, 29*(1), 151–169.

Smith, P., Martino, D., Cai, Z., Gwary, D., et al. (2007). Agriculture. In *Climate Change 2007: Mitigation. Contribution of Working Group III to the Fourth Assessment Report of the IPCC*. Cambridge, UK; New York: Cambridge University Press.

Smith, F. A., Smith, R. E. E., Lyons, K., & Payne, J. L. (2018). Body Size Downgrading of Mammals over the Late Quaternary. *Science, 360*, 310–313.

Spellberg, B., Guidos, R., Gilbert, D., Bradley, J., et al. (2008). The Epidemic of Antibiotic-Resistant Infections: A Call to Action for the Medical Community from the Infectious Diseases Society of America. *Clinical Infectious Diseases, 46*(2), 155–164.

Springmann, M., Clark, M., Mason-D'Croz, D., Wiebe, K., et al. (2018). Options for Keeping the Food System within Environmental Limits. *Nature, 562*(7728), 519–525.

Steinfeld, H., Gerber, P., Wassenaar, T., Castel, V., et al. (2006). *Livestock's Long Shadow: Environmental Issues and Options*. Rome: Food and Agricultural Organisation of the United Nations. 416 pp.

Stull, D. D., & Broadway, M. J. (2012). *Slaughterhouse Blues: The Meat and Poultry Industry in North America*. Belmont, CA: Wadsworth/Thompson.

Sullivan, K. (2018, March 7). Riverside Meats Closes Echuca Abattoir as Probe Continues. *The Weekly Times*.

Thomas, M. D. (2016, November 24). Secret Video Inside Victoria's Riverside Meats Abattoir Reveals Shocking Abuse. *The Age*. Online. February 2017.

Tone, E. (2014, August 26). Why I Stopped Being a Vegetarian: The Harm-Free Way to Eat Meat. *Elite Daily*.

UK Veterinary Medicines Directorate. (2019). *UK One Health Report – Joint Report on Antibiotic Use and Antibiotic Resistance, 2013–2017*. New Haw, Addlestone: Veterinary Medicines Directorate.

UNEP. (2012). Avoiding Future Famines: Strengthening the Ecological Foundation of Food Security through Sustainable Food Systems. *United Nations Environment Programme (UNEP)*. Nairobi, Kenya: UNEP.

UNEP. (2018, November 8). What's in Your Burger? More than You Think. *UN Environment*.

Vergnaud, A. C., Norat, T., Romaguera, D., Mouw, T., et al. (2010). Meat Consumption and Prospective Weight Change in Participants of the EPIC-PANACEA Study. *American Journal of Clinical Nutrition, 92*(2), 398–407.

Vidot, A., & Conifer, D. (2016, May 20). Investigation Launched into Claims Australian Cattle Were Slaughtered with Sledgehammer in Vietnam. *ABC News*. Online. July 2016.

Wahlquist, C. (2017, August 10). WA Seeks Powers to Prosecute Live Exporters after 3,000 Sheep Die on Ship. *The Guardian*. Online. April 2018.

Waite, R., & Vennard, D. (2018, October 17). Without Changing Diets, Agriculture Alone Could Produce Enough Emissions to Surpass 1.5°C of Global Warming. *World Resources Institute*.

Walker, P., Rhubart-Berg, P., McKenzie, S., Kelling, K., et al. (2007). Public Health Implications of Meat Production and Consumption. *Public Health Nutrition, 8*(4), 348–356.

Walls, J. (2016, November 24). Echuca Abattoir Should Be 'Shut Down Immediately'. *Bendigo Advertiser*. Online. February 2017.

Wang, Y., & Beydoun, M. A. (2009). Meat Consumption Is Associated with Obesity and Central Obesity among US Adults. *International Journal of Obesity, 33*(6), 621–628.

Watters, N. (2015). 16 Million People in the US Are Now Vegan or Vegetarian! *The Raw Food World*. Online. February 2017.

Wellesley, L., Happer, C., & Froggatt, A. (2015). *Changing Climate, Changing Diets*. p. 64.

Wheal, C. (2013, May 24). Animal Farm. *Chriswheal.com.* Online. February 2017.

Wilson, B. (2019, January 4). Protein Mania: The Rich World's New Diet Obsession. *The Guardian.*

Woginrich, J. (2011, January 20). My Beef Isn't with Beef: Why I Stopped Being a Vegetarian. *The Guardian.* Online. August 2016.

Worldwatch Institute. (2013). *Grain Harvest Sets Record, But Supplies Still Tight.* Washington, DC: Worldwatch Institute. Product No. VST 101. Online. December 2014.

Worldwatch Institute. (2015). *Global Meat Production and Consumption Continue to Rise.* Washington, DC: worldwatch.org. Online. June 2015.

WWF. (2017). Appetite for Destruction.

Younger, J. (2015). *Breaking Vegan.* Fair Winds Press.

3

Theoretical Framework: Advancing and Enacting a Critical Posthumanism

Critical posthumanism, as described by Cudworth (2005, 2011) and by Cudworth and Hobden (2014b), addresses what these and other authors perceive as a prevailing human-centrism in certain accounts of posthumanism (Giraud 2013; Weisberg 2009). Critical posthumanism provides the overarching, and ongoing, theoretical impulse for my research, while the practical theoretical and analytical framework is inspired primarily by Foucault in combination with social practice theories. For the remainder of this chapter, I provide a broadly contextualised explication of critical posthumanism before introducing Foucault's work and clarifying how it aligns with this explicitly emancipatory approach. I then articulate what Foucault's theorisations on power, knowledge, language, discipline, sexuality, and ethics contribute to a study of meat's persistence in social practices, and how I apply these to critically explore the persistent edibility of food animals among self-identified producers and consumers of ethical and sustainable meat.

© The Author(s) 2020   **55**
P. Arcari, *Making Sense of 'Food' Animals*,
https://doi.org/10.1007/978-981-13-9585-7_3

1   A More Robust Tool

Persistent constitutions of meat and food animals are firmly situated in networks of relations characterised by the systemic domination of nature by humans (Cudworth 2005: 45). The historical effects of this human-centric system of natured domination are of such magnitude that they have been accorded their own geological epoch, the Anthropocene, where "humans and our societies have become a global geophysical force" (Steffen et al. 2007: 614; also Rickards 2015). Indeed, archaeologists have proposed that the 'broiler' chicken represents a marker species for our reconfigured biosphere, based on their human-induced biological changes and sheer numbers (Bennett et al. 2018).

Using accounts of the Anthropocene as a theoretical fulcrum, I draw on Cudworth's (2005, 2011) unique perspective on the term, which she uses to underscore the contribution of ecofeminism to understanding "how *societies* are organized with respect to 'nature'" (2005: 1, emphasis added). In this light, while the Anthropocene has become widely accepted as a definition of our current geological location, and a conceptual frame through which to "understand the evolving human-environment relationship" (Biermann et al. 2016: 341), certain limitations have been noted. For one, Biermann et al. argue that the Anthropocene "risks being framed and understood in a way that is too 'global' and monolithic, neglecting persistent social inequalities and vast regional differences" (2016: 342). Consequently, the term "does not refer to a passage *out* of certain social conditions" (Alberts 2011: 5; see also Bai et al. 2016). In other words, the causes and effects of, and potential alternatives to, the Anthropocene still need to be clearly articulated. Lövbrand et al. similarly observe that a focus on environmental, rather than social, change has seen attention "diverted away from the social and cultural norms, practices and power relations that drive environmental problems in the first place" (2015: 212). It is for similar reasons that Haraway (2016: 47) commits to using the term "sparingly", proposing the Capitalocene as a more accurate designation of this period in our history.

Recognising the potential power of the Anthropocene as a narrative for change, these and other authors are challenging its tendency towards universality and emphasising its social and cultural dimensions (Hickmann et al. 2018). Without this level of critical analysis, it risks being subsumed in the 'non-political politics' of climate change (Swyngedouw 2013) and thereby remaining one step removed from the "radical change in our socio- political co-ordinates" (Swyngedouw 2015: 137) that is required to realise a future different from the one towards which we are heading.

For these reasons, the Anthropocene can be deployed in a way that seems descriptive and fairly neutral, and while its co-noun 'anthropocentrism' denotes the inherent prejudice (of human-centredness), it is perceived as being "rather weak for the capture of more direct aspects of human domination" (Cudworth 2005: 64). Both terms are therefore somewhat limited as robust conceptual tools to imagine and enact social change (ibid.: 64–70). As Cudworth later explains:

> 'anthropocentrism' has been used so broadly to describe human centred attitudes, beliefs and behaviours, that its link to a system of social organization has been lost. In addition, 'centrism' is inadequate to rise to the challenge of interrogating forms of exploitation, and in the case of some non-human animals, oppression. (2011: 68)

In response to the political weakness of 'centrism' (Cudworth 2008: 34), Cudworth proposes 'anthroparchy' as a more meaningful and useful concept, as it explicitly foregrounds the bases of human's dualistic orientation towards 'nature' (2005, 2008, 2011). Anthroparchy directly implicates "different forms and practices of power: oppression, exploitation and marginalization" (Cudworth 2008: 34) with the emphasis on 'the environment' as what is dominated in the anthroparchal society (ibid.):

> Anthroparchy is a formation of social relationships in which non-human nature is cast as a series of resources for human ends, and in which human interests inform the systemic ordering of social control over the environment. (2011: 68)

Anthroparchy therefore relies on a fundamental dualism between the human and the 'environment', and dualisms have long been identified as constituting the roots of domination and oppression, whether relating to gender, 'race', social class, religion, ability, or any kind of distinction that delineates an 'other'. Overcoming these and other dualisms is perceived as a way through and out of the environmental, social, political, and ethical challenges the human race faces today, and scholars (e.g. Foucault 1989; Deleuze and Guattari 1987; Latour 1993; Merleau-Ponty 2002; Carson 1962; Plumwood 1994; Braidotti 2013, to name just a few) have been applying themselves to this task ever since Nietzsche's seminal *Beyond Good and Evil*, written in 1886. However, with regard to more recent scholarly activity in fields relevant to my research, there are certain anti-dualistic approaches and theories that stand out for addressing questions relating to nonhuman animals and/or meat. These can be found especially in the work of ecofeminists such as Val Plumwood (1994, 2002, 2013a), Carolyn Merchant (1990, 2005) and Karen Warren (1997, 2000), and posthuman theorists such as Donna Haraway (1990, 2008), Sarah Whatmore (2002, 2013), and Cary Wolfe (2010). There are no dividing lines between my over-simplified divisions—ecofeminism can be posthumanist and vice versa, and the work of associated scholars similarly defies strict categorisation. However, the range of perspectives I am most interested in is unified by the shared intent to "productively trouble habitual ways of thinking and acting on both academia and everyday life" (Castree and Nash 2004: 1342).

This quote from Castree and Nash, which I also flagged in my introduction in relation to the role of heterotopia, serves well as the overriding intent of this book—**to trouble habitual ways of thinking and acting that involve food animals**. But before I explain the relevance of these anti-dualistic and posthuman perspectives to my research, I will first highlight their *potential* pitfalls. These are not conceived as faults of these perspectives per se, but are rather a question of how far they 'trouble habitual thinking'—the answer being, I argue, not far enough. To that end, I will demonstrate how non-dualistic and non-/posthuman perspectives often elide material conditions of inequality, thereby exposing "residual forms of essentialism lurking behind apparently nonessentialist forms of analysis" (Sedgwick 2003: 8).

2 An Obstinate Duality

As mentioned, dualistic and hierarchical conceptions of human/nature (nonhuman), culture/nature, male/female, and so on (see Plumwood 1994: 43; and Cudworth and Hobden 2014a: 757) have been soundly critiqued, and a range of anti- or non-dualistic and posthuman perspectives seek to move beyond such ultimately false binaries, at least in theory (Plumwood 2004; Haraway 1990, 2008; Whatmore 2002; Barad 2007). However, lived reality for food animals, and many 'others' who find themselves on the wrong (i.e. right hand) side of these binaries, reflects and reinforces profound dualisms that pervade associated practices and which need to be attended to empirically, before (or while also) thinking beyond them. Consequently, I agree with Weisberg (2009) and Cudworth (2014) in considering most anti- or non-dualistic and posthuman perspectives to be essentially idealist and therefore not sufficiently attentive to systemic dualisms to offer a useful framework for the exploration and analysis of unequal relations that are universally normalised, for example, as in power blocks or states of domination. It is the same sense in which Cudworth perceives studies of anthropocentrism to lack political teeth.

Some anti-/non-dualistic and posthuman perspectives elide any question of animals altogether, subsuming them within the term 'nature' (Warren 1997; Barad 2003). Similarly, Plumwood's overall oeuvre offers an invaluable critique of normative dualisms that underpin the colonisation of nature, women, and people, and yet she actively refutes similarly critical engagements with questions of human-animal relations such as those undertaken by ecofeminist scholars, including Carol Adams, Lori Gruen, Greta Gaard, and Marti Kheel (Plumwood 2004, 2013b; see also Alloun 2015). Still other posthuman and 'hybrid'[1] scholars, notably Whatmore (2002) and Haraway (2008), recognise what is unique and interesting about human relations with animals as opposed to generalised

---

[1] Citing Gillian Rose, Mei-Po Kwan explains, "Hybrids 'transgress and displace boundaries between binary divisions and in so doing produce something ontologically new,' [...] hybrid geographies are geographical practices (or 'boundary projects') that challenge the boundary and forge creative connections between social-cultural and spatial-analytical geographies" (2004: 758). Whatmore describes "hybrid mappings" that "emphasise the multiplicity of space-times generated in/by movements and rhythms of heterogeneous association" (2002: 6).

'nature'. The human-animal dualism is the main target in this body of literature, which argues for continuity rather than discontinuity and an appreciation of the porosity of borders, much as continental philosophers concerned with the 'question of the animal' have argued (Calarco 2008). However, I argue below that the agenda being advanced in these anti-/ non-dualistic and posthuman imaginings is one that is still decidedly human-centric, or, more accurately, anthroparchal.

Making an admittedly simplified generalisation across these works, while these enquiries into the human/animal distinction are framed as attempts to dismantle the universal anthropocentrism seen as pervading our discourses and institutions (Calarco 2008: 9), the main concerns invariably emerge as being about conceptions and definitions of 'animality', 'the Animal', 'humanness', 'the human subject', 'the self'. For example, the 'becoming animal' of the *human* being (Deleuze and Guattari 1987); discovering the animal that '*I*' am (Derrida 2008); the implications of the anthropological machine for *human beings* (Calarco 2008: 102, commenting on Agamben's writings; see also Chrulew 2012); or what *humans* have to gain from "becoming worldly" and embracing a "dance of encounters" (Haraway 2008: 296, 4). The (de/re)construction of what it means to be human seems to be the underlying focus of these projects with less, if any, attention paid to what this all might mean for the *animals* (as opposed to 'the Animal' as an aggregated and reified object) currently living a vividly and persistently anthropocentric, or anthroparchic, reality. Posthuman enquiries that focus on the *notion* of the animal, rather than animals themselves, thus take on a "radically passive dimension" (Chrulew, describing Agamben's The Open 2012: 58). In the case of Agamben, Calarco writes, "Agamben's writings…focus entirely and exclusively on the effects of the anthropological machine on *human beings* and never explore the impact the machines has on various forms of animal life" (2008: 102, emphasis in original). Sorenson also singles out the work of Derrida and Haraway for its questionable contribution and "bizarre" prominence within animal studies, arguing it offers little to animals, and that Haraway's is actually opposed to animal advocacy (2014: xix).

The anti-/non-dualistic and posthuman scholars I mention certainly vary in their level of critical engagement with different kinds of human-animal relations, Derrida (in Derrida and Roudinesco 2004) being notable here for at least confronting the issue of animal rights and the use of animals as food and entertainment. Yet there is a common reluctance to admit, let alone question, the fundamental circumstances and terms of human relations of animal use. As Giraud succinctly observes, "posthumanist relations towards animals end at the moment of consuming animal products" (2013: 50). Hence, although not all anti-/non-dualistic and posthuman perspectives are guilty of the "discursive wizardry" that Weisberg (2009: 37) and others[2] have identified in Haraway's oeuvre, I do agree with her assertion that "Haraway's work has become paradigmatic of a depoliticised approach within Animal Studies". To paraphrase Helena Pederson (2011: 75), this approach serves to obscure, dilute, or displace responsibility for animals' situation, reinforcing, rather than dismantling, their exploitation.

While anti-/non-dualistic and posthuman perspectives provide an interesting and productive line of theoretical enquiry, it is one that can, at least in the case of animals, appear premature and not sufficiently attentive to the multitude of dualisms that are still alive and well across social practices. However, unlike Weisberg, many critical animal studies (CAS) scholars, including Twine (2010b), Pederson (2011), Cudworth and Hobden (2014b), Giraud (2013), Deckha (2012), and Alloun (2015) take a more positive orientation towards posthumanism, seeing the recent, popular strands of thought as just one interpretation of a fundamentally useful intent to develop a non-anthropocentric ethic. Marrying this intention with the social and political critique that defines critical theory, and also CAS, generates a critical posthumanism that seeks to resolve theoretical tensions and potentially offers ways to materially enact and realise the anti-anthropocentric (or anti-anthroparchic) project that remains largely idealistic in posthuman theory (Giraud 2013; Cudworth and Hobden 2014a).

Cudworth is, to my knowledge, the first scholar to introduce critical posthumanism as an approach to social research that is more politically engaged than the cultural renditions of posthumanism that currently

---

[2] For example, Cudworth (2011), Pick (2012), Gaard (2013), Giraud (2013), Wadiwel (2015), and Sorenson (2014).

dominate this space (2011: 12).[3] In this respect, the intention behind Cudworth's proposed critical extension or realignment is similar to that which provoked her development of anthroparchy as a more accurate and necessary account of the mechanisms shaping the Anthropocene. As I have demonstrated, Cudworth and others identify a specifically "uncritical posthumanism" (2010: 79) that:

> lacks an understanding of species as an effect of social power in which non-human species and forms, and Other animals in particular, are marginalized, exploited and/or oppressed.

In its stead, Cudworth offers the following account of critical posthumanism, which includes a reference to Haraway's posthuman proposition that "we have never been human" (2008):

> A sufficiently critical posthumanism must draw in the insights about human centrism, human power and social justice provided by elements within political ecologism and critical animal studies. We may never have been human, but our social relations have been human exclusive, and latterly, also humanist. We need to understand the contingency of the categories of human and 'animal' whilst also hanging on to these concepts in our critiques of species relations. (2011: 13)

The mission to dissolve boundaries that are lived and experienced by sentient 'others' is not advanced simply by dissolving them conceptually or theoretically. As Cudworth goes on to assert, "We will not see justice for animals by deconstructing species" (2011: 13). Species and other differences (race, gender, etc.) need to be retained as agendas for critical social science for as long as they remain grounds for social exclusion, oppression, and domination. Therefore, while posthumanities have their

---

[3] The only other accounts of critical posthumanism that I have found are similarly inattentive to anything other than the human. The prefix 'critical' is used rather to describe a critical approach to theories and concepts of posthumanism itself, rather than society, social structures, and systems of power, as in critical theory and critical animal studies (see, for example, Herbrechter 2013a, b; Banerji and Paranjape 2016; forthcoming BRILL series titled Critical Posthumanisms: www.brill.com/products/series/critical-posthumanisms).

theoretical value, it is " 'the social' in social sciences" that Cudworth believes is in need of more critical problematisation (2011: 13).

Cudworth therefore outlines a posthumanism, and a science of the social, with an overtly emancipatory agenda, articulated more explicitly in a later work with Hobden (Cudworth and Hobden 2014b). Here, the authors additionally emphasise the embodied and "embedded character of human activity" (2014b: 146, also 144), seeing the inherent complexity of this embodiment and embedded-ness as something to be embraced and learned from/with rather than avoided with a sense of hands-thrown-in-the air futility. Delving into and through complexity is, they suggest, the only responsible way to approach global issues in a spirit that envisages the possibility for positive, emancipatory change rather than a disempowering fatalism (Cudworth and Hobden 2014b: 146).[4] With this in mind, across the breadth of Foucault's oeuvre exploring the interplay of power, knowledge, language, and bodies in diverse social 'locations', I recognise the blueprint for a resolutely non-dualistic, posthuman but also, importantly, a critical framework for embracing complexity that seems well suited to my intent to 'trouble habitual ways of thinking and acting' with regard to food animals.

3   A Foucauldian Lens

The framework for this book is therefore based on my identification of a complementary conceptual link between the critical posthumanism of Cudworth and colleagues—focused on foregrounding and articulating practices of power—and Foucault's accounts of power, knowledge, language, discipline, sexuality, and ethics. Effectively, Foucault's body of work on these topics provides me with the analytical tools with which to dissect and understand the practices of power that contribute to the ongoing 'oppression, exploitation, and marginalisation' of nonhuman animals, and the associated persistence of meat and food animals (Cudworth 2008: 34). Additionally, across his explorations of power,

---

[4] An emancipatory agenda is similarly absent in Haraway's (2016) more recent focus on complexity and 'staying with the trouble' (Hornborg 2017).

discipline, sexuality, and ethics, Foucault demonstrates intimate connections between knowledge, emotions, and the senses—all key players in the consumption of food—and how these operate co-constitutively, both reflecting and contributing to the exercise of power. Therefore, in this regard, his work provides a useful analytical framework with which to critically explore the persistent edibility of food animals, and thereby also trouble habitual ways of thinking about them.

Across the aforementioned accounts, the human that Foucault describes is a thinking *and* feeling body that is fundamentally both a product and productive of a broader social body or order. In other words, it is through this ordered social body that humans are at once shaped by their environment[5] and also implicated in the creation and operation of mechanisms that shape the terms and conditions of that environment for other bodies. Foucault most closely articulates these unifying aspects of his work in the following excerpt from *The Order of Things*:

> But to man's experience a body has been given, a body which is his body—a fragment of ambiguous space, whose peculiar and irreducible spatiality is nevertheless articulated upon the space of things; to this same experience, desire is given as a primordial appetite on the basis of which all things assume value, and *relative* value. (1989: 342, emphasis added)

Foucault's neglect of the bodies of nonhuman others (and also gendered bodies) has been noted by several authors (Palmer 2001; Cole 2011; Taylor 2013; Chrulew and Wadiwel 2016), some being more critical than others (e.g. Cavalieri 2008). However, this is not generally conceived as deliberate avoidance; "amnesia", as Cavalieri describes it (2008: 100), is not the same as purposive exclusion. Indeed, Foucault does not directly engage with other analytical approaches to overtly power-infused inter-human relations, such as feminism, postcolonialism, race studies, or disability studies (Chrulew and Wadiwel 2016: 4). And yet the value of his work in problematising conceptions of 'other-hood' and critiquing the systemic and normalising orders that bring about such

---

[5] In every way imaginable, including (but not limited to) materially, biologically, physically, emotionally, and intellectually.

'othering' has made it a prominent feature across all these fields, including critical animal studies. As Chrulew and Wadiwel comment, "the anti-dogmatic and provisional character of Foucault's infamous 'toolbox' not only tolerates but encourages such reinscriptions and intersections" (Chrulew and Wadiwel 2016: 4).

Indeed, there is a growing appreciation of the value of Foucault's ideas to the critical study of animals. Many CAS draw substantially from his body of work, particularly his conceptions of power (including disciplinary power, pastoral power, and sovereign power) (Palmer 2001; Twine 2010a; Cole 2011; Taylor 2013; Wadiwel 2002, 2015), and the associated power of knowledge (Johnson 2012; Cole and Stewart 2014), power of the gaze (Pachirat 2011; Freeman and Tulloch 2013; Acampora 2016), power of language (Peggs 2012); and ethics of self-care (Taylor 2010), to name just a few. An edited book dedicated to Foucault and Animals (Chrulew and Wadiwel 2016) confirms his contribution to this field.

Another scholar notable for his expansion of Foucault's conceptions of sovereignty and biopower to encompass not only human but all life, including 'the animal', is Agamben (1998, 2004). However, while Foucault might be charged with a certain 'amnesia' when it comes to nonhuman animals, the work of Agamben explicitly includes animals but has been described as decidedly anthropocentric in its register (Chrulew 2012). Thus, Chrulew asserts, Agamben's line of philosophical enquiry "does little to disrupt the thesis of the animal's captivation" (2012: 58).

To date, most of the aforementioned Foucauldian approaches to animals have dealt in a largely singular fashion with one or two of Foucault's concepts. None, to my knowledge, have drawn a direct connection across several, as I propose here, and applied that to the analysis of one specific aspect of animal use—the persistent consumption of their flesh. By focusing on a circumstance in which, (1) a distinctively 'valued' living 'other' is not only used, but also, in the final instance, consumed, and thereby made entirely 'other' through being extinguished and then assimilated; and (2) embodied human immaterialities are engaged via the emotions and all of the senses, there is a unique opportunity to apprehend "the tangibility of power, its texture and flavor" (Probyn 2000: 7). For even though, as Bertrand Russell observes, "forms of power are most nakedly and simply displayed in our dealings with animals" (1986: 20), those animals that are killed and eaten would have to represent the apogee of that power, for there is nowhere left to go in the expression of power beyond material embodiment.

Three schemas are variously emphasised across Foucault's body of work and these provide the overall architecture for my research design and this book. First, in *Power/Knowledge*, Foucault emphasises the integral relationship between knowledge and power: "The exercise of power perpetually creates knowledge and, conversely, knowledge constantly induces effects of power" (1980: 52). Cartographies of meat are permeated with knowledge, or "sets of knowledges" (82), and an examination of these knowledges, and their genealogy, will go some way to illuminating the domination of animals that is taken to be natural and unquestionable.

Second, in *The History of Sexuality* (1978), Foucault examines the pleasures invested by effects of power and their associated discourses, and the knowledges that emerge as a result of this interplay. This "regime of power-knowledge-pleasure", he argues, is what "sustains the discourse on human sexuality" (1978: 11). Exchanging sexuality for food is not a long bow to draw,[6] and leads me to enquire how emotions in general, not just limited to pleasure, might be invested by effects of power and associated discourses surrounding meat and food animals, and what knowledges and "truths" (1978: 123) this interplay contributes to.

Third, another of Foucault's key theorisations explores Jeremy Bentham's famous Panopticon[7] as a "technology of power" (1980: 148). Foucault describes how increased surveillance enables the effects of power to extend, "gaining access to individuals themselves, to their bodies, their gestures, and all their daily actions" (1980: 151–152). He calls this

---

[6] In Volume 2 of the *History of Sexuality: The Use of Pleasure*, Foucault notes "the long history of connection between alimentary ethics and sexual ethics" (1985: 51). These together relate to food, drink, and sex—the "three basic appetites", as Foucault (49) calls them, or, according to Aristotle, the "three common pleasures" (51). Many scholars since have also explored this connection (e.g. Probyn 2000; Cudworth 2011; Potts and Parry 2010), and it serves as a constant source of symbolic and narrative texture for works of art and literature across a variety of media (e.g. Margaret Atwood's *The Edible Woman* (1969); Jim Crace's *The Devil's Larder* (2002); Peter Greenaway's 1989 film *The Cook, the Thief, His Wife and Her Lover*, and the more recent *Raw* (2017) by Julie Ducournau, among others).

[7] The panopticon is an architectural form that Bentham proposed for the design of prisons. It comprises a central tower from which the many can be observed by the one, thereby reversing the principle of the dungeon. Instead, visibility becomes a trap (Foucault 1977: 200).

"subjection by illumination" or "power through transparency" (1980: 154). As well as using this lens of power to explore the increased visibility of 'happier' animals and 'better' meat production processes associated with ethical meat,[8] I link this visual-power association back to knowledge. For, as Foucault notes, those subjected to power through transparency exist "under the gaze of a permanent corpus of knowledge" (1977: 190) whereby the regarded 'other' is always "already encoded" (1989: xxii).

Completing the loop back to regimes of power/knowledge/pleasure, I also link transparency and visibility to pleasure, or in my schema, all emotions. Foucault draws implicit, rather than explicit, links between his theorisations on the gaze, or transparency, and his regime of power-knowledge-pleasure. In *Seeing and Knowing*, where he explores the clinical gaze, he describes the visible as being divided "within an already given conceptual configuration" (1973: 113), while in *Discipline and Punish* he describes a mechanics of forces behind the disciplinary gaze that "makes the penalty be feared" over the "pleasures" of the crime (1977: 106). Taken together with his account of 'knowledge-pleasure' in *The History of Sexuality Volume 1* (1978: 55), which he defines as "a knowledge of pleasure, a pleasure that comes of knowing pleasure", and where he also describes a "fundamental petition to know" linked to "a refusal to see" and "masking" of truth, the common threads between his philosophical enquiries in these areas are discernible, even if not explicit. This is especially so when this refusal to see, or masking of truth, is considered in light of his observation regarding the myth of the 'pure gaze' or what he calls the "speaking eye" (1973: 114). This 'pure' gaze, as a "servant of things and master of truth" (ibid.), is one that Foucault believes does not exist.

In *The Use of Pleasure* (1985), the second volume of the *History of Sexuality*, Foucault draws on classical Greek and Greco-Roman thought, Christian doctrine, and the pastoral ministry to conduct a genealogy of "the hermeneutics of desire" (5). He describes eating and drinking as 'natural pleasures' and notes "the moral problematisation of food, drink and sexual activity" (51) (the "three basic appetites") that characterised

---

[8] Which is an aspect of 'conscious' consumerism more broadly—the increased transparency and de/re-fetishisation of production processes.

Greek classical thought. It was Aristotle who identified the common principle that links all three, that being "the pleasures of contact and touch" (1985: 51). Throughout, Foucault also comments on the links between pleasure (and other emotions) and the senses, including sight, arguing, "there are sights capable of affecting the soul like venom" (1985: 41). Equally, I would add that there are touches, sounds, smells, and tastes, for example, relating to food, that would similarly 'affect the soul'.

It is from recognising, and explicitly connecting, these common threads in Foucault's work that I have arrived at a framework suited to my enquiry. A truly whole-of-body understanding of the mechanisms of power that keep food animals in a state of domination may therefore be undertaken by traversing rhizomatically through Foucault's accounts of:

1. Power/knowledge
2. The pleasure of knowing
3. The power of transparency or the gaze
4. The pleasure of looking

Integrating Laura Mulvey and others' work that more specifically addresses the pleasure of looking allows me to enhance my analysis in this fourth area, and reinforces the connections between power, knowledge, pleasure, and looking.

This Foucauldian framework also provides a way to empirically demonstrate why 'the critical' is needed in a posthuman approach, and determine what sort of embodied and embedded considerations need to be accounted for before a wholly posthuman territory can even be imagined or glimpsed. As Cudworth (2011) suggests, a constant and vigilant awareness of the prevalence of the anthroparchic gaze, even in the 'social' sciences, is imperative.

Relatedly, Foucault's observations regarding the potential for mechanisms of power to establish a formulaic and systemic block are especially relevant for my enquiry not only into mechanisms of power relating to food animals, but their persistence. There can be an overemphasis in some areas of the social sciences, including posthumanism, on a constant evolution, changeability, and transformation that is perceived to characterise the social world, which is in keeping with the rhizomatic assemblage of life described by Deleuze and Guattari whereby "everything changes" (1987: 21). Indeed, change is often conceived as a fundamental feature of all phenomena—the "uneven front of change" (Schatzki 2016: 18). However, Schatzki and other scholars of social practice theory (e.g. Shove 2012; Walker 2013; Halkier et al. 2011; Watson 2016) also recognise that social order, continuity, stability, habit, routine, and persistence are produced by everyday practices that are reproduced through performance. As Schatzki explains:

> Activities happen. Happening, however, is not equivalent to change … The performance of an action does not necessitate any more change than that the stock of events in the world has increased by one. In particular, a performance need not implicate further changes in social facts, phenomena, or events. In other words, an activity can just as easily maintain the world as alter it. In fact, this is the usual case. (2013: 82)

Hence, while certain features of Foucault's blocks or Schatzki's bundles of practices may change, these changes are limited and "occur amid general continuity in activities, arrangements, interwoven time-spaces, practice organizations, entwined practices, and links between practices and arrangements" (Schatzki 2013: 97). In this way, bundles can survive both small, isolated changes and larger, more extensive changes (ibid.). I suggest that radical social transformations require a better understanding of these nexuses of profound persistence and resistance to change.

Amidst the myriad changes, over time and across cultures, relating to food,[9] the consumption of animal flesh has remained a significant constant. This signals an opportunity to examine just such a persistent nexus of social practices in which taken-for-granted relations, between humans and food animals, are reflected and reinforced such that they represent a block or "context" of social life (Schatzki 2005: 467). A characteristic feature of these 'contexts', among which Schatzki includes agriculture, is a "persistence of structure from the past into the present" (Schatzki 2006: 1868). For those cows, pigs, chickens, sheep, goats, and other animals

---

[9] In terms of what is regarded as edible, how it is obtained, what purpose different foods are understood to serve both physically and symbolically, and the methods and tools used to prepare food for consumption.

cast at the base of this structure, as Horkheimer depicts in his 'Skyscraper',[10] this can have profound implications for their possibilities in life. But these implications also extend vertically, up the skyscraper. For as Gunderson notes, like more recent intersectional and critical animal scholarship (but "unlike many contemporary socio-environmental scholars"), Horkeimer and others of the Frankfurt School maintained that "the domination of nature is intimately tied up in the domination of the self and other human beings" (2014: 288). As indicated in the introduction, the aspect of this 'persistent structure' of domination I have chosen to examine, which I contend is key to more effectively understanding, and ideally changing, taken-for-granted human-animal relations, is how meat and food animals continue to be made sense of, or (re)constituted, as material objects of everyday practices. It is through unpacking these sense-making constitutions that relations and effects of power are revealed.

Emphasising that power is to be "comprehended as effect rather than object" (2016: 174), Watson identifies an affinity with Foucault in his practice-based approach to power as a social effect—one that can "direct or purposively influence the action of others" (170), rather than a force deliberately wielded by individuals. Recognising a similar affinity, Schatzki (2005) describes power as when "actions structure *others'* possibilities" (479, emphasis added). He continues:

the notion of power thus captures the responsibility that actions bear for the differential access that other people [and animals] have to the possibilities carried in practice-arrangement bundles. (479)

Though neither authors explicitly include animals in their designation of 'others', both acknowledge that nonhumans are part of practices (Watson 2016: 171; Schatzki 2005: 478). Whether conceived as blocks, bundles, or contexts, it is clear that unequal relations founded on dualistic

[10] In 'The Skyscraper', Horkheimer (1978: 66–67) paints a critical and vivid picture of the social order, at the base of which, "we encounter the actual foundation of misery on which this structure rises … Below the spaces where the coolies of the earth perish by the millions, the indescribable, unimaginable suffering of the animals, the animal hell in human society, would have to be depicted, the sweat, blood, despair of the animals … The basement of that house is a slaughterhouse, its roof a cathedral, but from the windows of the upper floors, it affords a really beautiful view of the starry heavens".

representations of an 'other' are reflected and reproduced through social practices, whereby they are experienced as systemically normalised, and normalising, aspects of everyday life. Drawing on Foucault (1973, 1977, 1978, 1980), I understand these unequal relations, specifically between humans and food animals, as expressions of invisible mechanisms of power supported by legitimated rights of entitlement and privilege. These unequal relations are in turn more tangibly expressed in persistent constitutions of meat and food animals, and it is therefore through these constitutions that the nature of these mechanisms of power can be apprehended.

4 Research Design and Methodology

Reflecting the theoretical approach of critical posthumanism described above, my methodology facilitates a critical, embodied, and emotionally engaged approach to understanding the persistence of meat consumption and the use of animals as food. Attention to language and discourse is considered a fundamental aspect of this. As Sedgwick comments, with reference to Foucault's account of the 'productive force of taxonomies', "[t]hat language itself can be productive of reality is a primary ground of antiessentialist inquiry" (2003: 5). Following Machin and Mayr, visual communication is considered part of discourse: "visual communication plays its part in shaping and maintaining a society's ideologies, and can also serve to create, maintain and legitimise certain kinds of social practices" (2012: 19). This becomes particularly significant when looking at the ways in which producers and suppliers promote their ethical and sustainable meat and how they represent animals.

My methodology is constructed around the collection, interpretation, and analysis of primarily discursive but also visual data relating to meat consumption and the use of animals as food. I am specifically interested in how food animals remain edible when their environmental and ethical legitimacy is challenged. Consequently, I focus on consumers and producer/consumers of so-called 'ethical' and 'sustainable' meat, and explore, through semi-structured interviews and an analysis of promotional materials, how my research participants make sense of, or (re)constitute, food animals and meat through their everyday 'meaty' practices.[11] In this way, by understanding how these sense-making constitutions are themselves constituted, I aim to identify what makes practices involving meat consumption and the use of animals as food so resilient to change, or what it is about this particular "edge of change" (Schatzki 2013) that does not change.

Based on this, the data component of my research entailed two main activities:

1. Analysing the data made via participants' conversations with me to explore how animals and meat are 'made sense of'.
2. Analysing the promotional material (textual and visual) of producers and retailers of ethical and sustainable meat in the greater Melbourne region of Victoria to examine another facet of how animals and meat are made to 'make sense'.

During interviews, producers and consumers were asked about their everyday practices relating to meat and food animals. This line of questioning helped determine when and how the ways in which meat and animals 'make sense' change, why they change, and how these changes affect the edibility of food animals. Exploring the interplay of knowledge, emotions, and the senses in different circumstances (of 'doings and sayings') in this way also foregrounded how the persistent edibility of these animals is constituted—that is, what tacit rules, knowledge, competencies, and skills remain consistent across the ways they and their flesh 'make sense'. This approach reflects Reckwitz's conception of social practices as constituted by "forms of bodily activities, forms of mental activities, 'things' and their use, a

---

[11] Meaty practices are conceived broadly as all practices that are in any way connected to the production and consumption of meat and the use of 'food' animals where their constitution as food is implicitly accepted. These include, for instance, any practices that involve the sourcing, buying, and eating of meat, practices relating to the breeding, purchase, feeding, rearing, trading, marketing, and general 'production' of meat, and other practices where 'food' animals might feature, for instance, as entertainment (e.g. petting zoos, rodeos), or objects of welfare efforts.

background knowledge in the form of understanding, know-how, states of emotion and motivational knowledge" (2002: 249). My analysis of promotional texts and images focused on the notions of 'address' and 'reception' as used in discourse analysis and cultural studies, denoting, respectively, the intended and the actual response of the reader or viewer (Sturken and Cartwright 2009: 72; also Machin and Mayr 2012).

As a point of note, I was not attempting to assess the accuracy or validity of any of my participants' knowledge and understanding regarding what is ethical or sustainable. Rather, my goal was to illustrate how fluid constructions of the ethical and the 'good' can be, and how these variously implicate one animal over another, one method of production over another, one way of acquiring it over another—all with tangible consequences on the lives of current and future food animals. It is the power of this albeit fluid knowledge that contributes to effacing a state of domination, as my analysis will demonstrate.

4.1 The Participants

Of the 26 consumers interviewed, 19 were female and the majority aged between 26 and 44 (Table 3.1). They lived mainly within the urban boundary of Melbourne, although five lived in regional areas one to three hours outside of Melbourne and one was from interstate, interviewed via Skype. The 15 producer/consumers that I interviewed were split more evenly between male and female, although often the person I interviewed was one member of a partnership. Table 3.1 provides basic demographic and dietary information for each participant as a point of reference for the data analysis chapters to follows.

As this table also illustrates, I ensured that I gathered data from producers of different types of meat, although beef and pork were the most common.

Table 3.1 Summary of participants including prior vegetarian or vegan practices, and current frequency of meat consumption

	'Gender'	Age	Past vegan (V) or Veg'n	Period vegan (V) or Veg'n	Frequency of meat consumption
Consumers					
Geoffrey	M	<25	–	–	1 or 2 times per day
Diane	F	26–34	Veg'n	5 years	Once per day
Beverley	F	26–34	Veg'n	12–18 months	Once per month
Julie	F	26–34	Veg'n	6 months	1 or 2 times per week
Ellis	F	26–34	–	–	Once per day
Dan	M	26–34	Veg'n & V	9 years	Once per day
Heather	F	26–34	Veg'n	2–3 years	2 or 3 times per week
Charlotte	F	26–34	Veg'n & V	2–3 years	1 or 2 times per week
Anthony	M	26–34	Veg'n	5 weeks	Once per day
Lisa	F	26–34	Veg'n	6 months	Once per day
Sophie	F	26–34	Veg'n	Current (1 year)	Veg'n
Damien	M	26–34	–	–	5 or 6 times per week
Fiona	F	35–44	–	–	3 or 4 times per week
Joyce	F	35–44	Veg'n & V	3 years	4 or 5 times per week
David	M	35–44	–	–	1 or 2 times per day
Maria	F	35–44	Veg'n	3 months	3–5 times per week
Anne	F	35–44	Veg'n	6 months	5 times per week
Natalie	F	35–44	Veg'n	12 years	Once per day
Helen	F	35–44	Veg'n	3–4 years	5 or 6 times per week
Michael	M	35–44	Veg'n	1 year	3 or 4 times per week
Henry	M	45–54	Veg'n	5 years	4 or 5 times per week
Sally	F	45–54	–	–	1 or 2 times per day
Tracey	F	45–54	Veg'n (+fish)	Current (23 years)	Once per day (for family)
Gillian	F	45–54	Veg'n & V	16 years	1–3 times per day
Lucy	F	45–54	–	–	2 or 3 times per day
Grace	F	55–64	–	–	5 or 6 times per week

Producers/Consumers[a]

Ian	C	M	26–34		—	1 or 2 times per day
Jennifer	P	F	26–34		—	3 times per week
Blake	C, P, Ch, S	M	26–34		—	6 times per week
Brigid	S	F	35–44		—	6 or 7 times per week
Florence	C	F	35–44	Veg'n	Current (30 years)	Veg'n
Shane	P	M	35–44		—	5 or 6 times per week
Finn	All-Retailer	M	35–44	Veg'n	1 year	1–3 times per day
Graham	Ch	M	45–54		—	1 or 2 times per day
Will	Ch	M	45–54		—	5 or 6 times per week
Trevor	C, P, Ch, S	M	45–54		—	5 times per week
Bella	C, P	F	45–54	Veg'n	7 years	6 or 7 times per week
Oliver	C, S	M	45–54		—	3 times per week
Alison	C	F	45–54	Veg'n	5–6 years	1 or 2 times per day
Melanie	P	F	55–64		—	1 or 2 times per day
James	C	M	55–64		—	5 or 6 times per week

[a]Animals used by producers are indicated as follows: C (cow), P (pig), Ch (chicken), and S (sheep)

References

Acampora, R. (2016). [Provocations from the Field] Epistemology of Ignorance and Human Privilege. *Animal Studies Journal, 5*(2), 1–20.

Agamben, G. (1998). *Homo Sacer: Sovereign Power and Bare Life*. Palo Alto: Stanford University Press.

Agamben, G. (2004). *The Open: Man and Animal*. Palo Alto: Stanford University Press.

Alberts, P. (2011). Responsibility Towards Life in the Early Anthropocene. *Angelaki: Journal of the Theoretical Humanities, 16*(4), 5–17.

Alloun, E. (2015). Ecofeminism and Animal Advocacy in Australia: Productive Encounters for an Integrative Ethics and Politics. *Animal Studies Journal, 4*(1), 148–173.

Bai, X., van der Leeuw, A., O'Brien, K., Berkhout, F., et al. (2016). Plausible and Desirable Futures in the Anthropocene: A New Research Agenda. *Global Environmental Change, 39*, 351–362.

Banerji, D., & Paranjape, M. R. (2016). *Critical Posthumanism and Planetary Futures*. New Delhi: Springer (India).

Barad, K. (2003). Posthumanist Performativity: Toward an Understanding of How Matter Comes to Matter. *Signs: Journal of Women in Culture and Society, 28*(3), 801–831.

Barad, K. (2007). *Meeting the Universe Halfway: Quantum Physics and the Entanglement of Matter and Meaning*. Durham; London: Duke University Press.

Bennett, C. E., Thomas, R., Williams, M., Zalasiewicz, J., et al. (2018). The Broiler Chicken as a Signal of a Human Reconfigured Biosphere. *Royal Society Open Science, 5*(12), 180325.

Biermann, F., Bai, X., Bondre, N., Broadgate, W., et al. (2016). Down to Earth: Contextualizing the Anthropocene. *Global Environmental Change, 39*, 341–350.

Braidotti, R. (2013). *The Posthuman*. Cambridge; Malden: Polity Press.

Calarco, M. (2008). Facing the Other Animal: Levinas. In *Zoographies: The Question of the Animal from Heidegger to Derrida* (pp. 55–78). New York; Chichester: Columbia University Press.

Carson, R. (1962). *Silent Spring*. Boston; New York: Houghten Mifflin.

Castree, N., & Nash, C. (2004). Mapping Posthumanism: An Exchange. *Environment and Planning A, 36*(8), 1341–1363.

Cavalieri, P. (2008). A Missed Opportunity: Humanism, Anti-humanism and the Animal Question. In J. Castricano (Ed.), *Animal Subjects: An Ethical Reader in a Posthuman World* (pp. 97–123). Waterloo, ON: Wilfrid Laurier University Press.

Chrulew, M. (2012). Animals in Biopolitical Theory: Between Agamben and Negri. *New Formations, 76,* 53–68.

Chrulew, M., & Wadiwel, D. (2016). Editors' Introduction: Foucault and Animals. In *Foucault and Animals* (pp. 1–18). Brill: Leden.

Cole, M. (2011). From "Animal Machines" to "Happy Meat"? Foucault's Ideas of Disciplinary and Pastoral Power Applied to 'Animal-Centred' Welfare Discourse. *Animals, 1*(4), 83–101.

Cole, M., & Stewart, K. (2014). *Our Children and Other Animals.* London; New York: Routledge.

Cudworth, E. (2005). *Developing Ecofeminist Theory: The Complexity of Difference.* Basingstoke; New York: Palgrave Macmillan.

Cudworth, E. (2008). Most Farmers Prefer Blondes: "The Dynamics of Anthroparchy in Animals" Becoming Meat. *Journal for Critical Animal Studies, 6*(1), 32–45.

Cudworth, E. (2010). 'The Recipe for Love'? Continuities and Changes in the Sexual Politics of Meat. *Journal for Critical Animal Studies, 8*(4), 78–100.

Cudworth, E. (2011). *Social Lives with Other Animals.* Basingstoke; New York: Palgrave Macmillan.

Cudworth, E. (2014). Beyond Speciesism: Intersectionality, Critical Sociology and the Human Domination of Other Animals. In N. Taylor & R. Twine (Eds.), *The Rise of Critical Animals Studies: From the Margins to the Centre* (pp. 19–35). Taylor & Francis.

Cudworth, E., & Hobden, S. (2014a). Civilisation and the Domination of the Animal. *Millennium - Journal of International Studies, 42*(3), 746–766.

Cudworth, E., & Hobden, S. (2014b). Liberation for Straw Dogs? Old Materialism, New Materialism, and the Challenge of an Emancipatory Posthumanism. *Globalizations, 12*(1), 134–148.

Deckha, M. (2012). Toward a Postcolonial, Posthumanist Feminist Theory: Centralizing Race and Culture in Feminist Work on Nonhuman Animals. *Hypatia, 27*(3), 527–545.

Deleuze, G., & Guattari, F. (1987). *A Thousand Plateaus: Capitalism and Schizophrenia.* London; New York: Continuum.

Derrida, J. (2008). *The Animal that Therefore I Am.* New York: Fordham University Press.

Derrida, J., & Roudinesco, E. (2004). Violence Against Animals. In *For What Tomorrow … A Dialogue* (pp. 62–76). Palo Alto: Stanford University Press.

Foucault, M. (1973). *The Birth of the Clinic: An Archaeology of Medical Perception* (R. D. Laing, Ed.). London and New York: Routledge.

Foucault, M. (1977). *Discipline and Punish: The Birth of the Prison*. New York: Vintage Books.

Foucault, M. (1978). *The History of Sexuality*. New York: Pantheon Books.

Foucault, M. (1980). *Power/Knowledge: Selected Interviews and Other Writings, 1972–1977*. New York: Pantheon Books.

Foucault, M. (1985). *The Use of Pleasure*. New York: Random House.

Foucault, M. (1989). *The Order of Things*. London; New York: Routledge.

Freeman, C. P., & Tulloch, S. (2013). Was Blind but Now I See: Animal Liberation Documentaries' Deconstruction of Barriers to Witnessing Injustice. In A. Pick & G. Narraway (Eds.), *Screening Nature: Cinema Beyond the Human* (pp. 110–126). New York; Oxford: Berghahn Books.

Gaard, G. (2013). Toward a Feminist Postcolonial Milk Studies. *American Quarterly, 65*(3), 595–618.

Giraud, E. (2013). 'Beasts of Burden': Productive Tensions between Haraway and Radical Animal Rights Activism. *Culture, Theory and Critique, 54*(1), 102–120.

Gunderson, R. (2014). The First-generation Frankfurt School on the Animal Question. *Sociological Perspectives, 57*(3), 285–300.

Halkier, B., Katz-Gerro, T., & Martens, L. (2011). Applying Practice Theory to the Study of Consumption: Theoretical and Methodological Considerations. *Journal of Consumer Culture, 11*(1), 3–13.

Haraway, D. (1990). *Simians, Cyborgs and Women: The Reinvention of Nature*. New York; London: Routledge.

Haraway, D. (2008). *When Species Meet*. Minneapolis; London: University of Minnesota Press.

Haraway, D. (2016). *Staying with the Trouble: Making Kin in the Chthulucene*. Combined Academic Publishers.

Herbrechter, S. (2013a). *Posthumanism – A Critical Analysis*. London; New Delhi; New York; Sydney: Bloomsbury Academic.

Herbrechter, S. (2013b, April). Rosi Braidotti (2013) The Posthuman. *Culture Machine: Reviews*, 1–13.

Hickmann, T., Partzsch, L., Pattberg, P., & Weiland, A. (2018). *The Anthropocene Debate and Political Science*. Abingdon: Routledge.

Horkheimer, M. (1978). *Der Wolkenkratzer.* Dawn and Decline: Notes 1926–1931 and 1950–1969, New York 1978, S. 66.

Hornborg, A. (2017). Dithering While the Planet Burns: Anthropologists' Approaches to the Anthropocene. *Reviews in Anthropology, 46*(2–3), 61–77.

Johnson, L. (2012). *Power, Knowledge, Animals.* Basingstoke; New York: Palgrave Macmillan.

Kwan, M. (2004). Beyond Difference: From Canonical Geography to Hybrid Geographies. *Annals of the Association of American Geographers, 94*(4), 756–763.

Latour, B. (1993). *We Have Never Been Modern.* Cambridge, MA; London: Harvard University Press.

Lövbrand, E., Beck, S., Chilvers, J., Forsyth, T., et al. (2015). Who Speaks for the Future of Earth? How Critical Social Science Can Extend the Conversation on the Anthropocene. *Global Environmental Change, 32*, 211–218.

Machin, D., & Mayr, A. (2012). *How to Do Critical Discourse Analysis.* London; Thousand Oaks; New Delhi: SAGE Publications Ltd.

Merchant, C. (1990). *The Death of Nature: Women, Ecology, and the Scientific Revolution.* New York: HarperCollins.

Merchant, C. (2005). *Radical Ecology: The Search for a Liveable World.* London; New York: Routledge.

Merleau-Ponty, M. (2002). *Phenomenology of Perception.* London; New York: Routledge.

Pachirat, T. (2011). *Every Twelve Seconds: Industrialized Slaughter and the Politics of Sight.* New Haven; London: Yale Agrarian Studies Series.

Palmer, C. (2001). 'Taming the Wild Profusion of Existing Things'? A Study of Foucault, Power, and Human/Animal Relationships. *Environmental Communication: A Journal of Nature and Culture, 23*, 339–358.

Pederson, H. (2011). Release the Moths: Critical Animal Studies and the Posthumanist Impulse. *Culture, Theory and Critique, 52*(1), 65–81.

Peggs, K. (2012). *Animals and Sociology.* Basingstoke; New York: Palgrave Macmillan.

Pick, A. (2012). Turning to Animals Between Love and Law. *New Formations, 76*, 68–86.

Plumwood, V. (1994). *Feminism and the Mastery of Nature.* London; New York: Routledge.

Plumwood, V. (2002). Decolonising Relationships with Nature. *PAN: Philosophy Activism Nature, 2*, 7–30.

Plumwood, V. (2004). Gender, Eco-Feminism and the Environment. In R. White (Ed.), *Controversies in Environmental Sociology* (pp. 43–60). Cambridge, UK; New York: Cambridge University Press.

Plumwood, V. (2013a). *The Eye of the Crocodile.* Canberra: ANU E Press.

Plumwood, V. (2013b). Animals and Ecology: Towards a Better Integration. In L. Shannon (Ed.), *The Eye of the Crocodile* (pp. 77–90). ANU E Press.

Potts, A., & Parry, J. (2010). Vegan Sexuality: Challenging Heteronormative Masculinity through Meat-free Sex. *Feminism & Psychology, 20*(1), 53–72.

Probyn, E. (2000). *Carnal Appetites: FoodSexIdentities.* London; New York: Routledge.

Reckwitz, A. (2002). Toward a Theory of Social Practices: A Development in Culturalist Theorizing. *European Journal of Social Theory, 5*(2), 243–263.

Rickards, L. (2015). Metaphor and the Anthropocene: Presenting Humans as a Geological Force. *Geographical Research, 53*(3), 280–287.

Russell, B. (1986). The Forms of Power. In S. Lukes (Ed.), *Power* (pp. 19–27). New York: New York University Press.

Schatzki, T. R. (2005). Peripheral Vision: The Sites of Organizations. *Organization Studies, 26*(3), 465–484.

Schatzki, T. R. (2006). On Organizations as They Happen. *Organization Studies, 27*(12), 1863–1873.

Schatzki, T. (2013). The Edge of Change: On the Emergence, Persistence, and Dissolution of Practices. In E. Shove & N. Spurling (Eds.), *Sustainable Practices: Social Theory and Climate Change* (pp. 79–110). London; New York: Routledge.

Schatzki, T. R. (2016). Practices, Governance, and Sustainability. In Y. Strengers & C. Maller (Eds.), *Social Practices, Intervention and Sustainability: Beyond Behaviour Change* (pp. 15–30). London; New York: Routledge.

Sedgwick, E. K. (2003). *Touching Feeling: Affect, Pedagogy, Performativity.* Durham; London: Duke University Press.

Shove, E. (2012). Habits and Their Creatures. In A. Warde & D. Southerton (Eds.), *The Habits of Consumption* (Collegium - Studies across Disciplines in the Humanities and Social Sciences) (pp. 100–112). University of Helsinki.

Sorenson, J. (2014). Thinking the Unthinkable. In J. Sorenson (Ed.), *Critical Animal Studies: Thinking the Unthinkable* (pp. xi–xxxiv). Toronto: Canadian Scholars' Press Inc.

Steffen, W., Crutzen, P. J., & McNeill, J. R. (2007). The Anthropocene: Are Humans Now Overwhelming the Great Forces of Nature? *Ambio, 36*(8), 614–621.

Sturken, M., & Cartwright, L. (2009). *Practices of Looking: An Introduction to Visual Culture*. New York; Oxford: Oxford University Press.

Swyngedouw, E. (2013). The Non-political Politics of Climate Change. *ACME: An International E-journal for Critical Geographies, 12*(1), 1–8.

Swyngedouw, E. (2015). Depoliticized Environments and the Promises of the Anthropocene. In R. L. Bryant (Ed.), *The International Handbook of Political Ecology*. Cheltenham, UK: Edward Elgar Publishing.

Taylor, C. (2010). Foucault and the Ethics of Eating. *Foucault Studies*, (9), 71–88.

Taylor, C. (2013). Foucault and Critical Animal Studies: Genealogies of Agricultural Power. *Philosophy Compass, 8*(6), 539–551.

Twine, R. (2010a). *Animals as Biotechnology: Ethics, Sustainability and Critical Animal Studies*. London; Washington, DC: Earthscan.

Twine, R. (2010b). Intersectional Disgust? Animals and (Eco)Feminism. *Feminism & Psychology, 20*(3), 397–406.

Wadiwel, D. J. (2002). Cows and Sovereignty: Biopower and Animal Life. *Borderlands e-Journal, 1*(2), 1–8.

Wadiwel, D. J. (2015). *The War Against Animals*. Leiden; Boston: Brill Rodopi.

Walker, G. (2013). Inequality, Sustainability and Capability: Locating Justice in Social Practice. In E. Shove & N. Spurling (Eds.), *Sustainable Practices: Social Theory and Climate Change* (pp. 181–196). London; New York: Routledge.

Warren, K. (1997). *Ecofeminism: Women, Culture, Nature*. Bloomington: Indiana University Press.

Warren, K. (2000). *Ecofeminist Philosophy: A Western Perspective on What It Is and Why It Matters*. Lanham; Boulder; New York; Oxford: Rowman & Littlefield.

Watson, M. (2016). Placing Power in Practice Theory. In A. Hui, E. Shove, & T. Schatzki (Eds.), *The Nexus of Practices: Connections, Constellations and Practitioners* (pp. 169–182). London; New York: Routledge.

Weisberg, Z. (2009). The Broken Promises of Monsters: Haraway, Animals and the Humanist Legacy. *Journal for Critical Animal Studies, 7*(2), 22–62.

Whatmore, S. (2002). *Hybrid Geographies: Natures Cultures Spaces*. London; Thousand Oaks; New Delhi: SAGE Publications Ltd.

Whatmore, S. (2013). Dissecting the Autonomous Self: Hybrid Cartographies for a Relational Ethics. *Environment and Planning D: Society and Space, 15*, 37–53.

Wolfe, C. (2010). *What Is Posthumanism?* Minneapolis; London: University of Minnesota Press.

Part II

Categories and Boundaries

Introduction to Part II

Categories and boundaries are the starting point for this book as they reveal the arbitrary and conventionalised nature of the ways in which animals and meat are classified. They also appear as more or less culturally and historically (and often legally) stable, serving as ready-made rubrics that can be both efficient and desensitising, as Douglas (2002: 37) explains:

> As learning proceeds objects are named. Their names then affect the way they are perceived next time: once labelled they are more speedily slotted into the pigeon-holes in future.

A well-known and oft-cited passage from the Book of Deuteronomy (14: 3–20) begins: "3. You shall not eat any abominable things. 4. These are the animals you may eat: …", and goes on to describe which animals are sanctioned for eating in terms of their being 'clean' or 'unclean' (Christian Old Testament 1250–1200 BC). This is one of the earliest examples of a dietary edict. Many such edicts, oral and written, have issued from religious, philosophical, scientific, medical, and various other 'authorities', from different cultures, since historical records began. For example, Hindu Dharma texts forbid the consumption of "five-nailed animals, except for the hedgehog, hare, porcupine, monitor, lizard, rhinoceros, and tortoise" (Gautama Dharma-Sutra 17.27, in Jamison 1998).

What these texts highlight is that the orders and 'pigeon holes' that humankind uses to make sense of experience do not exist "except in the grid created by a glance, an examination, a language" (Foucault 1989: xxi). Indeed, Melanie Bujok asserts that "the animal" as a term used to denote all species except humans "does not, in fact, exist", being rather a political device to "mask diversity and to place all other animals at a distance" (2013: 37). Thus, though we know nature only through such denominations, Foucault describes how it:

> glimmers far off beyond them, continuously present on the far side of this grid, which nevertheless presents it to our knowledge and renders it visible only when wholly spanned by language. (1989: 175)

It is this language that then "furnishes us with the rooms we dwell in" (Sanbonmatsu 2014: interview). What sort of possibilities would open up if we were able to see past this grid or "tabula" that we impose on the world, look into the gaps created by momentary breaks? Would we see absurdities in our current orders and classifications? Would we recognise other ontological pathways for animals, and all of nature, once the opacity of the grid imposed by our current values starts to dissolve? These questions relate to my location in vegan heterotopia and I use them to frame the discussion in this first empirical section of the book where I draw on my discussions with producers and consumers about their meat-related practices to address my first objective: **How does certain 'knowledge' of animals contribute to their edibility and non-edibility?** My intention is to draw the reader's attention to grids, breaks, gaps, absurdities, and the alternative pathways that might glimmer beyond the system of categories and boundaries that delineate various permutations of food animals and their flesh.

References

Bujok, M. (2013). Animals, Women and Social Hierarchies: Reflections on Power Relations. *Deportate, esuli, profughe, 23*, 32–48.

Douglas, M. (2002). *Purity and Danger: An Analysis of Concepts of Pollution and Taboo*. London; New York: Routledge Classics.

Foucault, M. (1989). *The Order of Things*. London; New York: Routledge.

Jamison, S. W. (1998). Rhinoceros Toes, Manu V.17–18, and the Development of the Dharma System. *Journal of the American Oriental Society, 118*(2), 249–256.

Sanbonmatsu, J. (2014). Interview with John Sanbonmatsu (S. Rodriguez, Ed., pp. 1–12). Online. Direct Action Everywhere.

4

Animal Categories and the Maintenance of Order

1 Taking Note of Categories

As I will demonstrate throughout this book, rather than being an outcome of individual attitudes, behaviours, and choice (Shove 2010), the ways my participants 'make sense' of food and other animals reflect historical practices of anthroparchal animal use that have become deeply normalised. Early critical theoretical accounts of these uses identified categories that have endured, and remain just as relevant today. One of the first documented formal classifications of animal use I am aware of is by Henry Salt who, in his 1894 volume on animals' rights, includes chapters on domestic animals, wild animals, animals slaughtered for food, sport, fashion, and finally experimental torture. More recently, this conception of use can be traced through the writings of animal rights scholars including Singer (1975), Regan (2004 [1983]), Benton (1993), Adams (1994), Dunayer (2001), and Cudworth (2003), among others. In *Natural Relations*, Benton (1993: 62–67) itemises nine admittedly overlapping categories of what he calls human-animal relations, or what I prefer to call human relations of animal use:

1. To augment human labour
2. To meet bodily needs
3. As entertainment
4. For education
5. For commercial exploitation (where profit-making is a fundamental element of their use as labour, food, entertainment, and education)
6. For physical protection
7. As domestic pets
8. As symbolic resources
9. Wild animals (Benton acknowledges that 'wildness' is a social relation)

That at least some of these categories were more tacitly recognised as part of everyday life well before these formal accounts is clear from Erica Fudge's archival studies of sixteenth- and seventeenth-century documents relating to livestock (e.g. 2002, 2013a, b). And before Benton, in 1965, author and social critic Brigid Brophy wrote regarding animals that:

> We employ their work; we eat and wear them. We exploit them to serve our superstitions…we sacrifice them to science … we could quite well enjoy marksmanship or cross-country galloping without requiring a real dead animal to show for it at the end. (3)

Several scholars have expanded on Benton's (1993) classification to consider species hierarchy within these systems of use (Cudworth 2003), how this hierarchy extends into legal protection instruments for nonhuman animals (O'Sullivan 2011), and how cultural representations of animals vary according to species and use (Morgan and Cole 2011).

By recognising the widely different animal attributes, constitutions, and ecologies that contribute to our equally varied relations with them, such categories provide some degree of grounding to explorations of human-animal relations in acknowledging that there is a diverse reality that "exists independently of our perceptions, theories, and constructions" (Maxwell 2012: 5). However, it is at that point of assignation, where differences are carved up and ordered, where one differently identified body is given more or less worth than another, that a reality of difference gets mistaken for a reality of value. Speaking of racial difference Claire Jean Kim (2015: 15) explains:

Nature, of course, has always offered up a great deal of visible human variation by way of skin color, hair texture, facial features, and so forth, but "race" is a historically and culturally mediated way of reading, classifying, and ranking bodies, of assigning some more worth than others on the basis of physical variation. It is a means of producing and disciplining different and inferior bodies.

The aim of this chapter is to expose the grid of surface differences, so that it might then be possible to glimpse beyond it, to where categories and boundaries become meaningless, recognising that "[a]nimal species differ at their peripheries, and resemble each other at their centres; they are connected by the inaccessible, and separated by the apparent" (Foucault 1989: 291).

1.1   Animal Categories in Practice

The usual suspects of designated food animals appear throughout my participants' accounts of their meaty practices and there is unanimous agreement that sheep, pigs, chickens, and cows are regarded first and foremost as "our food animals" (Natalie C). Therefore, for all of but one of my participants—producer Florence (explained below)—food animals are clearly edible. However, for a considerable number, this was not always the case. Regarding their current and previous dietary practices, and providing some justification for my questioning, in Chap. 3, of the longevity of vegetarian and vegan practices, 14 of the total 41 participants (11 consumers and three producers) had been vegetarian for between six months and 12 years when they were younger (Table 3.1). A further four consumers had also been vegan for between two and twelve years. An additional two consumers (self-described pescetarians Tracey and Sophie) do not currently eat meat (from terrestrial animals) but are happy to buy and cook it for their families or partner and do not consider using animals for food as inherently problematic. Florence, the vegetarian producer, does not believe animals should be eaten but still participates in producing and supplying it to others. In sum, 21 of the 41 participants involved in the production and consumption of ethical and sustainable meat also have at least six months' experience being vegetarian or vegan. How is it, then, that these food animals have remained fundamentally and persistently edible?

Reflecting the same understanding about food animals as former vegetarian Natalie (C) (above), Blake (P) emphasises that these animals are "existing for that". This determination is suffused with a sense of 'how could it ever not be thus', recalling what Foucault (1989) described as the impossibility of thinking otherwise, for as Natalie (C) says, "this is our culture", or, as Lisa (C) reasons, it is "inevitable, that's why it's ['things that have been farmed'] here I guess ... that's sort of why it exists is to be killed". These and other participants portray any alternate existence for sheep, pigs, chickens, and cows as unimaginable, because the implications would evidently be ridiculous, as former long-term vegan Gillian (C) implies: "they should just all live, that's a great idea, well where the hell are they going to go ... really?"

That categorisations of certain animals are understood as empirical truth rather than a "historically and culturally mediated way of reading, classifying, and ranking bodies" (Kim 2015: 15) is reflected in the seeming imperviousness of their boundaries—the edges hardened and stabilised over time (although individual animals are sometimes permitted to transgress, as I will demonstrate). To think of food animals otherwise is therefore deemed almost illogical because "chickens are chickens" as Charlotte (C) says, and "a pig is a pig is a pig, it should be valued for what it is [i.e. pork]" (Gillian C). As pescetarian Tracey (C) reminds us, "in the end they are animals, so I'm not confused about the difference between people and animals". To be so confused would, by inference, mean that you were wrong-headed, muddled, and not 'seeing' clearly. Joyce (C) explains you must instead be "clear-eyed about that", or open your eyes (Anne C) and "broaden [your] mind" (Gillian C). However, this is not the kind of clarity implied by Foucault's (1989) critique of order, based as it is on a fabricated and learned construction, or grid, of 'the Animal'. As Butler asks in *Gender Trouble*, "[w]ho devises the protocols of "clarity" and whose interests do they serve?" (1990: xx).

In Butler's seminal volume, she describes the conception of (human) gender categories as the result of repeated and sustained social performances (191–192). Later, in *Bodies That Matter* she extends this theory of performativity to describe how even before birth a girl is 'girled' or a boy is 'boyed', representing "at once the setting of a boundary, and also the repeated inculcation of a norm" (Butler 1993: 8).[1] That all animals are similarly 'done', as in performed from birth, being first, and most inescapably '*nonhuman* animal' and subsequently 'dogged', 'horsed', 'pigged', 'cowed', 'chickened' and so on is illustrated by my participants. Diane (C), a former vegetarian, has observed that some people are surprised when a duck or chicken walks up to them for a pat, but, she says, "it's because they don't remember that those animals are exactly the same as their dog and cat". Referring more explicitly to normalised categories of use, Michael (C) comments that his pets are companions, "they are never entertainment", and perceives animals variously "as part of food, visual enjoyment, as part of nature, companion animal, there's many ways of looking at it". This is the first process in animals' objectification, where their individuality is reduced to their species and use. These "processes of objectification", Foucault observes, "originate in the very tactics of power and of the arrangement of its exercise", and are one of its most obvious effects (1977: 102).

Hence, the edibility of food animals is broadly conceived, and maintained, as an indisputable and God-given truth, as Trevor (P) queries rhetorically, "if we weren't supposed to eat any meat at all, why would we, the Creator or whoever, why would we have had meat here in the first place?" Following Butler, the 'speech act' does indeed appear to exert a degree of agency through both its performative and linguistic dimensions, and play a role in the ongoing materialisation of these 'social kinds'—their constitution first as animals, then as meat, beef, steak, and so on (1990: xxvii, 1993: 13). However, despite this categorising of edible kinds and the distancing from human kinds that it accomplishes, my data show that in most cases people readily acknowledge themselves as being animals at some level.

---

[1] Butler's theory has been critiqued for failing to account for how 'human-ness' is similarly performed resulting in an inherent speciesism (Iveson 2014). The significance of this critique lies in how oppressions based on gender, gender identity, sexual orientation, 'race', religion, ability, economic status, and many subaltern 'others' draw repeatedly on tropes of 'more' and 'less' human, and 'more' and 'less' animal (Iveson 2014: 28; see also Kim 2015). However, while wholeheartedly accepting Iveson's objections, I see value in thinking how, like gender, animals can be similarly 'done' (Gherardi 1994; Martin 2003) by the performance and reproduction in social practices of pre-determined categories which are in turn re-inscribed on their bodies.

Reflecting on her animal-eating practices, Joyce (C), a former vegetarian and vegan, reasons that "a lot of animals kill animals to eat, and humans are animals". Anne (C), a previously short-term vegetarian, contends that we face "life and death like every other animal", while Diane (C) states clearly, "I am an animal, and I'm an omnivore" and "I'm comfortable eating animals"—that is, nonhumans.

1.2   The Human-Animal (Dis)continuity

There are times when humans make positive associations with certain animal attributes: 'strong as an ox'; 'brave as a lion', 'wise as an owl' (Dunayer 2001: 165), suggesting a desire to identify more closely with particular animals (in certain circumstances), to enhance human qualities they are also perceived to possess, but in greater measure. The mythic personification of the noble savage, living 'in a pure state of nature', is another aspect of these positive associations with animal nature, absent the corrupting forces of civilisation (Ellingson 2001). Having animal instincts is also viewed positively in situations construed (and constructed) as competitive and/or predatory, such as in business, sport, dating, and sex.

However, these symbolic gestures to a certain qualitative continuity are not reflected in social practices where associations with nonhuman animals are overwhelmingly negative and reflective of the human-animal binary (Adams 2010 [1990]; Dunayer 1995; Yates 2010; Kim 2015). As Dess and Chapman note, "in everyday parlance, *animals* means not, and less than, human" (1998: 156, original emphasis). References to animals are more widely used as derogatory epithets to 'explain' particularly distasteful human practices (Sergie 2018; Murdoch 2016). When discussing food, 'animal' becomes something of a misnomer for it is in fact species difference that defines the primary category being 'done' to these animals and which makes all 'other-than-human' animal species edible. Species difference is, as Iveson (2014: 25) says, the fundamental "unmarked-but-marking" filter through which all social 'norms' pass—the ultimate 'group' that is maintained through repeated and routine, everyday enactments of us-and-them boundaries (Bauman and May 2001).

As discussed in Chap. 3, some posthuman and ecofeminist scholars such as Haraway (1990), Whatmore (2002), Plumwood (2001), and Warren (1997) argue that the perception of boundaries, and the binaries or dualisms that they define—between culture/nature, human/animal, male/female and many more—needs to be "jettisoned" (Whatmore 2002: 159). Against all perceived manifestations of anthropocentrism, they advocate for the "deconstruction of symbolic, discursive, institutional and material arrangements that produce the category 'human' as something unique, distinct, and at the centre of the world" (Pederson 2011: 67). Minimising boundaries is often undertaken as a rejection of human exceptionalism, defined as "the idea that humankind is radically different and apart from the rest of nature and from other animals", in preference for a "human/animal continuity" (Plumwood 2007: 1) of "mutually affective" relations (Latimer and Miele 2013: 5) that emphasise "alongsideness" (Latimer 2013).

The understanding of a human-animal continuity, that is, the absence of a boundary, is used in relation to ethical and sustainable meat to support continued (and, in some cases, renewed) practices involving food animals and their flesh. For example, Maria (C), who had a brief encounter with vegetarianism as a teenager, asserts kinship with all animals and characterises humans as "in the food chain". Consumer Anne, who had a similarly brief stint eschewing meat, remarks on the "ethical dangers for humans in setting themselves above nature … divorcing ourselves from the food chain", or, as Joyce (C) puts it, being "in denial about humans being animals". These narratives directly reference the human exceptionalism argument notable in the work of Plumwood (2004) and Haraway (2010) (who are cited specifically by two of my participants), and also Bennett (2010), who theorise human-nonhuman relations as a mutual "becoming with" in an undifferentiated assemblage of all "vibrant matter" (Haraway 2008; Bennett 2010). As Maria (C) comments:

> In some ways, the same human exceptionalism that allows us to treat animals the way we do in the industrial system, saying, well I'm human, I don't need to eat meat, therefore I won't, is also a form of human exceptionalism because it says well, we are different in the food chain because we have morality and ethics … We are also animals and so

whilst we don't have to eat meat we are also food, you know, and we are being fed on all the time by all sorts of organisms that we would not like to think about.

This system of knowledge also informs Alison's (P) assertion that "I don't have a problem with eating something that would probably want to eat me". However, as explained above, there is an element of evasiveness in this understanding of human-animal continuity, especially in putting humans on par with other animals as prey—that "we are all here together, … we share what we do" (Gillian C). James (P) astutely points to the truth of the matter when he ponders, "what it'd be like if someone ate us, … if we actually … were farmed or had a predator". As he intimates, while we may have once been more embedded in the food chain, where the possibility of encounters with 'something that would probably want to eat you' (including other humans) would have been very real and more common, that has not been the case for centuries and we have never been farmed for our meat.

Arguments for theoretically 'jettisoning' the human-animal binary are problematic for the "detachment from the actual life conditions of animals" (Pederson 2011: 67) that it can foster, and its tendency to minimise the very real differences between humans and other animals in their capacities to affect the lives of others. This is a view supported by Pederson and other critical animal studies (CAS) scholars (e.g. Weisberg 2009; Twine 2010; Cudworth 2011; Deckha 2012; Giraud 2013). For animals that are genetically designed, bred, raised, and used for food, notions of a mutually affective continuity seem particularly untenable. Until such time as practice falls in line with theoretical ideals, there is an ongoing need to acknowledge dualisms, understand the mechanisms behind their construction and maintenance, and determine how they can be transcended practically and not just theoretically. My participants' accounts of their (ethical) relation to food animals and nature, which evince an implicit speciesism, and also anthroparchy, illustrate this need.

Contrary to its anti-dualistic intent, the notion of an idealised human-animal continuity, as advanced by Plumwood (2002), Haraway (2008), and others, and reflected in discourses of ethical and sustainable meat (Bost 2012; King 2017; Mark 2017), conceals, justifies, and reinforces persistently dualistic practices of animal use. A prevalent and enduring discontinuity between the categories 'human' and 'animal' supports the persistent use of animals as food. It is, therefore, the architecture of this discontinuity that I seek to understand.

2   Stable and Unstable 'Food' Animals

Given the open pantry to which speciesism permits humans access, what is regarded as acceptable to eat depends on a swathe of interconnected meanings, knowledges, skills, and materials. In this section, I provide a more detailed analysis of how the edibility of animals is constituted as part of practices involving ethical and sustainable meat, and, from this, discern what appear to be the most resilient connections. I am necessarily selective in the practices I examine; although I acknowledge that other practices support the edibility of animals, multiple studies are needed to explore these in depth.

Most of my participants are very assured in identifying what animals they will and will not eat. There are certain animals they would never consider eating. For instance, Henry (C) initially had trouble describing an animal that he would not eat, but was then quick to qualify: "within reason—I'm not going to eat a whale or a turtle", as if that would be 'of course' unthinkable. For Henry, who spent five years as a vegetarian, the understanding of an animal as endangered renders it non-edible, despite his gastronomic adventurousness. James (P) similarly says he would "certainly not [eat] anything that's endangered". The speculative implication here is that it is the endangered status of some animals that guards against their edibility (notwithstanding that the appellation 'endangered' is also largely socially constructed, albeit in relation to concrete conditions in 'nature'; see Sutton 2004: 65). Anne (C) alludes to "freaky" animals as being the only ones she would not eat, "like, I wouldn't go to Africa and eat a monkey or China and eat a dog or something like that [laughs]". And yet, these 'freaky' animals that are 'outside' reason might hold a different meaning for another person. For example, Oliver (P) describes himself as "pretty versatile, I've eaten monkeys, I've eaten lots of you know, different [animals/species]."

The meanings humans associate with these variably 'off-limit' animals can therefore change depending on the practice. Given that meanings are malleable, then within the universally normalised understanding of non-human animals as a natural resource (to be eaten, or otherwise used, as per Brophy 1965, Benton 1993, Cudworth 2003 and others) these endangered, 'freaky', weird, or companion animals could theoretically become edible, or, in the language of Deuteronomy, shift from being unclean to clean.

Mary Douglas' notions of dirt and purity come to mind here. For some, to think of a certain animal as food transgresses a boundary of edible purity. Given such a situation, the animal, or their flesh, would become "matter out of place" (Douglas 2002: 36). Hence the sense of revulsion and danger that is evoked by the thought of eating something that is considered 'dirty', that is, contaminated by virtue of being out of order. As Douglas elaborates, "[d]irt is the by-product of a systematic ordering and classification of matter, in so far as ordering involves rejecting inappropriate elements" (2002: 36). However, as indicated above, the constitution of certain animals as unclean or 'dirty' in terms of edibility does not negate the possibility of them becoming edible. Rather, the socially determined appropriateness of this edibility construes it merely as dangerous depending to what extent it is "likely to confuse or contradict cherished classifications" (Douglas 2002: 37). What this underscores is the implicit edibility of all animals and the arbitrariness of related classifications where the *degree*, rather than the *fact*, of their edibility is what is open to change.

To illustrate, animals generally regarded as pets, such as dogs, present greater resistance to being thought of as food. A lot of my participants housed a canine companion or two and they often volunteered that there was no question of their being eaten—"it's just not a meat that you'd eat" as Florence (P) explains, and as Anne (C) also indicated above. However, as with 'endangered', 'freaky', and weird, there is the implication and potential that dogs can be 'un-petted' to become edible. Dan (C), who also has dogs, and was both vegetarian and vegan for nine years, is a little ambiguous when he speculates that he would "really struggle eating a dog if I was in you know south east Asia where it was on the menu", while Maria (C), who also has a canine companion, says she "would probably be willing to try dog meat".

Eating horses is met with slightly less resistance, perhaps owing to their greater mobility and acceptance across multiple categories and practices—sport, entertainment, pets, labour, fashion (horse hair), food, and wild animals. Melanie (P) keeps and breeds ponies and says with astonishment, "I don't eat them [laughs]". But Alison (P), who also keeps horses, is less adamant—"I probably wouldn't eat horse". Blake (P) thinks "there's probably nothing wrong with horse meat", while Geoffrey (C), a committed carnivore, has eaten horse meat in Europe. These examples highlight the permeability of edible boundaries depending on negotiations between variable meanings associated with practices that involve food across time and space (negotiations I will explore further in Part III).

The transgression from dirty to pure, or from unclean (inedible) to clean (edible), can also work the other way, from edible to inedible, and is equally dangerous to the normal ordering of animals. Stepping away from my participants for a moment, a prominent, and revealingly problematic instance of this is illustrated by the following story of Baa the sheep. I include it here as I will make reference to Baa's situation at several points in the book.

In 2000, a Melbourne man purchased a three-year-old ewe from a nearby farm. After nearly 11 years, during which time Baa had become this man's beloved pet, he was forced to engage in a lengthy and expensive court battle to fight a council ruling that prohibited keeping 'livestock' on a suburban block. He fought the ruling for the next two years, costing him hundreds of thousands of dollars, but ultimately lost. By this time, at the age of 16, Baa had developed cancer. Her health steadily declined over the following year and she died at the age of 17 at home with her owner. He described his relationship with her as "true love" and that he had not expected to have such joy with her. On her last day, she was "still wagging her head" (Townsend 2014).

This story demonstrates the effect that the (taxonomical) naming, classifying, and 'ordering' of animals has on the stability of associated practices, and the ways both discourse and practices work to reproduce and reinforce these orders and limit opportunities for transgression. By contesting the incongruity and, more proscriptively, the illegality of treating a sheep— that is, livestock, natural resource, food—like a pet in Australia, Baa becomes a heterotopic site where normalised orders are disturbed and subverted.[2] However, for most of my participants, transgressions from edible to inedible are less radical.

Effecting a more ambiguous form of boundary blurring, the popularity of keeping pigs as pets has spread in recent years (Blakemore 2011; Major 2017), and the focus of breeding practices has expanded from improving their economic and gustatory value to humans, to include aesthetic values such as size, colouring, other physical traits, and temperament, in a similar way to dogs.[3] Whether this has led to the increased recognition of their intelligence, or vice versa, is unclear, and probably a little of both. However, this intelligence, especially when likened to dogs, causes pigs to be the typically farmed edible animal whose boundaries have become less solid and more easily transgressed than those for cows, sheep, and chickens.

On the one hand, pigs are celebrated for the variety of ways their flesh can be treated and the different types of meat that can be produced (Villas 2011; Pearsall 2015)—sentiments also shared by my participants. Gillian (C), a former vegetarian and vegan of 16 years, explains:

> I like pork, pork is really flexible in what you can do with it, preserved meat as well as lunch meats and cooking … I don't do a lot of pork cutlets or pork roasts but I do have bacon and I have lardons and lard [laughs], … and if I make soup it's got the bacon rinds in it, … so you can use everything of the animal, that's what I love about it.

Andrew (P) agrees, "pigs are just amazing animals, in that you can use so much of [them]". Bacon, pork belly, salamis, and even blood pudding feature prominently throughout my participants' accounts and pigs are commonly agreed to make a significant and prized 'contribution' to the available range of edible meat products, as Maria (C) explains, "pork can go into so many things, … pork belly, bacon, pancetta, … the flavour is

[2] Due in part to their constitution as livestock according to the National Livestock Identification System (NLIS). As such, in Australia, as in most Western nations, they are subject to a different set of welfare guidelines and codes of practice than 'companion animals'.

[3] For example, see: http://americanminipigassociation.com/owners/ready-mini-pig-owner/mini-pig-facts-myths/, www.orangegrovefarmstinypigs.com/specific-breedsize-standards- and also Cyranoski (2015).

in the fat". The pig is often the animal that provokes a (humorous) sense of the inevitability of its flesh being eating, and its power to 'break' vegetarians and vegans, as Gillian (C) recalls, "bacon was a problem, oh boy, [groans, laughs], and that was what broke me". Their 'efficiency' and usefulness are also admired. Considered the "tractors" of the farm, their talent for "digging up the earth" (Alison P) means they "go through [the land], turn it all over, get rid of all the weeds" (Gillian C). They are, moreover, "very efficient … because you can feed them anything" (Geoffrey C).

On the other hand, there are participants for whom a recognition of the intelligence and 'pet potential' of pigs unsettles normalised understandings of their edibility. Speaking of her sons' ambivalence about eating pork products, pescetarian Tracey (C) explains, "it's the pet thing… they've seen pigs as pets". Diane (C) simply says "pigs are smarter", and similarly for David (C), "pigs are much more highly intelligent". The straying of pigs from the traditional and comfortable conception of farm animals as just passive, "large bovine, ruminants" (Tracey C) presents a discernible challenge to 'normal' classifications and thus to practices. Finn, a producer who spent one year being vegetarian and now eats meat at least once a day, reveals, "*pork* breaks my heart cos they're just so smart, like they're really smart animals" (emphasis added).

The edible 'purity' of pigs is being challenged for many participants to a greater extent than it is for cows, sheep, and chickens. With new meanings being associated with 'pig' (intelligent and friendly) along with the traditional ones (efficient food, tractor), they are transgressing normalised boundaries, falling in and out of edibility depending on the practice. Nevertheless, in all cited cases, products made from pigs are still being eaten by these participants despite the chink in the food/pet boundary. This chink, or glimpse through at least one denominational grid (food animals), has only slightly increased the perceived level of danger in eating pigs. The resilience of pigs' general understanding as food is summed up by Graham (P), who keeps pigs, some of whom he names. He emphasises their difference from "a complete pet", saying "ones that are named, pigs, you know they're alright, their taste outweighs the…"— while trailing off here, it is clear that Graham means that their taste outweighs any level of attachment that comes from naming them.

While my data exhibit significant variation in whether certain species are perceived by producers and consumers solely as food or not, such determinations are clearly not purely reducible to simple personal attitudes and opinions about what is and is not edible. Rather they are an amalgam and manifestation of much broader shared understandings— about life, consciousness, nature, and human-animal relations—that are constantly evolving and being shaped through myriad connected discourses and practices. As Dan (C) indicates, his 'decisions' surrounding what he regards as edible stem from something much more complex:

> If an animal has a conscious awareness of self then I wouldn't eat it because I wouldn't eat another person, because they have a conscious sense of self, a storied sense of self that goes over time, a narrated sense of beginning, middle, end of who they are. I think possibly chimps have that and other higher kind[s] of primates so I wouldn't choose to eat a chimp.

Similarly, for Sophie (C), the possibility of some kind of mutual connection with an animal is enough to disturb common boundaries around what is and is not edible: "If I could have a relationship with that animal potentially, I don't really want to eat it". Incidentally, seven months after conducting the interview with Sophie, she informed me that she had become vegan.

Yet another way that the order of edible animals can be disturbed is their age. Ellis (C), a once-a-day meat eater, finds chicken "really problematic" because "most chicken you eat ... would only be about eight weeks old". As Will (P) says, "we call it chickens, we're eating chicks". Diane (C) feels the same about veal, which she "wouldn't eat", and, ever since breastfeeding her son, Natalie (C) avoids eating suckling animals because she "just related to it too much". However, as with animal classifications, determinations regarding what/who is 'too young' to eat are equally arbitrary. Furthermore, unlike intelligence or relatability, age does not seem to have the potential to exclude an entire species from being considered edible, but instead modifies under what circumstances it becomes edible.

For example, in response to the routine slaughter of approximately 800,000 'bobby' calves[4] per year in Australia (Voiceless 2015) at a few days old (the legal minimum being five days), a little complex of practices has emerged, involving producers from several states, to produce 'ethical' veal that has had "some time" (usually around a couple of months) to run around "really enjoying their life" (Alison P).[5] As Finn (P) explains on his website, "we sell grass fed veal…from fully weaned cattle, aged between four to six months. This may sound young, but *in agricultural terms* it is not. At least the animals have the opportunity to experience life" (emphasis added). The already arbitrary evaluation of an animal being 'too young' to eat can be ameliorated by adopting an *agricultural* perspective, and drawing on similarly arbitrary evaluations of the alternatives, such as being 'wasted' as a by-product of the dairy industry, or living only a few days as opposed to six months, at most, before being eaten. 'Waste' here is being defined in a way that prioritises a certain construction of 'ethical' value, for whether at five days or a few weeks/months old, male calves are still slaughtered and their carcasses profitably used in cheesemaking and canned export products, as well as "pet food, leather goods, the pharmaceutical industry or to be processed into pink veal for human consumption" (Voiceless 2015).

As well as illustrating how naming, order, and classification shape animals' edibility, the observations in this section, including the story of Baa, are suffused with emotions. They trace, through discourse and related practices, attempts to navigate and negotiate between pleasant and less pleasant feelings. This process is a key element in my analysis of the persistence of meat and will be the focus of Part III. In the final part of this first chapter of Part II, I turn to the ways in which edible orders are maintained so that glimpses beyond Foucault's grid do not constitute a significant threat to normalised classifications.

---

[4] Unwanted calves destined for slaughter. Most bobby calves are male, but they also include female calves who are, for whatever reason, considered unsuited to 'herd' replacement or milk production (Voiceless 2015).

[5] Similar practices are evident or emerging in the UK (Gordon and Baroke 2014), the US (Black 2009), Canada (Fiorucci 2018), and New Zealand (Heaton 2017).

3   How Discourse Can Inform Practice

From the preceding discussion, it is clear that the categories and boundaries of order, although flexible and continually evolving, require work to maintain them. This is because while being iteratively enforced and reinforced through everyday practices, they are also occasionally challenged. One way in which challenges are negotiated—and thereafter deflected or incorporated in some way—is with recourse to dominant, normalised (and normalising) discourses. In relation to ethical and sustainable meat, I identify three what I term validating discourses that are drawn on to mount a forceful and decisive defence of meat consumption and the use of animals as food—to maintain and, above all, validate the proper order when "aberration", "defilement", and "abomination" threaten "the margins of the lines" (Douglas 2002: 123 and various).

3.1   Validating Discourse No. 1: The Value of Contingent Life

This discourse explicitly prioritises 'contingent life' by maintaining that 'we' have an obligation to grant life, any life, because life in and of itself is worth more than the quality of that life. Contingent beings are defined as those who do not (yet) exist and whose existence depends on particular practices, such as the killing and/or eating of existing animals. Their lives are thus understood as 'contingent' rather than 'inevitable' existence (Kagan 2015: 141). The contingent life discourse plays a fundamental role in shaping the common or 'normal' regard towards food animals as a category, in policing its boundary, and therefore in reinforcing the persistence and resilience of associated practices. However, it is not a new discourse. In 1914, Henry Salt identified "the Logic of the Larder" as when a food animal's few hours of suffering are perceived to outweigh the "enormous benefit of life" (1914: 1).[6]

---

[6] That life lasts an average 5–7 weeks for 'broiler' chickens or 'meat' birds (natural lifespan 8 years); 5–7 months for pigs (lifespan 10–12 years); 18 months for 'beef' cattle (lifespan 15–20 years); and 6–8 months for lambs (lifespan 12–14 years). The lifespans of most of these animals are not too dissimilar from domestic cats and dogs and so there is the direct equivalence of considering them

Moves in support of 'ethical' veal, as mentioned previously, are especially illustrative of the contingent life discourse but it appears more broadly across my participants' accounts, underscoring their regard for animals and meat. For example, Alison (P) believes "there is something to do with the actual animals being born just as they would've been in nature, masses of animals being born and having a life". From this, it follows that killing animals for food becomes "morally permissible and even obligatory" (Visak 2013: 130) as long as the animals lead pleasant lives and would not have otherwise existed.

This logic has proven resilient and recurs as a consistent theme in my data. As Gillian (C) explains, "if we didn't eat them, they wouldn't exist". It is only by virtue of being eaten that they are 'granted' a life at all, for as Blake (P) exclaims, "if we all stopped eating meat, all the animals would die out, for sure". Producers James and Andrew respectively echo this understanding: "If we weren't eating him then he wouldn't have lived"; and "if we weren't here farming and eating the animals … they wouldn't actually exist". There is remarkable similarity in the ways this understanding about using animals as food is expressed, suggesting that it is a perspective that over time has taken on the unreflexive quality of accepted common lore. Some even construe this as a 'deal' that animals themselves have signed up for, as producer Alison explains, they "understand they're here for that, they're actually here for us to use … animals accept that". Grace (C) agrees "that's the bargain in a way, animals have made". (The attribution of subjectivity, thoughts and, feelings to animals that is evident here will be explored in Part III.)

However, the fallacy of this position is revealed by the "nonsense" of "predicating … happiness or unhappiness, of that of which we can predicate nothing" (Salt 1914). Put more simply, "non-existent animals cannot benefit from being caused to exist" (Visak 2013: 136). Hence, bargaining on the value of such contingent lives, and constructing the 'kindness' of bringing them into being to justify the killing of existing

---

to have lived 'enough' of a life to be acceptably killed for food at between 5 months and 18 months of age. Contributing to the perception of these food animals as 'old enough' are the genetic modifications that have been undertaken to make them grow bigger and faster than they would 'naturally'.

animals for food, even very happy animals, do not endure scrutiny. For "killing an animal which otherwise could have continued a happy life counts as a welfare loss" (Visak 2013: 135).

A pivotal tenet in the discourse of contingent life is the naturalness of this arrangement, as Alison (P) and Grace (C) illustrate above. Indeed, earlier in this chapter, in the section titled 'The human-animal (dis)continuity', I described how participants draw on understandings of humans being part of nature to legitimise their eating of animals. While naturalising the edibility of nonhuman animals in general, these understandings also foreground the social construction of nature and the targeted application of naturalisation this then allows for. This provides the basis of my second and third validating discourses.

3.2   Naturalised Discourses

In the context of consumption, Richard Wilk defines naturalisation as when wants are defined as needs, taken out of contention and embodied as (natural) taste, urge and impulse (2002: 10). They become so naturalised they are not given any conscious thought (Wilk 2009: 150). The goal of naturalisation (as a process), Wilk says, is "to make some practices unthinkable while making others seem 'normal'" (2002: 11)—a further iteration of boundaries and categories. The advantage of drawing on common understandings of what is 'natural' is that it allows associated discourses and practices to leverage the fallacy of an appeal to nature. This fallacy holds that what is 'natural' is inherently right or good and what is 'unnatural' is inherently wrong or bad.[7]

While naturalisation contributes to the general edibility of all animals, my focus here is where and how naturalisation is deployed to further

---

[7] Melanie Joy famously highlighted that the association of meat with being natural is one of three defence mechanisms that support the ideology of carnism. The other two are that it is also normal and necessary. According to Joy, these defence mechanisms rationalise a values-behaviour gap in relation to eating animals. As well as locating the constitution of carnism in individuals and their beliefs, rather than in social practices, Joy does not problematise how what is deemed natural is constituted more broadly, or, in other words, the *process* by which something is naturalised, which is through shared understandings and the social practices they are part of. Instead, carnism is conceived and approached as a reified ideology associated with 'meat' that is detached from the array of social practices and complexes that shape it.

distinguish *more* natural, that is, 'ethical' and 'better' (as opposed to just 'right' and 'good') animals and meat. I identify two aspects to naturalisation in this regard, which function as two further validating discourse. The first narrows uncritical invocations of the 'rightness' of 'nature' to more specific determinations of what is (more) 'natural' or ethical. The second construes all that is 'natural' as inherently benevolent, which adds an enhanced and more 'productive' (after Foucault) sense of 'goodness' to what is already considered ethical. As a note, when discussing their meaty practices, participants use a range of words to denote something less artificial and therefore 'good' or 'better'—closer to some (mythical) understanding of nature/natural. These include genuine, authentic, normal, pure, real, proper, and traditional. I therefore consider how the 'natural' is constituted through the use of all these words.

3.3 Validating Discourse No. 2: Invoking a Better, More Ethical 'Nature'

Above all other meats, the flesh of wild animals is regarded by many of my participants as the most ethical because, as Beverley (C) says, "[the animal] can live a natural life rather than a kind of weird farmed environment that's not natural for it". Beverley spent just over a year being vegetarian and now eats meat once a month on average. Pescetarian Sophie (C) is of the same mind. She not only regards these animals as "natural wildlife", but the act of killing them as "sort of how it would happen naturally"—or "natural hunting" as Beverley (C) describes it. This imbues practices that involve hunting and eating 'wild' animals with a sort of purity, less blemished by human intervention—matter more in place. However, unlike Beverley (C), most of my participants do not think a farmed environment is 'weird' and perceive 'traditional' farming as natural.

Speaking of his farm as a whole, Andrew (P) describes his efforts to "get a *more* natural system in place" (emphasis added), one that necessarily includes animals in a biological (i.e. natural) interaction between soil and plants. Graham (P) also appeals to the natural sciences, explaining how "the whole soil life, the biology and everything, it's all tied in together, it's tied in with different animals". "If you take animals out of that system", Andrew (P) explains, "you get woodlands or more complicated kinds of forests, so…"—the implication being that woodlands and complicated systems are undesirable in some way and grasslands are natural. In response to being asked to consider food systems without animals, Simon (P) asks, "what's going to happen to our grasslands?" This is redolent of the common discourse surrounding what are judged to be 'wastelands' with no intrinsic value and therefore best put to (typically economic) use (Baka 2013). As committed meat-eater David (C) says, referring to large tracts of Australia, it is "space that really is only suited to pasture". Hence the land is perceived and defined in terms of its 'natural' potential as a place to raise food animals.

'Nature' becomes almost reified as a living and even thinking entity, with Florence (P) describing sustainable farming as "the way it should be and that nature would like it to be". A distinct system of thought or knowledge becomes evident wherein 'the soil' and 'ecosystems' are understood as needing to be cared for by humans, and where ethical/sustainable farming systems are seen as the way to achieve this because they are "so beneficial to the soil … that's the way it should be" (Will P). Gillian (C) believes "healthy soil needs animals as part of it" (i.e. 'livestock'), and Bella (P), another formerly vegetarian producer (of seven years), similarly stresses the importance of "doing the right thing by the soil" by having animals as part of "a really nice system". As Florence (P) explains, the way "to return [the land] to its natural state" is by farming in the right way, an ethical way, where everything is "treated humanely", including the animals and the environment.

The naturalness of ethical and sustainable farming systems, as described by my participants, is constituted in opposition to 'unnatural' practices, or, more accurately, they are co-constitutive. For example, Florence (P) describes farming "more intensely", and creating "dusty, dirty paddocks", remarking "how awful that would be, and how unnatural". While the "high tech and unnecessary" process of producing in vitro meat prompts former vegetarian Heather (C) to declare her "preference for low tech natural processes", that is, farming animals. For ethical farming practices are 'known' to be 'good' and 'natural' not only for the soil, ecosystems, and the land, but also for the animals involved.[8]

---

[8] Note that the meat that is the product of practices perceived as unnatural is not necessarily inedible, as I will illustrate in Part II. The 'natural' edibility of animals in general remains.

An alternate system of knowledge, one that suspends the normalised boundaries and categories that divide up the 'natural' world according to anthroparchal constitutions of purpose and value, might offer a different view of this farming 'nature'. Several authors have remarked that the agricultural landscape so widely regarded as 'natural' is the result of centuries of environmental and habitat destruction (Redman 1999; Nibert 2013)—a destruction that has steadily intensified to the present day (Grigg and Halford 2013; Tanentzap et al. 2015). Some have even questioned the benefits of the Agricultural Revolution (Larsen 2006; Harari 2011), with Jared Diamond (1987) going so far as to describe it as the worst mistake in human history. Recent research from Harvard Law School has modelled repurposing UK land currently used for animal agriculture to achieve substantial carbon dioxide removal while increasing food security (Harwatt and Hayek 2019).

Naturalisation contributes to already naturalised animals (and flesh), that are part of *more* 'natural' farming practices, being constituted as 'essential' and unquestionably 'right'. Relatedly, but also distinctively, the discourse of a better or truer naturalness also fosters a sense of the benevolence of associated systems and practices—the goodness that they bestow on animals, communities, and individuals. This sense of synergistic benevolence and mutual reciprocity constitutes the third validating discourse.

3.4   Validating Discourse No. 3: The Benevolence of the 'Natural' Contract

My participants consistently draw on and articulate a common 'knowledge' that it is through ethical farming practices that, as Graham (P) says, "animals are grown properly" in ways that take their individual well-being into consideration. The farmers applying these ethical principles are in turn described in benevolent terms. Because they "pretty much go all out … to make it natural, healthy, happy for the animal" (Heather C), the farmers are seen as "doing the right thing … it's all very real and genuine" (Florence P). Jennifer (P) describes the lives of such farmers as "based around being authentic … with absolutely everything they do".

Authenticity is a common trope in studies of consumption (Beverland et al. 2008; Leigh 2006; Banet-Weiser 2012), and especially consumption of food (Beverland and Farrelly 2010; Weiss 2012; Simunaniemi et al. 2013; Pratt 2008). Jeffrey Pratt observes that 'the authentic' is related to other words, including "the natural, the organic, the local, the rooted, the distinctive" (Pratt 2008: 56) in ways that they "become synonymous, or at least immediately evoke each other" (ibid.). He links this "pre-set discursive field", which could also include the ethical, the good, the real, and the traditional, "precisely" with the romantic tradition. It is this discursive field of romantic "folksy charm" that Anne-Marie Todd (2009) identifies in her analysis of the construction of "Happy Cows and Passionate Beefscapes" in food advertising. 'Natural' landscapes and food animals are foregrounded amid allusions to what has been 'lost'— "solitude, wilderness, lush landscapes, free-flowing water, and clean air" (172, citing Corbett 2006). But the consumer can regain some part of this loss through the consumption and literal embodiment of these natural, happy animals. Table 4.1 illustrates how the producers that I interviewed endeavour to tap into this discursive field to create similar associations with their animals, farming practices, and meat in their promotional materials.

With these promotional materials, under so-called "old-fashioned", "natural", and "traditional farming methods", the paddocks and pastures required for raising food animals, and the practices of grazing and rotation, become 'natural' by association. They come to be understood as part of an ethical, holistic, and, above all, natural system. 'Farming' animals, or raising them for meat, is thereby conflated with "respecting", "regenerating", "nurturing", "enhancing", caring for, and generally supporting key elements of this 'nature' including its ecosystems, soils, and biodiversity.

The discursive field of authenticity and benevolence evident in these promotional materials contributes to constitutions of 'good' meat. In other words, it shapes how these producers' meat is made sense of. Geoffrey (C) explains that as long as an animal has been raised "correctly", meaning, "in the sun, on the pastures, eating grass, kind of just hanging around and being happy", then he regards that as "optimal" and therefore the more ethical meat to eat. Producer Simon succinctly sums

Table 4.1 Authentic, natural, and sustainable idylls within producers' promotional materials

Farm	Animals
• The farm is fully **sustainable**, a running creek and dams on the property	• Our animals are **happy** and **stress free** … grow in a **natural** and healthy environment
• **environmental** farming principles	• **Free-Range** Pigs are **happy** pigs.
• care of the **natural** biology and enhancing our **biodiversity— ecologically sustainable** farming practices	• Our pigs live as pigs are meant to
• **sustainable** philosophy	• Our cattle lead a contented low-stress life live a **happy life** out in the open
• **regenerating** land—a **holistic** farming system. **Natural** and ethical farming. Our philosophy of farming **naturally**	• '**Beyond**' Free range, pasture-raised, ethically grown … allowing the behavioural instincts be **real**.
• support **local** farmers—**local**, ethically grown chickens—operates within the **ecological** boundaries of our property, **regenerating** the land & soil.	• raised the **old-fashioned** way
• a truly **sustainable** way to farm.	• **true** free range, outside, cage free and **happy animals**
• **clean natural** farming practices.	• Our animals **roam free** across our **paddocks**, eating beautiful **green pasture**
• **respecting** the land that we farm	• Organic—Grass-fed—**Ethical**
• environmentally **sustainable** and ethical	• Our lovely animals graze on **native pastures** and live and grow the way **Nature intended**.
• **Environmentally sustainable** … sustainable farming practices	• **Natural** Beef.
• **sustainable**, organic/biodynamic, non-chemical/no-hormone practices— health of the land and water for future generations, build connections with local families	• **clean**, wholesome, **natural** beef and lamb… **genuine** seasonal lamb … **traditionally** aged

up the deceptive, normalising 'logic' of this discourse when he exclaims, "nature does it [i.e. produces meat] naturally".

The intrinsic benevolence of what is constituted as 'nature' and 'natural' processes, which include meat, and notions around the 'proper' and 'correct' way to treat edible animals in return, frequently segue into references to a natural contract between humans and these animals. This contract is understood as one that is willingly entered into by food animals, or is at least in their best interests. Joyce (C) describes some of the terms of this contract:

The idea of a social contract between the animal and the human … like the trade-off that an animal makes for a life of never having to forage, never having to … be exposed to danger is the fact that it dies before its time. But maybe it dies before its time [in the wild] or maybe it would have died a horrible death you know as a piglet, aged 13 days or something, who knows.

Grace (C) similarly explains how the contract benefits food animals:

Animals are sort of in the evolutionary decision making thing, got domesticated … not to be treated the way we treat them, you know like on factory farms and stuff like that. I don't think that's what they agreed to, but if you look say a subsistence villages, you see pigs and chooks and guinea pigs or whatever running around without fences. They're going to be eaten, but they've moved in with the people because it's a guarantee … that their genes will be passed on. So I think there's an element of, that's the bargain in a way, animals have made.

In support of the terms of this contract, there is a common narrative, illustrated in Joyce's (C) comment above, whereby the lives of animals in the wild are construed as dangerous and harsh, and more especially their deaths as brutal and much worse than the 'instant death' of the slaughterhouse. Daily meat-eater Geoffrey (C) places a lot of weight on this argument and explains at length:

If we don't eat the meat, how else will it end its life? Do we let them rot and die of old age? Do we allow foxes and other predators to tear them apart as they are living? … It's like instant death so it doesn't really affect them, they're never under stress or anything like that. The other option is if they just stayed on the farm and slowly withered away and died … I reckon it's actually a lot nicer for them just to have an instant death than if they were just roaming around on their own and died from breaking a leg or being attacked by another animal which would happen if they were just in the wild and us humans didn't eat them…Right now the meat I buy lives better than some humans.

Former long-term vegan and vegetarian Dan (C) expresses a similar point of view:

No-one dies peacefully, really, especially not in the wild. So, an animal living on a free range, healthy farm where they're provided with good food is probably living a better life than they would have lived necessarily out in the wild sometimes. Getting eaten alive by a lion is not as nice as getting shot with a bolt through the head in a slaughterhouse.

Mary Douglas describes a "ritually contracted death" as one that separates the personal life of the sacrificial subject from their public life—or constituted role in that public life (2002: 69). The purpose of the ritual narrative is to "control situations" and thereby modify or mediate the experience (68, 70). Drawing on Douglas' account, one could conceive of my participants' contractual narratives as supporting a kind of secular ritual of meat consumption. This ritual can even take on an animated life of its' own.

As Geoffrey (C) illustrates above, many of my participants conjure the thoughts and feelings of food animals at the point of slaughter to support these animals' 'preference' for the 'instant death' that humans provide in contrast to the constructed ravages of nature. So as well as being 'not really affected' and 'never under stress', Julie (C) maintains they are "not in a state of panic and fear", because, as Henry (C) says, they "don't see it coming, no idea, nothing"; they have "a few minutes of fear, potential fear, but minutes, not days", Nicki (C) explains, and even, as Sophie (C) describes, just "fall asleep with their friends".[9] Sometimes, both consumers and producers will ventriloquise food animals, play-acting little scenes as they describe them being led into a new paddock, loaded into transport for the slaughterhouse 'with their friends', or making amorous advances to their field-fellows. However, these animated and subjectified animals always at the same time remain resolutely objectified as meat—that is their socially determined sacrificial role. The seemingly contradictory positions of animal-as-subject and animal-as object can easily co-exist, as Morgan and Cole also observed in their analysis of the discursive representation of 'happy pigs' (2011: 125).

---

[9] This is in reference to the CO_2 gas chambers now widely used for the slaughter of pigs—a method of killing that is far removed from simply falling asleep (Akbar 2015). CO_2 stunning, as it is commonly referred to (or Controlled Atmosphere Stunning), is used in slaughterhouses in the UK, US, Europe, Canada, and New Zealand.

These animals' objectified flesh is also woven into the contractual narrative, being the 'gift' that is 'given' in return for the benefits of domestication or a quick(er) death. Alison, one of the formerly vegetarian producers, explains her production of meat from dairy cows:

> For me the ethical side is that … we've got these cows that would normally be slaughtered…it's our way of thanking them for what they've *given* us in their milk…then you're *given* these calves … animals understand they're here for us, they're actually here for us to use. (emphasis added)

In a similar vein, Jennifer (P) describes her relationship with the pigs that she raises for meat as one where "your energy that's gone into producing it is then *providing* you again with more energy back", and Oliver (P) emphasises the importance of consumers "giv[ing] dignity to the animal that *provided* them with meat". The conception of meat as a natural bounty that animals willingly 'give', 'provide', and bestow on humans suffuses my data. The state of domination in which food animals exist, evidenced by the fact that they are always destined, and liable at any time (e.g. due to inappropriate behaviour or insufficient worth), to 'fall off' the end of the power spectrum (after Palmer 2001), is concealed by a narrative of reciprocity. As Nicki (C) says of the sheep that she keeps, whose lambs are killed for home consumption, "they've given me joy looking after them, they're keeping down the grass, it's like an almost synergistic relationship". This implied 'self-mortification' is one of the features of pastoral power as described by Foucault (1979: 239), and one that Matthew Cole also observes in discourses of 'animal-centred' welfare in which food animals are seen as 'choosing' to self-mortify and submit their bodies in a "devotion of their lives to serving human appetites" (2011: 92).

In sum, the prioritisation of contingent lives, combined with the naturalisation of meaty practices and the constituted benevolence of the natural contract, is a key discursive mechanism by which the domination of food animals is substantially concealed. Drawing on Todd and Pratt, what is ultimately at work in how ethical meat, food animals, and practices are 'made sense of' is therefore a fusion of idealised notions of nature, pastoral life, local, organic, free range, the 'genuine', 'real', 'authentic', and 'natural'—a collective 'knowledge' that constitutes a "passion for [ethical] meat ... as a passion for nature" (Todd 2009: 175). However, this human-animal symbiosis and benevolence is constituted around a human-centric and naturalised want that requires the persistent regulation and control of these animals' lives until their unavoidable deaths. As Derrida remarks, "The appropriation, breaking-in, and domestication of tamed livestock are *human* socialization" (2008: 96, emphasis added).

4   Conclusion

Together, the three validating discourses I have identified in this chapter—the value of contingent life, invoking a 'better', more ethical nature, and the benevolence of the natural contract—act as boundary-forming devices, maintaining the overarching edibility of food animals. Given, then, that practices involving the use and consumption of food animals are constituted as natural, benevolent, and essential to their being alive, how are some animals and meat constituted as 'better' to eat than others? What other meanings are attached to food animals and their meat to make some more ethically edible than others? These are questions I attend to in the next chapter, where I demonstrate how these validating discourses can be seen to shape, alter, and modulate understandings of a range of meat-related practices, interacting with these further systems of distinction to perform even more variable and shifting dissections of the good, the bad, and the ethical.

References

Adams, C. J. (1994). *Neither Man Nor Beast: Feminism and the Defence of Animals*. New York: The Continuum Publishing Company.

Adams, C. J. (2010). *The Sexual Politics of Meat: A Feminist-vegetarian Critical Theory*. London; New York: Continuum International Publishing Group.

Akbar, J. (2015, February 17). Revealed: Disturbing Footage of Pigs Struggling to Breathe as They're Killed by CO_2 Stunning Method Being Used by Supermarket Abattoirs. *The Daily Mail*. Online. April 2018.

Baka, J. (2013). The Political Construction of Wasteland: Governmentality, Land Acquisition and Social Inequality in South India. *Development and Change*, *44*(2), 409–428.

Banet-Weiser, S. (2012). *AuthenticTM: The Politics of Ambivalence in a Brand Culture*. New York: NYU Press.

Bauman, Z., & May, T. (2001). *Thinking Sociologically*. Oxford; Malden: Wiley Blackwell.

Bennett, J. (2010). *Vibrant Matter: A Political Ecology of Things*. Durham; London: Duke University Press Books.

Benton, T. (1993). *Natural Relations: Ecology, Animal Rights and Social Justice*. London: Verso.

Beverland, M. B., & Farrelly, F. J. (2010). The Quest for Authenticity in Consumption: Consumers' Purposive Choice of Authentic Cues to Shape Experienced Outcomes. *Journal of Consumer Research*, *36*(5), 838–856.

Beverland, M. B., Lindgreen, A., & Vink, M. W. (2008). Projecting Authenticity Through Advertising: Consumer Judgments of Advertisers' Claims. *Journal of Advertising*, *37*(1), 5–15.

Black, J. (2009, October 28). The Kinder Side of Veal. *The Washington Post*.

Blakemore, S. (2011). Potbellied Pigs: The Right Pet for You? *petful.com*. Online. April 2018.

Bost, J. (2012, May 3). The Ethicist Contest Winner: Give Thanks for Meat. *The New York Times*. Online. May 2013.

Brophy, B. (1965, October 10). The Rights of Animals. *The Sunday Times*. Online. April 2015.

Butler, J. (1990). *Gender Trouble: Feminism and the Subversion of Identity*. London; New York: Routledge.

Butler, J. (1993). *Bodies that Matter: On the Discursive Limits of 'Sex'*. London; New York: Routledge.

Cole, M. (2011). From "Animal Machines" to "Happy Meat"? Foucault's Ideas of Disciplinary and Pastoral Power Applied to 'Animal-Centred' Welfare Discourse. *Animals*, *1*(4), 83–101.

Cudworth, E. (2003). *Environment and Society*. London: Routledge.

Cudworth, E. (2011). *Social Lives with Other Animals*. Basingstoke; New York: Palgrave Macmillan.

Cyranoski, D. (2015). Gene-Edited Pigs to Be Sold as Pets. *Nature*, *526*(7571), 18.

Deckha, M. (2012). Toward a Postcolonial, Posthumanist Feminist Theory: Centralizing Race and Culture in Feminist Work on Nonhuman Animals. *Hypatia*, *27*(3), 527–545.

Derrida, J. (2008). *The Animal that Therefore I Am*. New York: Fordham University Press.

Dess, N. K., & Chapman, C. D. (1998). 'Humans and Animals'? On Saying What We Mean. *Psychological Science, 9*(2), 156–157.

Diamond, J. (1987, May 1). The Worst Mistake in the History of the Human Race. *Discover Magazine*. Online. January 2018.

Douglas, M. (2002). *Purity and Danger: An Analysis of Concepts of Pollution and Taboo*. London; New York: Routledge Classics.

Dunayer, J. (1995). Sexist Words, Speciesist Roots. In C. J. Adams & J. Donovan (Eds.), *Animals and Women: Feminist Theoretical Explorations* (pp. 11–31). Durham; London: Duke University Press Books.

Dunayer, J. (2001). *Animal Equality*. Derwood, MD: Ryce Publishing.

Ellingson, T. (2001). *The Myth of the Noble Savage*. Berkeley; Los Angeles; London: University of California Press.

Foucault, M. (1977). *Discipline and Punish: The Birth of the Prison*. New York: Vintage Books.

Foucault, M. (1979, October 10 and 16). Omnes et Singulatim: Towards a Criticism of 'Political Reason'. *The Tanner Lectures on Human Values*. Stanford University, Palo Alto.

Foucault, M. (1989). *The Order of Things*. London; New York: Routledge.

Fiorucci, M. (2018, September 28). Veal: The Greener (and Rosier) Side. *The Healthy Butcher*. Online.

Fudge, E. (2002). *Animal*. New York: Reaktion Books.

Fudge, E. (2013a). Milking Other Men's Beasts. *History and Theory, 52*(December), 13–28.

Fudge, E. (2013b). The Animal Face of Early Modern England. *Theory, Culture and Society, 30*(7–8), 177–198.

Gherardi, S. (1994). The Gender We Think, the Gender We Do in Our everyday Organizational Lives. *Human Ecology Review, 47*(6), 591–610.

Giraud, E. (2013). 'Beasts of Burden': Productive Tensions between Haraway and Radical Animal Rights Activism. *Culture, Theory and Critique, 54*(1), 102–120.

Gordon, L., & Baroke, S. (2014, September 20). Veal: Evolving from 'Cruel Meat' to Ethical Choice. *Euromonitor International*.

Grigg, K., & Halford, J. (2013, May 31). Clearing More Land: We All Lose. *The Conversation*. Online. December 2016.

Harari, Y. N. (2011). *Sapiens: A Brief History of Humankind*. New York: HarperCollins.

Haraway, D. (1990). *Simians, Cyborgs and Women: The Reinvention of Nature.* New York; London: Routledge.

Haraway, D. (2008). *When Species Meet.* Minneapolis; London: University of Minnesota Press.

Haraway, D. (2010). When Species Meet: Staying with the Trouble. *Environment and Planning D: Society and Space, 28*(1), 53–55.

Harwatt, H., & Hayek, M. (2019). *Eating Away at Climate Change with Negative Emissions: Repurposing UK Agricultural Land to Meet Climate Goals.* Cambridge, MA: Harvard Law School, Animal Law and Policy Program.

Heaton, T. (2017, April 14). Should New Zealanders Eat More Veal? *Stuff.co.nz.*

Iveson, R. (2014). *Zoogenesis: Thinking Encounter with Animals.* London: Pavement Books.

Kagan, S. (2015). Singer on Killing Animals. In T. Visak & R. Garner (Eds.), *The Ethics of Killing Animals* (pp. 136–153). Oxford; New York: Oxford University Press.

Kim, C. J. (2015). *Dangerous Crossings: Race, Species and Nature in a Multicultural Age.* Cambridge, UK; New York: Cambridge University Press.

King, B. J. (2017). *Personalities on the Plate: The Lives and Minds of Animals We Eat.* Chicago; London: University of Chicago Press.

Larsen, C. S. (2006). The Agricultural Revolution as Environmental Catastrophe: Implications for Health and Lifestyle in the Holocene. *Quaternary International, 150*(1), 12–20.

Latimer, J. (2013). Being Alongside: Rethinking Relations amongst Different Kinds. *Theory, Culture and Society, 30*(7–8), 77–104.

Latimer, J., & Miele, M. (2013). Naturecultures? Science, Affect and the Non-human. *Theory, Culture and Society, 30*(7–8), 5–31.

Leigh, T. W. (2006). The Consumer Quest for Authenticity: The Multiplicity of Meanings within the MG Subculture of Consumption. *Journal of the Academy of Marketing Science, 34*(4), 481–493.

Major, T. (2017). *Mini Pigs with Big Appeal: Popularity Soars Despite Ownership Restrictions.* ABC Rural.

Mark, J. (2017, February 24). Toward a Moral Case for Meat Eating. *The National Magazine of the Sierra Club.* Online. April 2018.

Martin, P. Y. (2003). "Said and Done" Versus "Saying and Doing" Gendering Practices, Practicing Gender at Work. *Gender & Society, 17*(3), 342–366.

Maxwell, J. A. (2012). *A Realist Approach for Social Research.* London; Thousand Oaks; New Delhi: SAGE Publications.

Morgan, K., & Cole, M. (2011). The Discursive Representation of Nonhuman Animals in a Culture of Denial. In B. Carter & N. Charles (Eds.), *Human and Other Animals: Critical Perspectives* (pp. 112–132). Basingstoke; New York: Palgrave Macmillan.

Murdoch, L. (2016, September 26). Duterte's Drug Crackdown Hits Home as Man Shot Dead in Front of His Pregnant Wife, Family. *The Age*. Online. September 2016.

Nibert, D. A. (2013). *Animal Oppression and Human Violence: Domesecration, Capitalism, and Global Conflict*. New York; Chichester: Columbia University Press.

O'Sullivan, S. (2011). *Animals, Equality and Democracy*. New York: Palgrave Macmillan.

Palmer, C. (2001). 'Taming the Wild Profusion of Existing Things'? A Study of Foucault, Power, and Human/Animal Relationships. *Environmental Communication: A Journal of Nature and Culture, 23*, 339–358.

Pearsall, J. L. S. (2015). *Praise the Pig: Loin to Belly, Shoulder to Ham—Pork-Inspired Recipes for Every Meal*. New York: Skyhorse Publishing.

Pederson, H. (2011). Release the Moths: Critical Animal Studies and the Posthumanist Impulse. *Culture, Theory and Critique, 52*(1), 65–81.

Plumwood, V. (2001). *Environmental Culture: The Ecological Crisis of Reason*. London; New York: Routledge.

Plumwood, V. (2002). Decolonising Relationships with Nature. *PAN: Philosophy Activism Nature, 2*, 7–30.

Plumwood, V. (2004). Gender, Eco-Feminism and the Environment. In R. White (Ed.), *Controversies in Environmental Sociology* (pp. 43–60). Cambridge, UK; New York: Cambridge University Press.

Plumwood, V. (2007). Human Exceptionalism and the Limitations of Animals: A Review of Raimond Gaita's the Philosopher's Dog. *Australian Humanities Review, 42*(August), 1–7.

Pratt, J. (2008). Food Values: The Local and the Authentic. In G. De Neve, P. Luetchford, J. Pratt, & D. C. Wood (Eds.), *Research in Economic Anthropology* (pp. 53–70). Bingley: Emerald Group Publishing Limited.

Redman, C. L. (1999). *Human Impact on Ancient Environments*. Tucson: The University of Arizona Press.

Regan, T. (2004). *The Case for Animal Rights*. Berkeley; Los Angeles: University of California Press.

Salt, H. S. (1914). Logic of the Larder. *Henrysalt.co.uk*. Online. June 2016.

Sergie, M. A. (2018, April 8). Trump Calls Assad 'Animal', Blames Putin after Alleged Chemical Attack. *The Age*. Online. April 2018.

Shove, E. (2010). Beyond the ABC: Climate Change Policy and Theories of Social Change. *Environment and Planning A, 42*(6), 1273–1285.

Simunaniemi, A.-M., Sandberg, H., Andersson, A., & Nydahl, M. (2013). Normative, Authentic, and Altruistic Fruit and Vegetable Consumption as Weblog Discourses. *International Journal of Consumer Studies, 37*(1), 66–72.

Singer, P. (1975). *Animal Liberation.* New York: Random House.

Sutton, P. W. (2004). *Nature, Environment and Society.* Palgrave Macmillan.

Tanentzap, A. J., Lamb, A., Walker, S., & Farmer, A. (2015). Resolving Conflicts between Agriculture and the Natural Environment. *PLoS Biology, 13*(9), e1002242.

Todd, A. M. (2009). Happy Cows and Passionate Beefscapes: Nature as Landscape and Lifestyle in Food Advertisments. In J. A. Sandlin & P. McLaren (Eds.), *Critical Pedagogies of Consumption: Living and Learning in the Shadow of the 'Shopocalypse'* (pp. 169–179). New York; London: Routledge.

Townsend, M. (2014, August 18). Baa, the Springvale Sheep Whose Owner Fought a High Court Battle to Keep Her in His Backyard, Dies after a Lengthy Illness. *The Herald Sun.* Online. May 2016.

Twine, R. (2010). Intersectional Disgust? Animals and (Eco)Feminism. *Feminism & Psychology, 20*(3), 397–406.

Villas, J. (2011). *Pig: King of the Southern Table.* Hoboken, NJ: John Wiley & Sons Inc.

Visak, T. (2013). *Killing Happy Animals: Explorations in Utilitarian Ethics.* Basingstoke; New York: Palgrave Macmillan.

Voiceless. (2015). *The Life of the Dairy Cow: A Report on the Australian Dairy Industry.* Paddington, NSW: Voiceless. 92 pp.

Warren, K. (1997). *Ecofeminism: Women, Culture, Nature.* Bloomington: Indiana University Press.

Weisberg, Z. (2009). The Broken Promises of Monsters: Haraway, Animals and the Humanist Legacy. *Journal for Critical Animal Studies, 7*(2), 22–62.

Weiss, B. (2012). Configuring the Authentic Value of Real Food: Farm-to-fork, Snout-to-tail, and Local Food Movements. *American Ethnologist, 39*(3), 614–626.

Whatmore, S. (2002). *Hybrid Geographies: Natures Cultures Spaces.* London; Thousand Oaks; New Delhi: SAGE Publications Ltd.

Wilk, R. (2002). Consumption, Human Needs, and Global Environmental Change. *Global Environmental Change, 12*, 5–13.

Wilk, R. (2009). The Edge of Agency: Routines, Habits and Volition. In E. Shove, F. Trentmann, & R. Wilk (Eds.), *Time, Consumption and Everyday Life: Practice, Materiality and Culture* (pp. 143–155). London; New York; Delhi; Sydney: Bloomsbury.

Yates, R. (2010). Language, Power and Speciesism. *Critical Society, 3*(Summer), 11–19.

5

Negotiating Edibility

In 1964, Spanish poet and novelist Jorge Luis Borges cited an excerpt from a Chinese encyclopaedia, of "unknown (or apocryphal)" authorship, by which animals are divided into:

(a) those at belong to the Emperor, (b) embalmed ones, (c) those that are trained, (d) suckling pigs, (e) mermaids, (f) fabulous ones, (g) stray dogs, (h) those that are included in this class cation, (i) those that tremble as if they were mad, (j) innumerable ones, (k) those drawn with a very fine camel's hair brush, (I) others, (m) those at have just broken a flower vase, (n) those that resemble flies from a distance. (103–104)

Borges is deliberately poking fun at the ambiguities, redundancies, and deficiencies surrounding humans' arbitrary fragmentations of the universe. This 'classification' has in turn been widely cited, most notably by Foucault, to highlight how it is "system[s] of thought" that shape conceptions of 'Same' and 'Other' (1989: xvi)—and therefore of 'right' and 'wrong', 'good' and 'bad', 'clean' and 'unclean', 'edible' and 'inedible', and so on.

In this chapter, I explore how 'systems of thought', that make edible animals and meat *more* and *less* edible, are constituted as part of practices

© The Author(s) 2020
P. Arcari, *Making Sense of 'Food' Animals*,
https://doi.org/10.1007/978-981-13-9585-7_5

of ethical and sustainable meat. Delving deeper into my participants' accounts, I identify a variety of non-hierarchical distinctions that further carve animals and their flesh into diverse, overlapping, and fluid constructions of the ethical, less-ethical, and un-ethical. These distinctions include: (1) the species of animal; (2) the age and breed of the animal; (3) the type of 'production' system (organic, free range, wild, or farmed); (4) the type of operation (industrialised or not); (5) the size of operation; (6) the local-ness of production; (7) the type of outlet (supermarket, butcher farmers market, farm-gate) and, (8) the method of slaughter.

There are endless ways in which these eight categories (and possibly more) and the knowledges that constitute them can be put to work on an animal and its flesh. This fluidity creates a sense of the 'ethical' as a mobile and stretchy denomination that is pulled in one or other direction, resisting and contracting away from some categories while expanding to accommodate others. From this vantage point, the reader can perhaps begin to appreciate how well regulated the entire 'system of thought' that upholds nonhuman animal edibility is, and the extent to which it limits opportunities to glimpse any territory that might lie beyond, on 'the far side of this grid' (Foucault 1989: 175).

1   Separating the 'Bad' from the 'Good'

In this first section, I focus on the categorisation that takes place in association with attempts to determine the 'bad' which thereby (re)delineate the 'good'. For as Goodman et al. observe, food is not only good to eat, but "*good* food is even better to both think and feel with (and sometimes eat)" (2010: 1782, emphasis added). First, however, I provide a brief insight into my participants' meaty practices as context for their associated understandings.

In terms of their dietary and functional conceptions of meat, the majority of producers and consumers consider a 'main' meal without it 'not a big deal'. In fact, keenly aware and often troubled by some of the environmental, health, and welfare issues associated with practices of meat production and consumption, many were making an effort to eat less meat (apart from former vegetarians and vegans). Only five producers

Table 5.1 Summary of participants' reported frequency of meat consumption (n = 40)[a]

	Consumers	Producers[a]
1 to 3 times a day	5	5
Once a day only	7	
3 to 7 times a week	10	9
1 or 2 times a week	3	
Once a month (approx.)	1	

[a]One of the 15 producers I interviewed is vegetarian. They were a member of a family of producers and fully involved in the business

and five consumers reported that they resolutely ate meat at least once and sometimes two or three times a day (or cooked it for their family) and for these participants, meat was perceived to be a non-negotiable and necessary feature of their food-related practices (Table 5.1). A further seven consumers regularly ate it once a day. However, some of these consumers, and all of the remaining 23 producers and consumers, were variably ambivalent around the notion that a meal *had* to include meat. Yet despite this ambivalence, they still ate some meat, even if only once a month. So how does the meat they eat keep making sense to my participants?

1.1 Instrumental 'Meat' Categories

Meat can be categorised as more or less 'good' in different ways. It may be reduced to its chemical composition where it becomes nutritionally tagged as "a good source of protein" as Henry (C) calls it, or, according to Bella (P), a "high-nutrient food stuff", except when it is processed, at which point, according to nutritional guidelines, it becomes 'bad'. The flesh of food animals is also categorised according to the age, (genetically engineered) 'purpose', and breed of the animals, and by the different parts of their bodies that it comes from. These biologised subdivisions are co-constituted along with knowledge, skills, and tacit rules that describe the 'appropriate' or best use for each type of flesh—what it is understood to be for, how it ought to be prepared and cooked, and how it 'eats' (sensorially). In contrast to the ecological niches associated with nature, food animals have thus been adapted, both physically and taxonomically, to fit what Nick Fiddes (1992) describes as various economic niches based on entirely human-centric understandings of, and appetites for, their flesh. Thus, he continues, "[w]e arrogate to ourselves divine power not only of life and death but of evolution and destiny itself" (81).

However, rather than determinations of what is 'good' and 'bad' about the health, safety, tastiness, cut, or quality of the animal and/or their flesh, which have been the traditional measures of 'goodness', it is more recent determinations that I am interested in here. For these have arisen out of a need to identify what are ethically and environmentally 'good' animals and meats to eat in response to consistent challenges to fundamental practices of 'growing' animals to eat. These are challenges that many consumers, including my participants, have confronted and which have made them reassess how they approach eating meat. With respect to this reassessment, the next section highlights two associations with food animals and meat that are considered particularly 'bad' or unethical.

1.2   Bad Meat: Factory Farmed Animals and Their Flesh

According to Douglas, the process of rejecting or excluding as an "abomination" strengthens the delineation of what is 'good' (2002: 40), thus emphasising and maintaining the appropriate order. Factory farmed, industrial, and mass-produced animals and meat collectively comprise one of the most prominent and consistent 'bads' for my participants. Lisa (C), who eats meat daily, states firmly, "factory farmed *things* are both unethical and unsustainable" (emphasis added). Nicki (C) "can't handle … the thought of a factory produced piece of meat" and Damian (C) calls it simply "horrendous". Factory farming is commonly associated with industrialisation, as David (C) says, "it's the industrial side of the equation which is problematic", and as Bella (P), referring to a particular pork product, explains, "it's mostly not very good cos it's industrial pigs".

Chickens and pigs are the food animals most associated with industrial or factory farming—"chickens and *pork* are more your intensive *things*"

(Melanie P, emphasis added), they are "the extremes because growing indoors and never seeing the light of day, never seeing a blade of grass" (Will P). Natalie (C) believes, "chicken and pork is really priority... they've got it worst here", and they have indeed become the main focus of animal advocacy campaigns aiming to draw attention to and improve the conditions in which the industry raises, or 'grows' these animals. Increasing awareness of these conditions—sow stalls, cement floors, cages, crowding, and lack of daylight/fresh air—no doubt contributes to the perception among my participants that eating these animals is especially 'bad' and unethical, as Charlotte (C) explains, "I would never knowingly or willingly eat chicken or pork that I know is not organic or sustainable labelled or free range". Abhorrence of factory farming of chickens is especially strong. David (C) considers chickens (and dairy) "very much the most animal exploitative industries that there are", while Henry (C), who was previously vegetarian for five years, refers to it poignantly as "Armageddon for chickens". 'Disgusting' and 'gross' are some of the other adjectives used by my participants to describe industrial chicken products. The perceived 'badness' of this meat is generally compounded by its typical retail setting—a supermarket. As Simon (P) tells me, his family has "stopped eating supermarket chickens".

1.3 Animal Flesh from the Supermarket

Indeed, the pinnacle of 'bad' meat for most of my participants is "chicken from the supermarket in a bag" (Brigid P). As Brigid goes on to exclaim, "Oh my god, it just freaks me out, where that chicken's been or how it's been raised, even though it is *just* a chicken" (emphasis added). David (C) would "never go and buy a chook off the shelf at the supermarket, ever, just full stop". Describing a "supermarket bird" as a "dirty bird", and that she would rather not have meat at all than eat it, Maria (C), and many others, combine a rejection of all supermarket meat with a rejection of what it is understood to represent. This variously includes industrialised, homogenised, and/or environmentally harmful production practices, the mistreatment and abuse of animals, and anti-competitive corporate practices. In these, and many other ways, supermarkets represent a defilement of the "symbolic systems of purity" associated with ethical and sustainable consumption (Douglas 2002: 36; see also Lewis 2016; Micheletti et al. 2012).

Literally referred to as dirt, the supermarket chicken becomes emblematic of much larger processes of pollution. In response, my participants explicitly situate their meaty practices in direct opposition to the globalised, corporatised, impersonal model of mass production and market homogenisation, preferring "to support small independent producers" (Anne [C]). Large-scale industrial systems are therefore associated with 'bad', unethical, and unsustainable meat from factory farmed animals and, prefiguring my exploration of the role the senses and the emotions, there are in some cases very strong feelings attached to it. Grace (C) describes walking into the supermarket and being "*assaulted* with all that meat in plastic and it's all sort of red and it's just *depressing*" (emphasis added), while Maria (C) finds it simply "distasteful".

1.4   (De)Constructing Distaste

Looking more closely at what is depressing, distasteful, gross, horrendous, and generally problematic about factory farmed and supermarket meat for my participants, I suggest it is not simply that it is abhorrent and incompatible with a new system of knowledge relating to ethical and sustainable consumption. It is also that the mechanisms of power by which food animals are dominated in these systems have been made more obvious and harder to understand as normal or natural. Hence, it is the close confinement in harsh concrete and metal environments, and the force and harm applied to animals' bodies, that are identified as the 'problems' visibility has revealed rather than the actual use of these animals' use as food. Although visibility is contributing to why such meat is increasingly considered out of place, it does not generally trouble the perpetual state of domination in which these animals exist. This speaks to the 'power of transparency', which I will explore in Part IV.

Effectively, humans' domination of animals is revealed as a 'right' so normalised that it continues to remain invisible.

This observation supports one of Foucault's central points regarding power—that it is "tolerable only on condition that it mask a substantial part of itself. Its success is proportional to its ability to hide its own mechanisms" (1978: 86). In *The Subject and Power*, Foucault further explains how power is only possible under conditions of freedom—the freedom to resist, against which unequal relations come to light and power is seen, and felt, to be exercised. As Foucault explains, "without the possibility of recalcitrance, power would be equivalent to a physical determination" (1982: 790). The point at which disciplinary actions and techniques—the hidden mechanisms of power designed to counter resistance and maintain order—are no longer hidden and tip over into physical determination, or domination, seems to be when a theory of rights holds sway. The essential role of this theory, according to Foucault, is to fix the legitimacy, and, I would add, the truth of power (1980: 95). To this end, its essential function is then:

> to efface the domination intrinsic to power in order to present the latter at the level of appearance … as the legitimate rights of sovereignty, and … as the legal obligation to obey it. The system of right is … designed to eliminate the fact of domination and its consequences. (Foucault 1980: 95)

To illustrate his point, Foucault describes the situation of a married woman in the eighteenth and nineteenth centuries who may have had certain capacities for resistance within the institution of marriage but who was ultimately under a state of male domination in which all such acts of resistance were "ultimately only stratagems that never succeeded in reversing the situation" (1994: 292). A theory of male sovereignty and rights thereof, articulated in law, essentially rationalises mechanisms of domination (and still does today), thereby justifying or effacing their effects (Foucault 1980: 106).

Clare Palmer notes the relevance of Foucault's conception of power to the situation of animals, particularly his distinction between a 'power relationship', where resistance is always present, and 'domination', where the capacity for resistance has been eliminated (2001: 342). Referencing a range of wild, agricultural, urban, and domestic human-animal relations, she highlights the differences between them in terms of the extent to which the animals are "free to opt for other possibilities" or involved in relationships that are "persistent and nonreversible". Palmer's conclusion is that most of these are relations that no amount of "tricks" or "stratagems" can reverse:

> It is hard to think of many human/animal power relationships which contain within them the possibility of power reversal perhaps because of the ways in which humans affect the *constitution* of many animals and/or because of the probability that sufficient resistance on the part of an animal to human power will result in humans moving along the power spectrum to domination—and ultimately to physical violence or death which "drops off" the edge of the power spectrum (the trapped wild animal may be shot; the bucking horse can be sent to the knackers. (2001: 351, emphasis in original)

As Palmer observes, animals generally become "things which cannot resist" (2001: 358) although she guards against making universal claims by noting that the "specificity of particular contexts and environments in which they may be located" must be understood (358). In the case of food animals, however, I argue this relationship is one that always 'drops off' the spectrum and is therefore always one of domination—a fact that is elided in discourses and practices of (human) rights relating to food, whether juridical, religious, secular, or 'natural'.

Hence, the real 'problem' or distaste associated with factory farmed animals and supermarket meat is that the appearance of a benevolent human-animal relationship sanctioned by rights of juridical, religious, secular, and 'natural' law can no longer be easily upheld. The state of domination that these rights actually support is made visible in the obvious repression of the lives and bodies of these animals so that the perception of a 'relationship' (in the sense of mutuality and two-way connection that this word implies) becomes untenable.

Meat that is the product of these more undeniable mechanisms of power is therefore condemned by my participants. It has invaded the artifice of 'possible resistances' that allows human relations with food animals to be characterised as 'husbandry' or as following the model of 'pastoral care' (Foucault 1979; Cole 2011). Under such characterisations, disciplinary measures are undertaken only in so far as they mutually benefit both humans *and* animals; 'for their own good' as it were, as Nibert explains:

> The term widely used to refer to this practice, 'domestication', has come to reflect what is largely regarded as the 'providential inevitability', the much-touted human-animal 'partnership' ... A benign partnership. (2013: 11–12)[1]

Concealment is thus essential to the normalisation and ongoing success of relationships of power. Once the *mechanisms* of domination become evident (or acknowledged), resistance intensifies, 'proper' order is destabilised, and matter becomes increasingly out of place until a new order is established (consider, for example, the suffrage, anti-slavery, and, more recently, MeToo and BLM movements). I argue that this is what is happening with my participants' demonisation of factory farmed and supermarket meat (and movements for 'better' meat more generally). In fact, Grace (C) makes an explicit comparison (with union movements) when she exclaims, "you know it's like having workers and things like that, I mean you don't abuse them".

However, Grace (C) highlights that it is not edible animals per se that are 'out of place', but rather the practices of domination and 'abuse', which in turn become inscribed on their bodies and flesh. These practises symbolically, but also literally, make the meat distasteful. Ethical and sustainable animals and meat are inscribed with different signs that restore a pastoral rhetoric of 'tender fondness' for the care and welfare of individual animals. Absent obvious mechanisms of power, the creative, productive, and positive aspects of a sovereign power, as described by Foucault (1980: 105), that bestows and protects the rights of its animal subjects, come to the fore and the negative aspects of a state of domination are minimised. As Foucault explains:

---

[1] Jeffory Clymer (2012: 84) identifies a similar "rhetoric of tender fondness" in his exploration of family, property, and race in the nineteenth century whereby institutionalised violence is cloaked in the "warm embrace" of interracial "affection" and "companionate marriage" (84). Still relevant today, a study of modern-day compassionate or benevolent sexism reveals that it both "masks" and reinforces gender inequality and thus undermines efforts towards equality (Hideg and Ferris 2016).

> What makes power hold good, what makes it accepted, is simply the fact that it doesn't only weigh on us as a force that says no, but that it traverses and produces things, it induces pleasure, forms of knowledge, produces discourse. It needs to be considered as a productive network which runs through the whole social body, much more than as a negative instance whose function is repression. (Foucault 1980: 119)

It is to this 'productive network' that I turn in the next section where I explore the positive aspects of power that traverse my participants' accounts of their meaty practices, helping power 'hold good' and supporting the continued use of edible animals in ways that are not 'distasteful'.

2   Making It All 'Better'

So, if, as producer Bella says, "you shouldn't eat factory farmed meat", or "random generic supermarket meat" according to Natalie (C), what meat 'should' you eat? Answers to this question reveal some broadly shared, but by no means unanimous, understandings of what is 'better' and 'best'. These understandings, and their associated discourses, contribute to what might be imagined as different (but co-constituted) layers of permeable and shifting boundaries which variously intersect, align, complement, or conflict. The most prominent of these are discussed in the following sections.

2.1   Cows and Sheep Versus Pigs and Chickens

In terms of preferred or 'better' animals, former 12-year vegetarian Natalie (C) believes that in general "[beef and lamb] seem to have better welfare outcomes than chicken and pork". The perception that these animals spend most of their lives outside in open paddocks is a common element in many of my participants' reasoning. "Cows, are more likely to have a kind of *genuinely* free-range system or a grass-fed system", says

Maria (C) (emphasis added), and Anthony (C) reasons that "cows and sheep I suppose, I feel like generally [they] have a better life because they're outside, they're eating grass ... I think in general they probably live a happier life [than pigs]".

However, for Finn (P), and also Diane (C), cows are actually more unethical than chickens because of "the amount of water that goes into creating a kilo of beef compared to the water, the grass, the methane emissions, all that CO_2 stuff that's coming out of the cows" (Finn). Considered by many the apex of 'bad' meat, chickens' "fast turnaround compared to pigs, cattle or sheep" (Will P) leads Diane (C) to view them as "more sustainable than beef" and "the meat of the future" according to Finn (P) due to their capacity to "pump[ing] out lots of *protein* for minimum resources (emphasis added).

Chickens may also be more easily construed as ethical because theirs tends to be the meat more commonly labelled as such—free range, organic, biodynamic, welfare approved—distinguishing them and their flesh more clearly and authoritatively from their factory farmed, industrial cousins. As Charlotte (C) says, "it's easier to get. It's probably the most common cheapest kind of meat that is more ethical or more free range or whatever". Similarly, "it's a lot easier to access the information on the chicken than it is on the beef", explains Geoffrey (C).

The availability of clearly labelled meat that has already had much of the work done for the consumer in terms of its construction as ethical foregrounds the hierarchies of meat that can emerge, converge, and conflict. Even though they endeavour to avoid supermarkets in principle, free range and organic labels can overcome that resistance for some participants like Julie (C) who says, "at a real push, I would buy the chicken that's free range", and Lucy (C), who is "happy to go supermarket free range or whatever, those sorts of things, that's fine". Such labelling embodies a broad set of knowledge apparatuses, including measurable criteria, system compliance, legislative support, and government oversight, that appears to engender at least a basic level of trust, even as a least worst option. Charlotte (C) succinctly captures the flow and negotiation between her different understandings of what is ethical and 'good':

It is much better to buy organic, free range chicken than not buy organic free range chicken. If I had a choice I'd buy it from a massive industrial company but it wouldn't be my first choice to buy from them, I'd rather buy it from a small non-organic, non-certified free-range chicken farm that I would visit, that is local to Victoria.

After being classified as an edible animal, Charlotte's chicken undergoes a further four potential levels of classification depending on the interplay between her prioritised sub-categories of how it was raised, the size of the operation, the type of operation, and its location. This further illustrates both the flexibility and mobility of boundaries between categories of 'good', 'bad', and also ethical.

2.2   'Local'

As Charlotte (C) indicates, where the animal comes from creates another opportunity for distinction, though, of course, location, type of operation, size of operation, and how it was raised are all inextricably related. Supermarkets are almost universally placed at the bottom of this hierarchy and their "bad", "crappy", "dirty" meat is mostly to be avoided. This leaves local butchers, farmers markets, farm-gates, and personal connections as the preferred avenues for obtaining "good meat" (Maria C) that is known to come "from a good source" (Geoffrey C). However, it is not necessarily these outlets themselves that possess the ethical credentials, but again, as with supermarkets, it is what they are understood to represent and the qualities, and knowledge, with which they then imbue the animals and the meat they provide. It is this embodied knowledge that is being offered and in turn valued by my participants.

Animals and meat from these 'good sources' become imbricated in what local food is understood to mean more broadly, which prominently includes supporting a community, and nourishing a sense of personal connection as opposed to the disconnection attributed to the industrial system. As Jennifer (P) illustrates, "to me they're linked, the eating the meat and having a flourishing community". Ellis (C) thinks there are "so many benefits" of eating ethical and local meat; "it's better for the planet, it's better for the community, you know community building, people being connected with their producers". Hence, it is a quality of being local, and the meaning that accompanies it, that my participants seem to particularly want to associate with their ethical meat practices. As Damian (C) proclaims, "the meat we eat *should* be from farmers markets and it *should* be sourced locally" (emphasis added). However, the definition of local is slippery and stretchy, and its ethical credibility can be revised depending on the circumstances and other ways in which 'ethical' is being understood.

Sometimes animals and meat that are from Australia in general could be construed as local and therefore 'better' regardless of how it was raised and produced. As Melanie (P) reasons, "it might be ethically grown overseas to what the overseas people class as ethical, but it won't be Australian probably", while Anthony (C) further condemns "the crap that is in the supermarkets" for not being "Australian made and owned". Different Australian states can also be associated with different kinds of production system, adding another potential distinction to understandings of what is local, ethical, and 'better'. Observing that "meat that was produced in Victoria is a lot less likely to have feedlots associated with it", Charlotte (C) concludes that buying meat from Victoria is therefore "quite different from getting it from Western Australia".

Local can therefore mean many things, including local to your neighbourhood, your region, your state, or your nation, and the degree to which animals and meat from any of these locales is considered 'better' will depend on how their construction as 'ethical' is shaped by other considerations such as the way it was produced. For instance, Anne (C) would "rather have the production standards than the low food miles", whereas Lisa (C) "would much rather have something that was local but maybe not organic".

Stănescu (2010) and Hinrichs and Allen (2008) have critiqued this kind of discourse surrounding local food, highlighting its inherent vagaries and inconsistencies. On the one hand, claims that local food is more environmentally sustainable, in terms of its ecological or carbon footprint, cannot be upheld when different foods, production systems, and associated energy usages are taken into account, which too often result in local options performing least well (Stănescu 2010: 12–13). On the other hand, there is the tendency for local food movements to draw on idealised depictions of a pastoral idyll that is "at times, distressingly sexist and xenophobic" (Stănescu 2010: 10). By "gloss[ing] over the issues of sexism, racism, speciesism, homophobia and anti-immigration sentiments" (Stănescu 2010: 8), a focus on the local, as opposed to the global, encourages what Hinrich and Allen's define as "selective patronage" that favours the social justice needs of designated groups while excluding other disadvantaged groups (2008: 331). These authors argue that 'local' needs to be 'retired' as a designation of more ethical and sustainable foods to encompass a wider range of social justice issues and avoid the potential for "nationalistic regionalism" that it contains. Nevertheless, the (unsubstantiated) 'knowledge' that local is better has been popularised to the point of an unquestionable truth, fuelled by the bucolic imagery of a certain kind of farming, one that is generally, but not always, small-scale and that is assumed to benefit the wider community or the nation as a whole.

2.3   Wild 'Meat'

Moving away from the idea of any kind of formalised farming system involving traditional food animals, but maintaining the focus on local, some of my participants consider it ethically preferable if the animal is instead "in the wild" at the time of death, or "wild meat" as Heather (C) calls it. As once-a-month meat-eater Beverley (C) explains, "the animal's had a happy life, and as painless as possible death". Pescetarian Sophie (C) reasons that "they're just running around in the wild and it's sort of how it would happen naturally".

Wild animals mostly eaten in Australia—kangaroos, rabbits, and deer—are also regarded as 'pests', and this additionally informs their construction as ethically edible. Gillian (C) will eat "wild shot venison, because that's a pest", and for Beverley (C), it is "especially overpopulated ones" that she is "happy" to eat, "ones that are causing damage to the environment or native animals". How what is classified as a native animal is negotiated against what is also regarded as a pest is not clear from my data, but this introduces another area of ambiguous and shifting boundaries around what is ethically edible.[2] Kangaroo to some extent exemplifies that conflict.

For Beverley (C), "if they say it's wild kangaroo, then I'm ok with buying that", and Lisa (C) and Sophie (C) are equally unequivocal: "Kangaroo meat is like all wild caught I think, which really sits well with me", says Lisa, while Sophie "will eat kangaroo because it's a natural wildlife and it's just been hunted and killed. It's natural, it's fine". Presumably, however, not all wild animals would be considered natural and fine to eat; therefore, the 'knowledge' of kangaroo, and to some extent also wallaby and possums, as pests makes their flesh edible while those of other native animals remains non-edible. Drawing an additional distinction between sustainable and ethical, Diane (C) surmises, "a sustainable meat would be you know, kangaroo or whatever, but kangaroo's not ethical". For Diane, and also Finn (P), who agrees that "roo … is not ethical", wild is not a sufficient criterion for being ethical, as hunting methods must also be considered. As Finn explains, "there's a lot of foul shots … shot in the neck, shot in the shoulder … joeys they just, you know, chicken them [*sic*] or smash them on the ground. It's … not a very well regulated, professional, streamlined industry".

Wild meat, as another category of 'ethical meat' is therefore further distinguished according to how the animal is hunted and, more especially, killed. As Bella (P) says, "a hunted one I probably would prefer, if it was a good hunter". She is therefore more ambivalent about wild, hunted meat being considered more ethical than farmed—"I think there are problems with both", she says.

2.4   Knowing

In an effort to negotiate these multi-faceted and shifting criteria of ethical-ness, participants often rely on a more general sense of knowing (as opposed to 'fact'-based knowledge) to inform their meaty prac-

---

[2] If the negotiation simply involves a perceived tipping point in populations whereupon an animal becomes designated as a 'pest', it is theoretically possible that koalas, wombats, and other native herbivores would similarly become edible if their habitats were to recover sufficiently to allow their populations to grow.

tices. Also highlighting the role of the imagination in the constitution of practices and their elements (see Chap. 7), this 'knowing' and the associations it evokes—especially sensory and emotional (explored in Part III)—shape the perceived edibility of meat and food animals and therefore contribute to their persistence as constituent elements of ethical practices.

For example, Maria (C) recalls having bought meat from a different producer than she would normally buy from; however, "I know the region their meat comes from, I don't know them personally. I know that it's free range, it's rare breed. I have less of an attachment to the producer but I feel comfortable with the choice". Will (P) also suggests that knowing is directly related to an animal's edibility: "you feel better about it if you know how it was grown and what sort of methods are used ... and even how it was killed". Such 'knowing' is now *de rigeur* as part of a new food ethics, as Diane (C) illustrates:

> Everyone's obsessed with you know, 'what's the story', 'where did it come from'? Nobody wants anything unless they know [where it comes from]. In a Virgin magazine, there was this big article on it, the new luxury is knowing where something came from.

Knowing (as opposed to seeing) can thus be enough in and of itself to lend certain objects some of the 'goodness' that comes from being authentic. As Ellis (C) explains, "if I know how the animals were raised and kind of where, that makes me comfortable enough to eat them". And as Anne (C) further illustrates:

I need to know the story in quite specific detail. So land management practices that are kind to the landscape and animal treatment methods that are about where and how an animal is processed and finished off basically, so minimum stress possible production. I don't need a certification of those things, but I do need to know, I do need to know the story of the animal basically.

This extends to 'knowing' about how these food animals are killed. Joyce (C) describes asking a producer to 'walk her through the process' in detail:

I said to her, 'tell me about how you kill your pigs?', and so she told me and that was the first time that I'd asked a farmer to walk me through the process. I didn't anticipate that I would find it upsetting and I didn't find it upsetting. I'm very matter of fact about it. It was good to know I suppose that there were farmers out there who had thought about it.

Making a more explicitly emotional association, Michael (C) says he "wouldn't be enjoying it if I didn't know where it came from". Some participants prefer to also see—a certain mediated visibility playing a significant role in the (re)constitution of ethical consumption more broadly, as I will show in Part IV. However, this is not always necessary. If it is simply somebody that you know, Graham (P) explains, "you can just tell, if people are right or not", and similarly Florence (P), "They've got the right thinking going on … they look and feel right".

Thus far, all of the considerations, negotiations, and orders of edibility that my participants express quite clearly and assertively describe various systems of knowledge that both shape their practices with regard to ethical and sustainable meat and simultaneously reinforce the domination of animals, in ethicalised ways, as a normalised part of those practices. However, the production practices that pose the greatest challenge to this (ethicalised) normalisation are the ones that tip the power relationship undeniably off the spectrum, that is, causing the animal's death. How ethical and sustainable meat withstands this ultimate challenge—how killing is constituted as 'good', 'bad', and 'ethical', thereby adding another layer to the shaping of boundaries—is the focus of the next section.

3 Kill-Ability and 'Better' Killing

Animals generally have to be deliberately killed in order for people to have access to all the types of meat that they consider edible. Hence, the capacity to kill animals and the way they are killed provide further opportunities (or necessities) for distinguishing between what is considered ethically 'good' and 'bad' while maintaining Foucault's 'productive' aspects of the obvious exercise of power this requires. This leads to the notion of ethical killing.

The capacity and skills to take an animal's life shape understandings of their edibility. Discussing which animals she might realistically raise for meat, Joyce (C) reasons that "feathers is one thing, I'm not sure I could thread a little bunny through … and yank and kill it with my bare hands". In the context of her conviction that "if you eat meat you should be able to kill it yourself", Lisa (C), who eats meat every day, goes on to attest that she "could definitely kill a chicken"; however, she envisages having "a pretty hard time killing … a pig or something really big like a cow".

In addition to species, age, the way it was raised, where it came from, and its perceived kill-ability, *how* an animal is killed is at least as, and sometimes more, important in constructions of the ethical-ness and edibility of its flesh, as Finn (P), Diane (C), and Bella (P) indicated in the previous section with respect to hunting. In confronting the *how* of killing, my participants clearly reveal their efforts to resolve the consequences of the increased visibility of factory farming and industrialised processes of animal use. Gillian (C) describes 'how an animal dies' as a "central tenet to ethical and sustainable production" and asserts ending the "mechanisation and the industrialisation particularly of the slaughtering process" is key to improving the meat that is available. Similarly, for Nicki (C), "ethical is more to do with the actual killing and processing of them".

Echoing Gillian's (C) understanding of industrialised slaughter processes, anything that avoids the use of abattoirs is generally regarded as more ethical. "Abattoirs are pretty horrific places", says Charlotte (C), and "any way to take out that process is good". David (C), who eats meat once or twice a day, sums up his own confusion in trying to attach further distinctions to the meat that he eats based on how it is killed:

> Should we use captive bolt stunning, should we use CO_2, should we do this, should we do that? Halal [is bad] because they have to be conscious when their throats are slit and this sort of stuff.

Anne (C) believes, "the only ethical slaughter is backyard stuff". Others concur, and "get[ting] to a point where we buy meat that is killed on the farm" is one of Diane's (C) main criteria in her vision for ethical meat, and also Alison's (P), who describes the "ultimate of having the non-aware slaughter on the property in their own paddock".

Understandings of on-farm-killed meat certainly add another dimension to the ethical-ness of meat for some of my participants, drawing once again on a collective 'knowledge' and discourse associated with local, small-scale, family-run farms. Speaking of her local farmers, Charlotte (C) relates that she would "have a lot less issues with eating the meat that they killed and produced on their farm than eating a chicken from a factory somewhere". Referring to an occasion where she was eating the flesh of an animal killed on-farm by friends, Diane (C) says, "they had given it a happy life and slaughtered it themselves so I was happy to eat that". David (C) alludes to what he considers more 'natural' when he compares the abattoir, where "you're putting them under that stress of the transport and whatever and taking them out and killing them" with the 'better' alternative "where they're just in their *normal* environment" (emphasis added).

As David (C) demonstrates, direct comparisons with industrial slaughter processes cast on-farm or home slaughter in an even more positive light. Brigid (P) describes the process "when they get on a big truck and go to a big abattoir and they get manhandled by people other than me", where instead she prefers "to do *right* by my [animals]" (emphasis added). Graham (P) illustrates the same comparative formulation: "to take them into a processing facility where there's pigs roaring and screaming, they can't not be stressed. If you just walk them up somewhere and gave them a bucket of feed and then shot them, humanely". Graham increases the stakes of his comparison by making denigrating assumptions about industrial facilities, the people who work there and where he imagines they come from. In a triple 'othering' of slaughterhouse workers, ex-prisoners, and animals, he states: "it's far better than being loaded into a vehicle and transported halfway across the state and then in the charge of somebody who probably just got out of prison, you know, certainly not compassionate".[3]

---

[3] Graham may be referring to prisoners being employed at abattoirs in the Northern Territory since 2014 as part of the state government's 'Sentenced to a Job' Program. However, questions of cause in the mistreatment of animals in slaughterhouses are a more complex social issue, as Richards et al. (2013) and Fitzgerald et al. (2009) have highlighted.

Drawing more explicitly on the notion of having a duty of pastoral care and the responsibility for individual life that accompanies it, and also supported implicitly by the universal understanding of humans' sovereign right over animals, killing the animal themselves is for some of my participants even better. Therefore, while the readiness and capacity to kill certain animals can render the species as a whole more and less edible, it is through killing an animal first-hand that that individual's flesh in particular becomes more ethically edible. Although Tracey (C) is pescetarian, she does not "mind the idea of eating an animal if I've killed it". Similarly, producer Finn, who was vegetarian for a year, would "feel a lot better about it if we ... actually did it ourselves". There is a tacit understanding, almost a rule, among these producers and consumers whereby the meat of animals that have been killed first-hand is bestowed with an extra quality, beyond ethical 'goodness'. That quality is a sense of that meat having been earned through the act of killing. It becomes a reward, something deserved in return for the physical and emotional labour of taking the life of the animal that it once was. This is a theme I will explore further in Part III. For now, I bring the focus back to negotiations around ethical killing, moving on from where the killing is performed, and by whom, to explore *how* food animals are (ethically) killed.

Killing animals is acknowledged as requiring a certain set of skills and material infrastructure. Joyce (C) has signed up for a "workshop on how to kill a chicken for backyard poultry"; while Nicki (C) has already attended such workshops and developed a skill set that enables her to regularly kill her own backyard chickens. She explains that of all the different methods, the "best" involves "a broomstick on the back of the neck and you just pull and it just snaps the vertebrae". There is further knowledge and skill required after this step to do it 'properly', as she explains: "the chicken needs to flap around to actually expel the blood, so if you keep them in the cone, the blood just sits there ... so you just let them roll around". Graham (P) concurs: "You've got to do it right ... do it properly. There's spots that you've got to hit them, and then they don't notice".

In light of this, much as a large number of my participants support an increase in backyard rearing and killing of food animals, and changes in regulations to permit the distribution of this meat, others are more circumspect. A friend once asked Natalie (C), "why don't you just kill it yourself?" She responded with, "I have no skill and I would cause harm". Concerned that animals might be killed by people who "don't necessarily have training or the background", Natalie prefers the process to be handled by "skilled slaughterhouse staff … skilled butchers". Melanie (P) likewise thinks, "all animals should be slaughtered properly by a professional slaughterman"; for as Will (P) emphasises, "it does require skills and also the right sort of equipment and restraining devices".

There is a sense in which recourse to a set of knowledges that describe, document, and thereby reify 'proper' and 'good' killing, much of which is highly formalised through manuals, equipment, training, and certification, legitimates the act of killing while creating distance from the living subject. Although the physical distance between human and killable animal may be reduced, a focus on the act and art of killing puts their individuality and life to one side. The animal's death becomes a technical exercise, which can be done the right or wrong way. At this point, the "thingification" (Palmer 2001: 358) of the animal is almost complete as the (short-lived) living component is, expertly and less expertly, excised to leave the edible flesh that was always the end game. Pachirat notes a similar strategy at work in the industrial slaughterhouse where:

> a focus on food safety deflects the attention away from the work of killing onto the technical realm of hygiene. The possibility of perceiving and experiencing what happens in industrialized killing is diverted into elaborate performances and deceptions generated by the focus on food safety. (2011: 206)

Technical expertise and know-how aside, after careful negotiation and ordering of their 'better' meat based on species, breed, age, location, production processes, and other criteria, it is still at the point of death that my participants perceive a potentially troubling (or disordering) loss of control in what is finally inscribed on their meat through the manner of slaughter. Backyard and on-farm slaughtered meat comprises only a tiny proportion of total meat consumed by the general population. For the majority, including consumers of all commercially produced ethical and sustainable meat, abattoirs and slaughtermen are the final arbiters of a more ethical outcome and so the focus turns to identifying more and less ethical methods of industrial slaughter. Abattoirs rarely permit visitors,[4] let alone allow the entire slaughter process to be witnessed. Therefore, for most of my participants, with few other means available, it is primarily hope and imagination that ensures meat does not become matter out of place due to unethical slaughter.

Natalie (C) *"hopes* [the animals] don't have to travel too far, and that they can go to a fairly small abattoir where they can get processed quickly and quietly…that they can't see what's going on and that it's just quick" (emphasis added). Referring to the use of CO_2 gas chambers for rendering pigs unconscious prior to slaughter, Melanie (P) says that to her knowledge, they "all go into a little room as mates, and they're gone to sleep, so to me that's not bad". Sophie (C) has a similar conception of the process: "they walk into this chamber with their friends and then they just gas them, and they just fall asleep and I thought that sounded about the most ideal way you could do it". The notion that being with their 'mates' or 'friends' makes slaughter more benign is a common one.

To emphasise, it is not individual attitudes or judgments regarding the 'best' or most ethical way to kill animals that I wish to highlight or critique here, but rather the persistence of socially, geographically, and historically normalised practices involving the acceptable, or at least justifiable, killing of food animals. In other words, how questions regarding the most ethical way to kill animals come to be constituted as legitimate and unproblematic questions in the first place.

Reflecting more generalised understandings among my participants of how 'properly' (i.e. ethically) slaughtered animals are imaginatively distinguished from the non- or less-ethically slaughtered ones, Joyce (C) remarks, "the magic phrase I want to hear is, they're killed first thing in the morning before the other animals, like for me that says those animals are being handled differently to how the rest of the animals are being processed". These hopeful and idealised constructions of ethical abattoir

---

[4] There is a shift in this space as many abattoirs now permit visitors by arrangement, while some in Europe and the US are open to the public and conduct daily tours. This development will be discussed in more detail in Part IV.

practices rely to a large extent on constructions of the animals' experiences and also feelings—a topic I will address in Part IV. As long as commercially sold meat has to be killed at a certified slaughterhouse, and as along as these businesses are not open to the public, how and whether the 'most ethical way' imagined by these participants is ever, or can be, achieved, will always be left to their imagination.

That there is clearly broad agreement on there being a 'proper' or 'right' way to kill animals—notwithstanding these understandings of 'proper' can vary widely—implies that there is also a wrong way to kill animals. Besides common assertions that it should *ideally* be quick and stress-free, there is generally no further meaningful explanation provided, and indeed these words—proper and right—do not demand it. Whenever they are used, they elicit a sense that you tacitly agree about what they mean even if it has not been made explicit. Moreover, in not having to qualify by what measure something is deemed 'proper' and 'right' or by whom, there is the sense of appeal to a higher authority that has decreed such a thing to be truth—proper and right according to some natural law. Yet, in and of themselves, these ethical distinctions that my participants describe are only meaningful with the a priori acceptance of humans' right to exercise power over food animals.

Interestingly, much of the concern expressed by my participants around how animals are raised and killed focuses on the effect this can have on the sensory qualities of the resulting meat, both actual and imagined—a finding I explore in Chap. 7. This is conceived as an added bonus, or another good reason for killing 'humanely'. 'Ethical' producers reflect and reinforce this expanded consumer knowledge in order to promote their meat. The website of one producer emphasises that "ethical considerations are foremost in our minds—not only is this kind to the animals but it results in tender meat" (Trevor). Indeed, consumers and producers frequently associate the quality and sensory attributes of ethical meat with quite specific or variably vague notions of an ethical life *and* death—highlighting how certain practices are understood to enhance the gustatory experience (see Chap. 7). As Nicki (C) suggests, "to know the process and to be familiar with something dying makes a big difference to how you appreciate meat".

No amount of stratagems or tricks can alter the lived reality for food animals—a reality that reflects their state of domination and can only end one way. Understandings relating to the various benefits that are transmuted to humans when the flesh of certain animals who have been treated in certain ways is consumed can be considered part of Foucault's 'productive network' that makes this normalised domination "hold good" (1980: 119). As Marilyn Frye aptly comments with regard to enslavement:

> Although the slave is not engaged in 'surpassing herself,' she *is* engaged in surpassing: she is engaged in the master's 'surpassing' *him*self. Her substance is organized toward his 'transcendence'. (1983: 66, emphasis in original)

However constituted, the categories, knowledges, and associated discourses discussed thus far support determinations of the 'ethical' that help to maintain meat as matter in place in the midst of potentially unsettling challenges, especially associated with factory farming and industrialisation.

4　Conclusion

To imagine a different order where cows, sheep, pigs, and chickens are relieved of the productive obligation that we assign as their birthright so they are no longer considered meat; and to also imagine a range of alternate knowledges and 'truths' that support these orders that are as 'unalterable and unquestionable' as our current ones, seems almost as fantastical as the Chinese encyclopaedic classification of animal types, cited at the beginning of this chapter, that so amused Borges (1964). Regarding this classification of uncertain authorship, Foucault remarked that it shattered "all the familiar landmarks of my thought … the thought that bears the stamp of our age and our geography—breaking up all the ordered surfaces and all the planes with which we are accustomed to tame the wild profusion of existing things" (1989: xvi).

The current normalised order is not more true or closer to 'nature' than any other order humans might construct. Looking back, Berger (1992: 4) notes that, before domestication, "[a]nimals first entered the imagination as messengers and promises", and looking forward, at least one of my participants entertains the possibility of a world where:

> cows and whatever are seen, not as pets, but as something that should be viewed, in nature … or something to paint, maybe to pat, maybe to ride on … Hopefully it will reach a point where it goes, the norm is not to kill. Michael (C)

Our present order simply reflects human's want to consume animals, reformulated and naturalised (and also materially, politically, and economically systematised) as a need. This 'need' restricts animals understood as food from "freely proceeding in their bodies or to use them as a means of acting … a total temporal and spatial control of their behaviour up to oppression and death" (Bujok 2013: 43–44). The normative and "productive force" (Sedgwick 2003: 5) of taxonomies means "it is no longer their identity that beings manifest in representation, but the external relation they establish with the human being" (Foucault 1989: 341). These relations are encoded in knowledge, truths, and validating discourses; in the systems, infrastructures, and practices they support; and are permanently inscribed on physical bodies, both human and nonhuman. As the protagonist of Emma Geen's novel observes, "Human language developed around human bodies, it never quite fits other ways of beings" (2016: 66).

It is to these bodies that I now turn in Part III. Following Foucault, I am interested in "how deployments of power are directly connected to the body—to bodies, functions, physiological processes, sensations, and pleasures" (Foucault 1978: 151–152). Therefore, it is the immaterial dimensions of our human bodies that I am primarily concerned with: the links between desire and power (Foucault 1981: 52–53), and between knowledge and pleasure through the pleasure of knowing—"a knowledge-pleasure" that Foucault articulated in relation to discourse and sexuality (1978: 77). In other words, how knowledge and power are also emotionally and sensorially embodied.

References

Berger, J. (1992). Why Look At Animals? In *About Looking* (pp. 1–14). London: Bloomsbury.

Borges, J. L. (1964). *The Analytical Language of John Wilkins. In Other Inquisitions 1937–1952* (pp. 101–105). Austin: University of Texas Press.

Bujok, M. (2013). Animals, Women and Social Hierarchies: Reflections on Power Relations. *Deportate, esuli, profughe, 23*, 32–48.

Clymer, J. (2012). *Family Money: Property, Race, and Literature in the Nineteenth Century*. Oxford; New York: Oxford University Press.

Cole, M. (2011). From "Animal Machines" to "Happy Meat"? Foucault's Ideas of Disciplinary and Pastoral Power Applied to 'Animal-Centred' Welfare Discourse. *Animals, 1*(4), 83–101.

Douglas, M. (2002). *Purity and Danger: An Analysis of Concepts of Pollution and Taboo*. London; New York: Routledge Classics.

Fiddes, N. (1992). *Meat: A Natural Symbol*. London; New York: Routledge.

Fitzgerald, A. J., Kalof, L., & Dietz, T. (2009). Slaughterhouses and Increased Crime Rates: An Empirical Analysis of the Spillover from 'The Jungle' into the Surrounding Community. *Organization and Environment, 22*(2), 158–184.

Foucault, M. (1978). *The History of Sexuality*. New York: Pantheon Books.

Foucault, M. (1979, October 10 and 16). Omnes et Singulatim: Towards a Criticism of 'Political Reason'. *The Tanner Lectures on Human Values*. Stanford University, Palo Alto.

Foucault, M. (1980). *Power/Knowledge: Selected Interviews and Other Writings, 1972–1977*. New York: Pantheon Books.

Foucault, M. (1981). *The Order of Discourse* (R. Young, Ed.). London and New York: Routledge.

Foucault, M. (1982). The Subject and Power. *Critical Inquiry, 8*(4), 777–795.

Foucault, M. (1989). *The Order of Things*. London; New York: Routledge.

Foucault, M. (1994). *Ethics, Subjectivity and Truth* (P. Rabinow, Ed.). New York: The New Press.

Frye, M. (1983). In and Out of Harm's Way: Arrogance and Love. In *The Politics of Reality: Essays in Feminist Theory* (pp. 52–83). Santa Cruz: Crossing Press.

Geen, E. (2016). *The Many Selves of Katherine North*. London: Bloomsbury Publishing.

Goodman, M. K., Maye, D., & Holloway, L. (2010). Ethical Foodscapes?: Premises, Promises, and Possibilities. *Environment and Planning A, 42*(8), 1782–1796.

Hideg, I., & Ferris, D. L. (2016). The Compassionate Sexist? How Benevolent Sexism Promotes and Undermines Gender Equality in the Workplace. *Journal of Personality and Social Psychology, 111*(5), 706–727.

Hinrichs, C. C., & Allen, P. (2008). Selective Patronage and Social Justice: Local Food Consumer Campaigns in Historical Context. *Journal of Agricultural and Environmental Ethics, 21*(4), 329–352.

Lewis, T. (2016). Ethical Consumers and Sustainability Citizenship. In R. Horne, J. Fien, B. Beza, & A. Nelson (Eds.), *Sustainability Citizenship in Cities: Theory and Practices* (pp. 199–208). London: Routledge.

Micheletti, M., Cheng, S.-L., Stolle, D., Olsen, W., et al. (2012). Habits of Sustainable Citizenship: The Example of Political Consumerism. In A. Warde & D. Southerton (Eds.), *The Habits of Consumption* (Collegium - Studies across Disciplines in the Humanities and Social Sciences) (pp. 141–163). Helsinki: Helsinki Collegium for Advanced Studies.

Nibert, D. A. (2013). *Animal Oppression and Human Violence: Domesecration, Capitalism, and Global Conflict.* New York; Chichester: Columbia University Press.

Pachirat, T. (2011). *Every Twelve Seconds: Industrialized Slaughter and the Politics of Sight.* New Haven; London: Yale Agrarian Studies Series.

Palmer, C. (2001). 'Taming the Wild Profusion of Existing Things'? A Study of Foucault, Power, and Human/Animal Relationships. *Environmental Communication: A Journal of Nature and Culture, 23*, 339–358.

Richards, E., Signal, T., & Taylor, N. (2013). A Different Cut? Comparing Attitudes toward Animals and Propensity for Aggression within Two Primary Industry Cohorts—Farmers and Meatworkers. *Society and Animals, 21*(4), 395–413.

Sedgwick, E. K. (2003). *Touching Feeling: Affect, Pedagogy, Performativity.* Durham; London: Duke University Press.

Stănescu, V. (2010). 'Green' Eggs and Ham? The Myth of Sustainable Meat and the Danger of the Local. *Journal for Critical Animal Studies.* VIII, (1/2), 8–32.

Part III

The (Dis)Pleasure of Knowing (About Animals and Meat)

Introduction to Part III

What is evident throughout producers' and consumers' accounts of their meaty practices that I have drawn on thus far, especially when they deal with which animals are considered acceptable to eat (e.g. locally farmed, free range, or wild food animals but not 'baby' animals or 'pets'), which animals can be more problematic (e.g. pigs), and the enactment of distinctions between 'good' and 'bad' meat is that a gamut of emotions contributes to these constructions. In Part III, I foreground these emotions by examining how Foucault's regime of power/knowledge/pleasure draws attention to the positive and productive aspects of power, in contrast to the negative mechanisms of exclusion, repression, and domination. It is these positive mechanisms, and more specifically "the conditions of their emergence and operation", that Foucault argues ought to direct analyses of power, "insofar as they produce knowledge, multiply discourse, induce pleasure, and generate power" (1978: 73). Negative views of power tend to ignore the emergent and dispersed nature of power, with all its positive and productive aspects linked to bodily pleasures, and are therefore insufficient to explain its persistence (Westwood 2002).

Of course, productive mechanisms, and the positive emotions associated with them, are given form and definition by what evokes negative emotions, and vice versa. In what follows, my participants' accounts of *dis*pleasure are therefore given equal weight in the constitution of productive mechanisms of power as their accounts of pleasure. Indeed, it is my contention that it is through the identification of how displeasure (e.g. with factory farming) is negotiated and reconciled to maintain the edibility of (ethical) animals that the most effective and persistent positive mechanisms are foregrounded. In light of this, the primary focus of Part III is demonstrating how emotions as a whole (associated with both pleasure and displeasure) shape the constitution of meat and food animals, becoming an integral part of associated knowledges and truths that contribute to the normalisation of animals as food, and thereby helping to maintain mechanisms of domination. The senses provide a secondary focus, being inevitably intertwined with knowledge and emotions. As Merleau-Ponty asserts, the notion that the senses somehow preceded knowledge, performing their functions in a purely objectively fashion, is an illusion. They belong "to the domain of the constituted and not to the constituting mind. [...] Perception is thus thought about perceiving" (2002: 43–44). I will show that this is particularly the case in the multisensory world of food and eating.

Part III therefore addresses my second and third objectives by exploring **how sensory and emotional associations with animals and meat shape their edibility or non-edibility**, and **where and how an embodied knowledge of 'food' animals is challenged and how their edibility is maintained.**

References

Foucault, M. (1978). *The History of Sexuality*. New York: Pantheon Books.
Merleau-Ponty, M. (2002). *Phenomenology of Perception*. London; New York: Routledge.
Westwood, S. (2002). *Power and the Social*. London; New York: Routledge.

6

Sensory Connections and Emotional Knowledge

I begin this chapter with an account of the senses and their role in Foucault's nexus of power/knowledge/pleasure—the 'regime' that he identifies as being what "sustains the discourse on human sexuality" (1978: 11) and which I am arguing similarly sustains the discourse on human's use and consumption of food animals. Being conventionally thought of as belonging to individuals, the senses are a tricky area to tackle from the perspective of socially constituted orders and practices, which is the overall orientation of this book. I confront this problem with specific reference to the knowledge-pleasure aspect of Foucault's nexus which he describes as "the pleasure of knowing" (1978: 71), identifying the senses as providing a connection between pleasure/displeasure (or emotions) and knowing (or knowledges and truths).

© The Author(s) 2020
P. Arcari, *Making Sense of 'Food' Animals*,
https://doi.org/10.1007/978-981-13-9585-7_6

1   Brought to Your Senses

1.1   Sensible Theory

The senses are the *permeable* interface by which places, people, 'things' (living and non-living), and situations are seen, heard, touched, smelled, and tasted.[1] I stress permeable to emphasise that this is not a dualistic framing that posits a purely biological bridge between 'inner' and 'outer' worlds, between mind and body. Recognising that this has been the prevailing model of (anthropological, sociological, and cultural) sensory scholarship over at least the past 30 to 40 years, scholars such as Chau (2008), Hsu (2008), Hayes-Conroy and Hayes-Conroy (2008), and Vannini et al. (2013) counter with the inseparability of sensation and perception, body and mind. As Elisabeth Hsu (2008: 437) argues, "one cannot overemphasize the social and contextual nature of sensory experience".

Adam Chau goes even further and posits that "the sensory event is located not in the individual body but rather in the social collectivity itself" (2008: 288). He calls this the 'sensory-production model' to highlight the role of this social collectivity, through the 'sensory orders' they constitute, in shaping "ways of perceiving, knowing, and being in the world" (490). For Chau, every culture has its own 'sensory order' and this notion of order is common across the scholarship that promotes a sociology of the senses. For example, Vannini et al. (2013) draw on the work of sociologist Eviatar Zerubavel to describe how sensory experience is understood via "mental clusters" or distinctions produced by "lumping" and "splitting"—analogous to the delineation of boundaries, categories, and 'orders' discussed in Part II. The authors explain that these distinctions are not the outcome of individual thought processes, but rather constituted through "social and cultural *sensory communities*": groups of people who share common ways of using their senses and making sense of sensations" (2013: 7, emphasis in original). Again, as with the categories and boundaries discussed previously, though these sensory distinctions might seem

---

[1] Acknowledging these are the senses most associated with human experience and that other animals may have many others.

natural, in fact "we have been socialized to 'see' them" (Zerubavel 1996, cited in Vannini et al. 2013: 7). In this way, "sensing is always already a 'making sense' because it always involves *an act of creation/ordering*" (Evans and Miele 2012: 301, emphasis added). What enables this socialised sense-making to cohere temporally and spatially to varying degrees is what Eva Illouz (after Scarry 1999) terms "perceptual mimesis" (2009: 404). This describes the capacity of the senses to be evoked in real time, through imagination, memory, and nostalgia (Illouz 2009; Appadurai 1996), and the more enduring of these perceptual bonds create sensory expectations about certain practices and encounters, including those involving food.

What this preamble on the sociology of the senses illustrates is that boundaries and categories, and 'orders of things', shaped by systems of knowledge, 'truth', and understanding that 'lump' and 'split', constitute not only our understandings of 'things' like meat and food animals, but also our sensory perceptions, or experience,[2] of them. Foucault makes similar reference to "the already 'encoded' eye" (1989: xxii). The senses thus become "part of old memories, new intensities, triggers, aches, tempers, commotions, tranquilities ... The visceral body feels them as intensities that have an impact on tasting", or more accurately how tasting (or smelling, touching, hearing, seeing) is interpreted (Hayes-Conroy and Hayes-Conroy 2008: 467). Distinctions between 'good' and 'bad' meat are therefore as much sensorially and viscerally constituted, as through the knowledges, 'truths', and emotions that surround different food animals, meats, and practices. An account from 'ethical' butcher Berlin Reed of one his most memorable post-vegetarian experiences serves to illustrate this melding in sense-making:

> I joined him in cutting. Gorgeous, juicy slabs of pork seemed to melt off the carcass ... I stopped to admire the marbling of the belly, which is of special interest to me for bacon curing, and pointed out the heavenly striations of fat and meat in mouthwateringly perfect proportion to one another. (2013: 95)

---

[2] Elisabeth Hsu (2008: 436–437) notes that 'sensory experience', rather than 'sensory perception', is often used to avoid the traditional conception of the mind-matter dualism that originated in the natural scientific framework.

This evokes Levi-Strauss' observation that "natural species are chosen not because they are 'good to eat' but because they are 'good to think'" (1991: 89). However, my data indicate that the shift towards 'better' meat demands going further, for it is commonly understood that such meat must not only be 'good to think', but 'also good to know'.

1.2   Knowing Taste

As discussed in Part II, whether via direct experience, imagination or memory, knowing the story of a piece of meat—the breed of the animal that it came from, how the animal was raised, where the farm was, what type of farm, where and how the animal was killed, what age the animal was when it was killed, where the meat was sourced, and many more distinguishing features—contributes to the degree to which it is understood as 'good' or 'bad'. But what I highlight here is how this more detailed 'knowing' also contributes to how 'good' or 'bad' it is perceived to taste.

Rather than more knowledge of production processes detracting from sensory pleasure, as many assume it will, including animal advocacy organisations (as I discuss in Part IV), this knowing actually enhances it. Producers are aware of this as an added marketing opportunity. One advertises a special spit service on their website where "the meat" can take "centre stage". The blurb continues:

> The background story behind the animal's origin and how the animals are grown adds to the drama and appreciation of enjoying a whole beast from nose to tail. At the same time you can feel good about directly supporting farmers and their small businesses.

The "pleasure of knowing", which Foucault originally articulated in reference to the scientification of sex and sexuality in the nineteenth century (1978: 70–71), is especially relevant today considering the "current hazing" of the twin carnal pleasures of food and sex (Probyn 2000: 9). Indeed, Hayes-Conroy and Hayes-Conroy use the phrase to describe the visceral politics of the Slow Food movement where "knowing where and how food was produced" contributes to the pleasures of eating (2010: 467).[3] As Carolyn Korsmeyer observes, "the aesthetic is cognitive; the sensuous and emotional layers of response have meaning" (2002: 117). This interplay of knowing/senses/emotions connections is prominent in my discussions with participants in relation to meat, food animals, and their related practices, characterised by a rhizomatic and almost synesthetic circulation of associations.

For instance, Finn (P) recalls the appearance of an 'unknown' steak served to him at a restaurant, saying, "I knew it was an unhappy animal … it just wasn't appealing to me, so I left it". Similarly, Jennifer (C) explains that the animals "take on the flavour and the taste and the condition of the environment that they're raised in". As Julie (C) says, "you can taste a well-bred animal". These participants' sensory and emotional expectations, experiences, and final assessments of meat are being shaped by their prior knowledge and understandings of how and where the animals were raised.

As well as how an animal was raised, my participants demonstrate that his/her breed and age can similarly evoke quite specific sensory associations based on collective, socially mediated understandings of flesh. Trevor (P) comments that Brahman cows "aren't really the most tenderest animals to consume", while Bella (P) remarks "rare breed pork is amazing … it's just a totally different colour because they're working their muscles all the time so they develop all this strength and colour". Julie (C) remarks that "if an animal is older when it's slaughtered … you can definitely taste the difference". This leads to the final factor that appears to shape my participants' sensory experience of meat—the slaughter.

As noted in Chap. 6, the edibility of certain meat and animals is constituted in part by knowledge about the way they were killed and this also translates to its perceived sensory qualities. Contrasting "farmed venison" with its wild counterpart, Gillian (C) explains, "it doesn't have the same beautiful texture that I suppose in my mind a humanely slaughtered animal has". Geoffrey (C) is of the understanding that the quality of the

---

[3] Importantly for qualitative research, Beekman highlights that the reverse is also true—that there is also knowing in pleasure, and argues for using emotional perception as a source of knowledge about food by "listen[ing] to the aesthetic or cultural knowledge embedded in people's emotional responses" (2006: 309).

meat he eats is directly related to the way the animal was killed in terms of the stress it experienced: "If they send one cow off on its own, it stresses the animal which in turn destroys the meat, and the meat's not as tender". Having "read a little bit about the different ways of killing the animal", Heather (C) now feels she has an appreciation of the impact of "first the speed, the time in which it takes the animal to die … on the quality of the meat that you *harvest* from the animal" (emphasis added).

What becomes evident is that knowledges and accepted 'truths' regarding species, breed, age, and method of slaughter are engaged in constituting particular sensory associations with specific edible animals as 'good' or 'bad'. By these accounts, a happy, rare breed, older, and 'humanely' killed animal actually tastes better.

Some participants indicate awareness of how verbal, textual, and pictorial descriptions can shape their food experiences, predicting that in a blind taste test, they would be unable to tell the difference between meat that came from a factory farmed animal and an ethically raised animal. This suggests that when the more symbolic (Korsmeyer 2002) components of meaning are removed, the constitution of expectations and experiences is somewhat simplified, encompassing aesthetic meanings and distinctions of what Korsmeyer (2002: 129) terms "taste qualities". These include flavour, texture, smell, and colour, and are equally socially and culturally mediated (Sutton 2010; Hayes-Conroy and Hayes-Conroy 2010). Indeed, once an animal is made meat, my participants demonstrate their familiarity with these 'taste qualities', and the collective knowledge and expertise that shapes associated distinctions—an "ethical-aesthetic expertise and experience" that Evans and Miele also identified across their 48 focus groups (2012: 308).

As Anthony (C) explains, "chefs have taught me to pick good steaks … I'll be looking for marbling through the steak and you just don't see that in the *cows* in the meat or deli section of the supermarkets" (emphasis added). James (P) maintains that he can distinguish between breeds based on the "grain" of the animal's flesh, and both Gillian (C) and Simon (P) make clear distinctions between male, female, and heritage pigs based on smell and taste. As Simon (P) comments, "if you cut [a] boar open, when he's working, you can smell him … but female pork, it is a slight difference, you can taste it … I can taste it slightly when it's a male".

The way this specialist knowledge is expressed implies that it derives from a body of reliable, objective facts about meat. There is no appeal to a particular source or authority but rather the knowledge is conveyed as common truth regarding 'good' and 'bad' meat. However, it is also understood to be quite specialist knowledge not shared by all. As Gillian (C) illustrates, "you can't sell it as normal steak because it's doesn't look like a normal steak. [It's] beautiful, yellowed fat, rich red meat … so it has to go to a specialty retailer", presumably where those who share this knowledge will better understand the meat to be 'good'. As I illustrated previously, many of my participants relate their knowledge around the tenderness (mouth feel) and taste of meat to the way in which the animal was killed, making a single sensory connection from a practice, through to the animal and the final meat. Effectively, for the 'knowing consumer', the meat embodies all that happened to 'it' prior to that point, becoming sensorially and emotionally inscribed with its own history.

By introducing further categories of distinction that are, as discussed, also emotional and sensory, the emergence and increasing mainstreaming of ethical and sustainable meat has enhanced this specialist sensory knowledge and disseminated it more widely. This has created the opportunity for visceral connections with different orders of edibility. The level of engagement and intensity with which my participants describe these connections, often tinged with excitement, pleasure, or disgust, also highlights that this nexus of knowing/senses/emotions constitutes a noteworthy dimension of meaty practices that likely contributes significantly to the persistence of meat and the use of animals for food more generally. It is a dimension that is perhaps moving these practices in a more sensorially aware and reflective direction—re-invigorating and re-fetishising the consumption of meat and use of food animals which is under fire on several fronts.

1.3 Sensing Pleasure (and Pain)

Competency in sensing 'good' and 'bad' meat can sometimes outweigh what is known about the way the animal was raised. Despite being labelled free range, Maria (C) is suspicious of a particular producer's packaged chicken because "the fat is still quite white … I know that a grain-fed or corn-fed bird will be quite yellow". More typically, however, sensory associations tend to align with knowledge. Grace (C) notices the chicken she buys has "that firmer, denser meat than the ones that are in the supermarket that are sort of blown up, and grown too quickly". Chicken wrapped in plastic makes Natalie (C) "feel quite squeamish" whereas the ones that she understands to have "built up their muscles being outside" are considered edible. Once again, meat from a supermarket is widely understood, sensorially as well as intellectually, to be 'bad'. It is variously "tougher" (Melanie P), "pumped with water" (Florence P), "flabby" (Gillian C), "dulled and less flavourful" (Lisa C), and "slimy and sweaty ... pallid and less appealing" Natalie (C).

These participants' depictions of their sensory associations with meat, animals, and associated practices emphasise how important the senses are to the constitution and ordering of food animals and meat discussed in Part II—helping to delineate what is 'known' to be edible/inedible and good/bad to eat. As previous studies have also demonstrated (Anderson and Barrett 2016; Evans and Miele 2012; Piqueras-Fiszman and Spence 2015) associations with meat that is 'known' as 'good' are resoundingly positive, emphasising that all these sensory associations derive as much from a pleasure of knowing as a pleasure of thinking. As Tracey (C) explains, "the meat tastes better because it's happier".

Furthermore, as Illouz (2009) and Appadurai (1996) indicate with reference to memory, imagination and the ignition of "triggers, aches and tempers", sensory and visceral associations with objects are particularly enduring over time. This suggests that the rhizomatic connections they form between knowledge and emotions (or Foucault's pleasures), here specifically in relation to meat and food animals, play a major role in the persistence of associated practices. Stepping back from this focus on the senses, but remaining aware of how their constant presence informs how meat and edible animals make sense in associated practices, I now turn to the emotions. I propose that they occupy a central but under recognised role in constituting meat and food animals, and hence in the performance, defection from, and persistence of meaty practices.

A note here on my use of the term 'object' or 'material object'. While reflecting the use of these terms in practice theories to describe the material components and infrastructures of a practice, I am also emphasising the objectification and 'thingification' of living beings who are used in practices. Nonhuman animals, 'other' humans, and other living entities are therefore included under these terms. However, I will occasionally refer explicitly to nonhuman animals when I wish to emphasise the distinction.

2 Locating Emotions in Practices

How something smells, tastes, sounds, looks, or feels tactilely in the present can act as a powerful emotional as well as perceptual trigger—or emotional mimesis, to use Illouz's (2009) terminology—causing past emotional associations to surface in the present. Like sensory associations, emotional associations can be persistent across space and time, held together by the intensity of the emotion and further 'charged', or made more vivid and visceral, by the senses. These associations have a specific 'object'—either the practice(s) that delivered the experience, such as a social event or meeting, or certain elements of it, for instance a particular food animal or meat. It is through these associations that an emotion may carry over to the same 'object' encountered in the present, in effect 'preloading' them with a legacy of emotional value that shapes how 'good' or 'bad' they are understood and sensed to be. However, these emotional associations are not static. They are at the same time open to being reconstituted at every newly unfolding moment of a practice. Conceiving social practices in this way, as emotionally and sensorially enlivened, and enlivening, goes some way to putting the visceral back in touch with the social, as Wetherell enjoins (2012: 10).

Emotions are not conceived as "interior properties" belonging to or expressed by individuals but rather "properties of the specific affective 'attunement' or mood of the respective practice" (Reckwitz 2016: 119). Sara Ahmed similarly describes how emotions can 'stick' to objects and move sideways, creating sticky associations with other objects and also backwards so that "'what sticks' is also bound up with the 'absent presence' of historicity", that is, what has over time become normalised (2004a: 120). However, it is less the object that an emotion attaches to than knowledge about that object—how it is being understood, for as

Martha Nussbaum (2003 [1997]) explains, "emotions embody not simply ways of seeing an object but beliefs ... about the object" (in Beekman 2006: 307). Hence, Ahmed describes "the work of emotions" as involving "the 'sticking' of signs to bodies" (2004b: 13). Respectively, Foucault (1989: 44) observes that, "to know an animal or a plant, or any terrestrial thing whatever, is to gather together the whole dense layer of signs with which it or they may have been covered". In this way, the evocation of an emotion does not depend on the physical presence of the object with which it is associated. It may also be an imagined object, knowledge of which, in the imagining of its 'signs', evokes the real emotion in the present and reinforces the association.

Emotions are thus perceived to circulate within and between practices of everyday life, along with knowledge and the senses, in a co-constitutive relation. Together these 'create' objects (human and nonhuman), and these objects continue "accumulating ... affective value as social goods" as they move around between the practices of which they are part (Ahmed 2010: 21). However, practices are dynamic and constantly evolving and so an object's emotional value may also change. In this, I draw also on Ahmed's conceptualisation of 'affective economies' in which emotions can become attached to, and contained by (but do not reside in) objects, and also slide between them. Thus, her focus is not on emotions per se, but the work that they do (2004a, b: 14–15). According to this, emotions (in terms of Foucault's knowledge/pleasure) can be conceived as shaping objects in certain ways. Hence, Ahmed posits, "we could ... ask how the circulation of signs of affect [understood here as knowledge and emotion] shapes the materialisation of collective bodies" (2004a: 121). This is an especially pertinent way of considering how emotions contribute to the collective materialisation of animals' bodies and their flesh in everyday practices, and the persistent ways in which these meaty materialisations 'make sense'.

In the same way that persistent notions of what is natural, good, and bad can be associated with different practices, animals, and types of meat, as I demonstrated in Part II, 'good' emotions can become associated with a practice, and 'bad' or less 'good' emotions with the animal or meat that is a material component of that practice. The emotional value that a practice or material component accumulates thus exerts its own agency within and between practices in a push/pull negotiation in terms of being attracted to what 'feels' good and repelled by what 'feels' bad. As Anthony (C) says, he wants to eat "meat that makes me feel happy to be eating it". Here, Anthony's happiness is clearly associated with a specific meat that, in being associated with his knowledge of what is 'good', is conceived as having the capacity to make him happy. He implies that happiness is a quality of the meat—the cumulative product of a bundle of 'good' practices—that he assimilates when he eats it. It becomes, as Matthew Cole explains in his analysis of 'Happy Meat', "something to be consumed along with the muscle fibres, fat and blood", producing a "morally satiated" consumer (Cole 2011: 94).

I have so far described the ways in which, following Reckwitz (2002, 2016), I conceive emotions as being part of and constituted through social practices—attached to the material objects of practices (living and non-living) and to practices as a whole—rather than emotions as individualised states. Indeed, the sociology of emotions has been a focus of scholarship and debate for an almost equal period of time as the sociology of the senses (Turner and Stets 2005). Therefore, drawing on and paraphrasing Vannini et al.'s (2013) conception of sensory communities, and supported by a significant body of literature on cultural differences in emotions,[4] I suggest emotions can be similarly framed in terms of "social and cultural [*emotional*] communities", defined as groups of people who share common ways of expressing their emotions and making sense of emotions (2013: 7, emphasis in original). Some scholars frame discussions of culture and emotions in terms of emotional practices (De Leersnyder et al. 2015; Mesquita and Walker 2003). However, they also describe cultural emotions as "a function of the interactions and relationships in which they take place" (De Leersnyder et al. 2015: 10), and as aligning with and reinforcing "the distinct cultural models (i.e. goals and practices) of self and relationship" (Mesquita and Walker 2003: 777). Thus, while there are clearly instances where the expression of an emotion could be perceived as a practice in itself—and I will allude to at least one of these instances later—these collectively expressed and understood emotions are seen as being constituted primarily by and through social practices.

---

[4] For example, Lim (2016); Mesquita and Frijda (1992) and De Leersnyder et al. (2015), to name a few.

Indeed, my participants illustrate just how central their emotions are to how meat and food animals make sense in their meaty practices. I recall here Florence's (P) assessment of herself, the only vegetarian in a family of meat producers, as being "a bit soft, or what I call soft I suppose, for animals". She appears to feel that she ought *not* to be soft, that it is an inappropriate feeling to have towards animals constituted and understood as food. She has a regulatory feeling *about* her feeling—that it is, in the context of meat-related practices, the wrong one. On the one hand, Florence's (P) comment highlights the enduring historicity of the valuing of reason over passion, or indeed any 'irrational' emotion, the latter being seen as biological, feminine, primitive, and therefore lesser, while the former is cultivated, masculine, rational, and therefore superior (Wetherell 2012: 95). On the other hand, it also indicates how emotions could be viewed as practices in and of themselves, as in being soft or sentimental. However, in the context of different practices, for example of caring, being soft or sentimental would be appropriate. It is, therefore, the meanings constituted in relation to practices and their elements that determine the (in)appropriateness of an emotion, not that the emotion is of itself right or wrong.

However, as noted previously, cultural differences in the expression and understanding of emotions have been recognised wherein high arousal emotions (such as joy or anger) are more generally disparaged in Eastern (collectivist) cultures compared with low arousal emotions (being solemn or reserved), while the reverse holds true in Western (individualist) cultures (Lim 2016). In the same way, 'softness' or 'sentimentality' is popularly understood in Westernised and some other cultures to be undesirable (Narula 2014; Knight 1999). Not only is it considered an inappropriate emotion to associate with certain practices or objects (such as slaughter or animals), but it is typically identified with women and femininity, and so tends to be construed also as weakness, eliciting associations with a range of similarly 'soft', sentimental, and feminised practices such as caring for others, or not eating animals. An extreme example is the Rudalis of Rajasthan, a group of low caste women who hire themselves out as professional wailers for funeral processions (Singh et al. 2014; Devi 1997). Singh et al. explain that a "triple oppression of class, caste and gender" is in operation here because, for the Brahmins—the high caste group, to sing or wail at a funeral is considered low class, or at least questionable, and not reflective of their high status (Hurlstone 2011).

Identifying appropriate 'affective' or emotional practices that "confer 'distinction' " while stigmatising others, Wetherell draws on Ahmed to describe how emotions intertwine with cultural circuits of meaning and value (and their objects) so that "some get marked out as disgusting and others as exemplifying moral virtue" (Wetherell 2012: 16). However, as Wetherell hints at here, what are perceived as emotional practices (such as the Rudali's performance of grief) are still being constituted within a broader nexus of social and cultural practices out of which they derive their distinct forms of expression and meaning. Emotions are therefore still being constituted through social practices.

Yet viewing emotions as also culturally constituted provides another way of thinking about how they shape practices involving food animals and meat. It highlights that not only are certain emotions associated with these practices and their elements, but in certain circumstances there can also be a 'preferred' cultural emotion. This is one that is widely considered *more* appropriate or 'better' than others, and which can consistently subdue conflicting or contradictory intra-practice emotions (when different emotions are attached to different material objects of a practice). The socially and culturally constituted knowledge and 'truth' of a preferred emotion, which can vary across situations, can in turn shape and alter how practices and their various elements are constituted and known.

It is therefore my contention that longstanding meaty practices in Westernised cultures shape, and are shaped by, the constitution of certain preferred emotions. These emotions are now so tightly associated with meat and food animals that they tend to dominate any practice of which these material(ised) objects are part. At the same time, associations with these dominant, or distinctive, emotions, which remain largely stable and persistent across bundles of practices, denigrate and stigmatise any contradictory emotions that may arise (such as 'softness' or sentimentality), along with their associated practices. In support of this contention, scholars have repeatedly highlighted "gendered norms of emotional expression" (Menely 2007), including the feminisation (and by association denigration) of sentimentality (Solomon 2004; Campbell 1994), particularly in relation to animals (Menely 2007). According to Menely (2007: 249):

> Sentimentality's devaluation has accomplished the crucial cultural work of guarding the border of human community, a border disrupted by the cross-species sympathies widely promoted within sentimental culture.

This offers an alternate and more nuanced view of what psychology-based accounts of human behaviour would identify as cognitive dissonance. Recognising the dynamic flow of emotional associations within and between practices and their objects, shaped further by social and cultural norms and expectations, explodes the utility of this popularised interpretation and demands that individuals are no longer blamed for so-called dissonant thinking. Interventions can then be more productively directed at the broader discourses and practices that materialised and instrumentalised animals are part of.

The next section explores the *kinds* of emotions that my participants' associate with meat, food animals, and meaty practices, and how these can be indicative of potential disturbances, or challenges, in their socially and culturally normalised sense-making.

3   Negotiating Challenges: Emotional (Dis) Comfort with Meaty Practices

Supporting my proposition of different and sometimes conflicting emotional associations within and between meaty practices, my participants variously indicated feeling happy about eating some animals and 'meats', more ambivalent about others, and overtly uncomfortable or disgusted about others. The most prevalent emotions associated either with particular food animals and meat, or with the practices these are part of—or, more accurately, with knowledge about them—are a sense of comfort and discomfort.[5] Some participants use the terms 'comfortable' or

---

[5] In relation to my third objective, I conceive emotional discomfort to be indicative of challenges to an embodied knowledge of animals' normalised edibility. My explorations of emotional associa-

'uncomfortable' (or variations thereof) in relation to having confidence or being confident in something or someone, and there is, of course, the physical understanding of these terms (Shove 2003). And while emotions could arguably be said to contribute to these meanings, and there is some overlap between them, here I focus on *emotional* comfort/discomfort as describing the extent to which something is sensed or understood as being as it 'should' or is preferred to be.

3.1   Theorising Comfort

Following Williams and Irurita, I define emotional comfort as "pleasant positive feelings, a state of relaxation", and its negative dimension, emotional discomfort, as "unpleasant negative feelings, a state of tension" (2006: 408). Emotional comfort is theorised in the health literature where it is recognised as one of three integrated aspects of patient care, along with physical and spiritual comfort—also conceived as (emotional) ease, (physical) relief, and (spiritual) transcendence (Kolcaba 2003; Williams et al. 2017). It also appears in explorations of comfort food, where it is associated primarily with physiological and psychological responses (Wansink et al. 2003) but also understood in terms of the social conditions that shape its constitution (Locher et al. 2005). Both approaches to emotional comfort reflect different temporal and spatial experiences, expressions, and understandings of wellbeing. Acknowledging that divisions between physical, spiritual, emotional, and psychological aspects of human experience are academic constructions, my conception of emotional comfort is best thought of as a temporal shift in these perspectives defined by a more ontological sense of comfort or balance shaped less by immediate biophysical responses (though these are part of it) and more by the world beyond one's physical body and an understanding of one's place in it.

---

tions are thus framed in terms of comfort, which maintains, and discomfort, which challenges this edibility. As well as avoiding the binary perspective of positive and negative emotions, these terms reflect the terminology used most frequently by my participants. I conceive them as my empirically supported variation on Foucault's conception of 'pleasure'.

As Dan (C) says with regard to his meaty practices, "it's about you making sense of life and meaning and right and wrong and where you fit into that and what you're comfortable with". It can be associated with a particular 'object' (e.g. meat) that "sits well with me" (Lisa C), or with a person, entity, or practice, for instance, "a source that I kind of feel comfortable buying from" (Andrew P). If a balance (within or across practices) between what is perceived as 'right and wrong' cannot be reached or is upset, discomfort may result, so that there may be practices that no longer feel comfortable, or foods that "I don't quite feel right about eating" (Sophie C).

3.2   Finding Comfort Through Discomfort

Accepting animals as food, there are certain practices surrounding the ways these animals are treated that are more commonly 'known' to be problematic and cause discomfort to those seeking 'ethical' and 'good' meat. Factory farming is generally regarded negatively, as I have illustrated. However, it is in the imagining of the devalued lives of the animals that are part of such systems, coupled with a knowledge and understanding of this as unethical, that an emotional association emerges that appears to reinforce this orientation. This could be thought of as a regime of power/knowledge/*dis*pleasure, or, in my framing, discomfort.

As Ellis (C) illustrates, "factory farming is pretty horrendous, that *thought* of feedlots and animals living their lives in kind of confined spaces is pretty, I dunno, I just think that's abhorrent" (emphasis added). Exhibiting both emotional and perceptual mimesis, David (C) recalls a visit to an industrial pig operation where there was "crate after crate after crate after crate and they smelt disgusting". Anne (C) relates an even more embodied association, asserting, "[factory farming] is just about trauma to animals and producers. And I don't want to eat that, I don't want to eat that violence"—alluding again to the treatment of animals being materially inscribed on the meat and, on ingestion, becoming part of her own body at a cellular level, just as Anthony (C), earlier in this section, was able to eat the happiness of a well-treated animal. Violence is reified and considered inedible, whereas 'good' treatment of animals is much more digestible, "[the meat] just sits better in you, knowing that the animal has lived well, and hopefully has died well", affirms Nicki (C). Damien (C) captures this sense of the embodiment of the treatment of animals more eloquently when, speaking of the meat he prefers to buy, he says, "I feel like there's an energy to it, it's had this love and it's been treated in a beautiful way and I just feel so much more nourished eating it rather than this factory farming". As Gillian (C) explains, factory farmed meat "doesn't feed your soul".

It is in my participants' responses to a final supplementary interview question, where they express discomfort in association with in vitro meat,[6] that an enhanced 'sense' of nourishment emerges as a particular association with the practice of consuming, or, more accurately, embodying the flesh, energy, or soul of a living, happy animal. For not only is in vitro meat considered "repellent" (Maria C) because it is a product of industrial processes and therefore 'unnatural' and not 'real', concepts addressed in Part II, but also because it does not derive directly and wholly from a once living animal. As Joyce (C) explains, "there's something extremely gut level and visceral and that's why that animal-ness, I think there's something in that". And Maria (C) similarly asserts, "the difference in the meat is about the life of the animal and if there's been no life, no animal, then it can't taste the same, I just know that it can't". Hence an even more basic and visceral emotional distinction emerges where meat (and presumably any meat, ethical or not) from living animals is digestible, and to a degree comforting, simply because it was once life, while meat from a laboratory is indigestible, because "there's no life force in that meat" (Gillian C). After all, if not violence or happiness or life itself, what is inscribed on in vitro meat? What is ingested when one eats it? The emotional sterility of a laboratory is perhaps not so appetising.

---

[6] Also known as cultured meat, Miller defines in vitro meat as "meat that is grown by proliferating cells in a nutrient-rich medium without the necessity of an animal's slaughter" (2012: 42). Some production methods rely on 'donor' animals and animal products, notably calf serum, although animal-free media are starting to be used. Much debate and contention surrounds this developing technological field (Pluhar 2009; Stephens 2010; Chiles 2013), and for many, the perpetuation of a 'carniculture', where 'meat', even animal-free meat, is prioritised "remains in need of careful thought" (Miller 2012: 43).

The visceral is "the realm of internally-felt sensations, moods and states of being, which are born from the sensory engagement with the material world" (Hayes-Conroy and Hayes-Conroy 2008: 462). In a 2011 article for the *LA Times*, the Pulitzer Prize–winning food critic Jonathan Gold refers to the exquisite rite of eating a particular food that was "heightened by danger, flavoured with death". This sentiment echoes my participants' accounts of emotionally inscribed meat and animals, and also presages associations of meaty practices with novelty and cultural omnivorousness (being the pursuit of anything defined as novel and exciting), which I discuss in more detail in the next chapter. Put simply, to ingest something is to also ingest everything socially constituted about it. Capturing the sentiments of many of my participants, Anne (C) says, "personally, I like my *food* to have had its feet on the ground" (emphasis added). In other words, life is appetising and affirming, while lifelessness is not.

However, while participants associate a degree of relative comfort with eating an animal who has lived and died (as compared to in vitro meat), there is still evident discomfort associated with certain meat, animals, and meaty practices, most especially in relation to practices of killing. For even meat that has had its feet on the ground is necessarily inscribed with some degree of violence and there is widespread agreement that the "violence" of the slaughterhouse is the one discomforting process of producing meat that even producers of ethical and sustainable meat "haven't got any control over" (Graham P) (unless they practice on-farm killing, in which case the resulting meat cannot legally be sold to the public in Australia).

Describing a vivid childhood memory of visiting an abattoir with her father, Julie (C) recalls that she "found it revolting and I remember leaving and saying thank you for killing our own *meat*, to my father, because it was so revolting, so revolting, I was disgusted". This memory strengthens her commitment to buying "as ethical as I can", being the response to her discomfort that is more relatable in the context of normalised meaty practices than not eating meat. A more recent familiarity with abattoirs both in Australia and overseas has imprinted itself on Henry's (C) memory, as he recounts:

it's certainly distressing, it's very confronting as well, to watch, to watch the whole live animals getting the chop, you know … less developed setting, where it's much more confronting, cos it's manual, and manual tends to be more grisly, that's quite confronting, yeah. Some of the conditions that they're killed in … there's lots of crap, lots of shit, they're all pissing, they're all upset.

Others who have visited abattoirs feel discomfort from imagining the emotions they ascribe to the animals in these situations. Finn (P) imagines (and perhaps somewhat feels) "just the stress of the process and pain, like stress and pain and confinement", and Alison (P) thinks of the "cows jammed in there … this real feeling of sadness". Those who have not experienced killing first-hand express sorrow for the animals, imagining what it would be like for them because "death has a smell, blood you know … it's death and blood. It does play on my mind" (Michael C); and also cling to the hope that an animal "would not be watching all of those animals die before them and knowing that they're coming next, that's horrible, that's just horrible" (Julie C). As Henry and Michael indicate, the visceral effect of memories is enhanced by multi-sensory (olfactory and aural) mimeses. Similarly, Sally (C) recalls seeing "big trucks going past with all the pink pigs in, screaming before they go off to the slaughterhouse", while Helen (C) would hear the "horrific" sound of "the pig squealing" during slaughter.

Yet this discomfort with killing is weighed in balance with other aspects of eating animals that are overwhelmingly associated with comfort[7] (or even more intense positive emotions). Moreover, as shown in Part II, there are normalised and appropriate ways to compensate for this discomfort—more normalised than forgoing meat. Hence, while most participants grapple with discomforting emotional and sensory associations

---

[7] I conceive this comfort as associated not only with other meaty practices, or elements thereof, but also with, and constituted by, the worldwide nexus of practices—environmental, social, economic, medical, educational, and so on—by which meat consumption and the use of 'food' animals have been universally, and systemically, normalised. This aggregate comfort is set against, and enhanced by, the discomfort—the "unpleasant negative feelings, a state of tension" (Williams and Irurita 2006: 408)—associated with plant-based dietary practices situated within or alongside practices where meat has been normalised. Admittedly, at the time of writing, this discomfort is being unsettled to a greater degree in Westernised nations than at any other time in recent history.

with food animals, meat, and meaty practices, including the most discomforting practices involving killing, these are negotiated against other, more 'productive', associations (e.g. with cooking, entertaining, eating). The outcome of these negotiations for all but one of my participants,[8] and including the 17 participants who had previously been vegetarian or vegan, is their continued participation in meaty practices.[9]

Amidst the often intense and visceral associations with the ways in which food animals are treated and especially killed, there are a few participants for whom the discomfort lingers—they do not feel quite right and evidently have more difficulty resolving the negotiation between multiple knowledges and emotions around their meat eating. Julie (C) reveals "those ethical issues around meat consumption have been really getting to me", while for Joyce (C), "they're the things I still feel pretty shit about and I'm still working through in my own head and may be the reasons why I may yet end up being a vegan".[10] This suggests that, theoretically at least, my participants' meaty practices can change, and they may even defect from them altogether. However, almost half have already defected *from* plant-based diets *to* meaty practices. Additionally shaping these negotiations is the stigmatisation of emotions commonly associated with practices that involve *not* eating animals, which can act as an added deterrent to defecting from those involving meat. Such stigmatisation is not only associated with non-meaty practices and the threat of social exclusion they carry (Twine 2014), but with the opposite of every positive meaning that meaty practices are also conflated with. Thus, a person risks taking up 'deviant' practices uncomfortably construed and understood as un-Australian, anti-farmer, impolite, and unhealthy, among others (as I illustrate later in this chapter). The transference of Australian nationalistic sentiment onto meat or food animals is noted also by Singer (2016: 191) and Dalziell and Wadiwel (2016).

---

[8] One participant informed me six months after our interview that she no longer ate meat. Having not followed up with the rest of my participants, I cannot say with any authority that they all continue to eat meat, or even that this one participant continues to *not* eat meat.

[9] I acknowledge here that the production and/or consumption of ethical and sustainable meat were my main recruitment criteria and therefore this is not surprising.

[10] As a teenager, Joyce spent two years following a vegetarian diet and 6–12 months following a vegan diet. With respect to her current practices, she explains, "my body seems to be a body that needs to eat meat".

It is understandable, in the context of the purposive sampling of my research design, that my data suggest defection from the normalised, animal-based diet is not common.[11] Indeed conversely, as discussed in Chap. 3, some reports highlight the "mainstreaming" of veganism (Crawford 2015) and its rise as a "lifestyle movement" (de Boo 2016). However, others have found that a large proportion (84%) of dietary veg*ns return to meat (Asher et al. 2014; ACE 2017).[12] While the available data is inconclusive, ongoing media reports would seem to support these findings. Furthermore (and perhaps contributing to these reversals), veganism is promoted as a dietary choice and something you can even do part-time—one day a week or for part of the day (Bittman 2013; Walker 2016; Flower 2018), divesting it of its broader philosophy in relation to animal exploitation. In sum, it is not possible to foresee how persistent these apparent dietary changes will turn out to be, especially considering that more than half of my participants (two thirds of consumers and a quarter of producers) are former (or current) vegetarians or vegans.

Through exploring the discomfort associated with certain practices of using animals for food, especially practices of killing, and the common (perhaps cultural) understanding that emerges about how this discomfort ought to be approached, I identify the first of two emotions of distinction that I will highlight in this book. As well as directly referencing Wetherell's (2012) meaning of distinction, I also intend 'emotions of distinction' as a play on the many ways in which my participants 'create' distinctions between 'bad', 'good', or 'better' meat. In this way, these emotions are

---

[11] I do not count reducing meat consumption as defection because it still involves the persistence of practices relating to eating animals and their constitution as food.

[12] The findings of this US study, involving 11,429 people over 17 years, have recently been critiqued by Lockwood (2019), specifically the aggregation of vegetarians and vegans (veg*ns) into one group, the question of whether *dietary* veganism is a 'plant-based diet' as opposed to veganism (i.e. an ethical commitment to oppose all animal exploitation), the relatively small sample size, and the underlying research design. A much larger UK study, conducted by EPIC-Oxford between the 1990s and 2010 with approximately 65,000 people over 35, found much smaller rates of recidivism over time (15%–27%). Further questions could be posed of both studies regarding the reliability of data from self-reporting participants, the risk of virtue signalling, and the contribution of age and ageing populations to the findings (Massow et al. 2019), all of which point to a need for more qualitative research to explore understandings and everyday practices surrounding both dietary and ethical veganism.

doubly distinctive. The first brings together common emotional associations that participants make in response to certain, and especially intense, discomforts. Collectively, it summons what I term requisite bravery.

3.3   Requisite Bravery: An Emotion of Distinction

I refer to these previously outlined emotional associations collectively as requisite bravery because, variously expressed, they resemble a social edict or precept of ethical meat eating that, I have observed, exists 'out there', like a law of nature that must be followed simply because "that's the reality of it" (Henry C). Requisite bravery is an association with practices involving ethical meat that, as it is characterised in quite definitive terms by my participants, 'has to' be experienced and almost performed not only to be considered ethical, but to earn the right to eat meat at all.

Just as Foucault (1982) argues that power cannot exist without the possibility for resistance, in the psychology literature it is said that bravery or courage cannot exist without fear, or what I have conceived as discomfort—it is a prerequisite; that is, they are mutually co-constitutive (Goud 2005; Seltzer 2015). Discomforting practices, or elements thereof, thus offer the best opportunity to exercise requisite bravery, for example, by confronting the 'reality' of killing an animal whose flesh you will eat because, "it's part of eating meat, and if you can't do it, you *should* really be questioning how you could be eating meat" (Ellis C). Ellis's (C) tone is scolding, implying that there is something shameful about not being able to face this discomfort and kill an animal but still eating meat, an observation supported by Turner and Stets who describe the "embarrassment and shame" that comes from not following the "rules of feeling and display" (2006: 26).

Exploring my data in more detail, the precept around bravery is sometimes applied to participants' own practices, such as Dan (C), who began eating meat again after many years of being both vegetarian and vegan yet still feels uncomfortable with the associated killing—"death", he says, "is a horrible thing". Dan (C) mitigates this discomfort by approaching this element of his renewed flesh eating in an 'ethical' way. Accordingly, he "*had* to be comfortable with [killing] to be comfortable then eating an animal … if that was something I was going to do then I *had* to be able to do it myself". David (C) acknowledges that killing an animal is "an intensely emotional sort of experience", and one that "I *need* to emotionally prepare myself to do beforehand", evoking a sense of steeling himself to undertake something that he does not find pleasant but is nevertheless 'known' to be necessary. Berating herself for not being "braver" about "the life and the death of things", Anne (C) decides, "if I'm going to eat this stuff I *need* to step into that space and take control of it … I feel like I *need* to do that". Tracey's (C) is more self-judgemental, "if I can't kill it, I *don't* deserve to eat it".

What these participants reveal is an implicit understanding that bravery, constituted in relation to ethical meat, earns them a greater right to eat meat than others. This contributes to the persistence of using animals for food more generally as a number of misgivings and challenges levelled at it can be aggregated and addressed through this ethical re-constitution, which at the same time promises a certain social distinction to its participants. This is why I identify this emotion of distinction, and one other identified in the next chapter, as mechanisms of power. These are in addition to, and work in association with, the validating discourses.

This sense of social distinction emerges also when the bravery precept is applied more broadly, decreeing how others ought to approach their meaty practices. As Lisa (C) very plainly states, "if you eat meat you *should* be able to kill it yourself, you *should* be able to watch it die and butcher it". If you cannot, she continues, "then you *shouldn't* be eating meat". There are many ways in which this requirement is expressed, but the meaning is very clear—"if you can't do it then perhaps you *shouldn't* be eating it" (Sally C). Producers are equally unequivocal: "people say, I don't want to see the face of whatever I'm eating … you've *got* to" (Blake, P), and Finn declares even more prescriptively, "no you can't watch it, you *have* to do it yourself" (Finn, P). Echoing Tracey's (C) self-judgement, Natalie (C) questions whether those who cannot watch the process "*should* be allowed to get away with that". Implicit in these understandings is that some kind of exclusion—from the 'ethical' community?— would, or should, follow from a lack bravery in relation to these practices.

Presumably there is in contrast an associated feeling of inclusive pride, or at least comfort, if you can follow the socially constituted emotional rules 'of feeling and display' in response to discomfort. The association with bravery appears repeatedly in my data. However, emphasising how its constitution is practice(s)-dependent, there is an additional and related discomfort, associated with how practices of *not* eating meat are broadly constituted and understood, which illustrates how any discomfort associated with eating food animals is the lesser of the two to confront. In other words, there is less social approval to be gained from exhibiting bravery in the face of this discomfort (associated with non-meat practices), and more likely the opposite—the discomfort associated with occupying heterotopic sites of vegetarian or vegan practices.

To clarify, by itself, bravery is "morally ambiguous … neither virtuous nor vicious" (Seltzer 2015). The constitution of bravery as a 'good' emotion depends on whether it is considered appropriate to a practice, and also how the material objects of that practice (such as food animals and meat) are constituted. It also depends on how fear or discomfort is constituted in relation to other practices, which represents the "fear of what people might say about you if you don't act courageously" (Seltzer 2015; also Stallen and Sanfey 2015).[13] Finally, while emotional associations are constituted within and also across social practices, these practices are also cultural.

The Constitution of Bravery

At least in Western cultures, bravery is commonly constituted as an emotion of distinction, and almost, at times, an emotional practice in and of itself. The capacity to 'steel' yourself, stand strong in the face of fear, and have 'the courage of your convictions' (rather than change them) is universally elevated and there is a tacit understanding of what bravery or courage looks like. Practices involving health care, emergency services, and the armed forces come to mind. But there are others where discomfort appears to be more purposefully sought in order to create the opportunity for bravery. The rise of extreme sports (Brymer and Schweitzer 2013) and adventure tourism (Carnicelli-Filho et al. 2010) can be seen as

---

[13] This point relates to a large body of literature on social conformity.

part of this quest, where "the search for new emotions and sensations different to those in daily routine has become a decisive factor" (Carnicelli-Filho et al. 2010: 953)—particularly, as these authors note, the emotion of fear and, it follows, what it takes to overcome it. There is also the rise of dark tourism, or thanatourism, which provides travellers the opportunity to seek out what were once sites of others' suffering and death (Stone and Sharpley 2008). Here, "a strange combination of empathy and excitement" turns these sites of tragedy into tourist attractions (Tarlow 2007: 51). As Beedie and Hudson (2003) note, these new tourist practices have successfully commodified adventure, or rather the emotions associated with it, much like, as I will be discussing in Part IV, visibility of meat production is being commodified, perhaps in part provoking (and selling) a similar excitation of associated emotions.

What these practices share is an element of fear or discomfort in "confronting death and dying" (Stone and Sharpley 2008: 576) in some form, with a concomitant element of social distinction that ensues from participating in them because "achieving an adventurous objective requires some kind of social validation to be meaningful" (Beedie and Hudson 2003: 637). With the term 'requisite bravery', I argue that it is the fortitude associated with facing and overcoming discomfort that invites social validation—the discomfort is expected (solicited even), but bravery is required to 'complete' the practice. The "capital potential of participation" (Beedie and Hudson 2003: 637) in practices associated with heightened emotions, especially fear, emphasises that these emotions are very much socially constituted and associated with practices in different ways. As Theodore Kemper observes, "a large class of emotions results from real, imagined or anticipated outcomes in social relationships" (1978: 43). These observations support Ahmed's (2004a) notion of an affective economy and are also reflected in dramaturgical theories of emotion where one emotion (e.g. bravery) is offered in exchange for another of higher emotional value (e.g. respect) (Turner and Stets 2006: 26).

In terms of food-related practices, bravery has already been associated with food tourism, also known as culinary tourism or food adventures (Molz 2007; Mykletun and Gyimothy 2010). Jennie Molz notes the double meaning of the 'Intrepid' traveller in denoting the traveller's character as well as the nature of their practices. She comments that at the same time as demonstrating their fearlessness in encountering the food of Others, the traveller's consumption of what may otherwise be considered inedible also "becomes a way of performing the self as adventurous, curious and open" (2007: 86). However, bravery is required not only to overcome a fear of the Other's food, but, as Mykletun and Gyimothy note, to confront a most vivid reminder "about death and dying" and the fact that animals are killed for our food (2010: 435). This echoes my participants' preferences for assimilating the life of a living animal, with its feet on the ground, rather than the non-life of in vitro meat—confrontation with the life and death of meat is affirming in a way not otherwise possible.

An encounter with freaky, scary, or frightening food experiences does not have to involve travel. Cultural omnivorousness (discussed in more detail in the next chapter) elevates the freaky and the low-brow in the context of "quality, rarity, locality, organic, hand-made, creativity, and simplicity" (Johnston and Baumann, in Kirkwood 2016). Echoing Jonathan Gold's excitement over food 'flavoured with death', Deborah Lupton observes "a machismo of eating, an almost inverse food snobbery, in which the more repulsive the food, the more points are won for appearing gastronomically brave and adventurous" (1996: 128). This bravery, or 'machismo', Joanne Hollows adds, "can extend to other acts of cultural omnivorousness that transgress boundaries within, as well as between, national cultures" (2003: 242).

Bravery, is perhaps, then, the price of inclusion. Maria (C) implies that being brave is the ethical price that you must pay in order to be more connected with your food: "you *have* to learn how to do these things, so there's an ethical cost in trying to develop a less substantiated relationship with the animal … to develop a real intimacy". This of course presupposes the constitution and 'truth' of animals as food, and implies that a less substantiated relationship with non-animal foods does not incur the same ethical cost, presumably because the emotions surrounding plant production, especially of fear or discomfort, are far less heightened, present less of a challenge, and therefore require less bravery—with less reward. The ethical cost is therefore outweighed by the social gains that come from performing a more 'appropriate' connection with meat and animals, especially those of the ethical order.

A note on masculinity is called for here. There is a substantial body of literature on gendered norms of emotional expression, or, more accurately, the gendering of emotions (Menely 2007; Citrin et al. 2004; Lewis and Simpson 2007), with bravery typically associated with masculinity, and with notions of power and dominance (Kinsella et al. 2017; Burnett 2001). Some authors have framed the surge in uptake of a more confrontational approach to the realities of killing and butchering food animals, particularly from women, as illustrative of a feminist approach to empowerment through the co-option of practices understood to be in the 'male' sphere. In reference to a text that exemplifies this trend, which the media have dubbed the New Carnivore movement, Parry (2010: 383) observes that "violence towards animal bodies [is presented] as a bold revision of traditional feminine gender norms, as well as performance of female empowerment".

I have not set out to explore how gender shapes the (re)constitution of ethical meat and food animals, primarily because this would have required a different approach to my interview questions, one that included 'feminised' foods, and also this question would deserve a book of its own dedicated to its proper treatment. I do, however, flag it in my conclusion as an important topic for further research that would complement and extend the substantial body of existing scholarship on the links between meat, masculinities, and also socioeconomic status (e.g. Adams 2010 [1990]; Sobal 2005; Ruby and Heine 2011; Emma Roe and the Manfood Project Bristol: https://man-food.org; Chan and Zlatevska 2019).

Hard-Won Gains

The precept surrounding killing animals for ethical meat also comes with the added understanding that it should not be easy. In Berlin Reed's *The Ethical Butcher*, the story of the author's personal journey from vegetarianism to high-profile butcher, Reed observes, "[Meat] is meant to be a hard won prize" (2013: 52). Graham (P) conveys a similar sense of trial when, referring to the killing that meat requires, he says, "you wouldn't want to enjoy that, that'd be a terrible situation ... people in their right mind don't enjoy that sort of thing". Some of my participants do express excitement at the prospect of killing animals themselves, but in performative and less conventional or mundane ways, such as pig shooting or "boar hunting with bows and arrows" in France, which Sally (C) thought "sounded quite exciting". Michael (C) has identified a couple of archery schools and explains that it is the increased connection with the act of killing that he finds appealing:

> Maybe it's a romantic notion, or the idea that you're not disconnected from miles away. From hundreds of metres away, it's fairly intimate.

Michael echoes Maria's (C) earlier notion of there being an ethical cost in being more 'intimately' connected with how animals become meat, an allusion Tracey (C) also makes when expressing her disappointment at her son's decision to shoot a kangaroo:

> We just said, 'how could you do that, how could you shoot it?', like that's just the worst form of … abrogation of responsibility. You're not even killing it directly, you're killing it through a gun which is just pathetic.

Tracey (C) implies that a more connected, intimate form of killing would have been more difficult, but (and therefore) more 'responsible' and ethical—the associated effort being a better way to show "a bit of humility and respect towards the animal" (Natalie C).

As discussed, it is the trial involved in the confrontation, and the bravery associated with pushing through the resistance of discomfort, which generates the rewards and their deservedness. Sally (C) describes the process as "very exhausting emotionally", while Gillian (C), acknowledging that she is not going to enjoy it, adds, "but I'm not supposed to enjoy it". Once again, a degree of sombre helplessness is implied in the face of an inevitable and foregone conclusion, "a humbling experience … [that] people should have" (Anne C), and that you "have to come to terms with" (Natalie C). Animals on Graham's (P) farm receive an apology before being 'sent off—"you can't do much more", he opines. Andrew (P) describes 'sending off' cows that have been with him for 12 years to be slaughtered, saying, "that is a bit harder I would say, it gets you in here [pointing to his heart], of course it does". What psychology and behaviour-based studies would construe as emotional dissonance or a "meat paradox" (Bastian and Loughnan 2016: 278, also Bastian et al. 2012; Loughnan et al. 2014), I therefore conceive as being very much part of the constitution, maintenance, and performance of the practice.

Consistent with the knowledge that it is the emotional labour involved in killing that 'earns' the meat, a sense of detachment is not to be prized. Referring to the emotional difficulty that he endures, David (C) says, "I really want to keep that". He goes on to explain:

> I don't want it to be a point where it's a production line … that just got to be the routine. I'd find it difficult to see how that could not take away from that connection that I would have and that way that I want to approach it emotionally.

Discomforting associations, which I explore in more depth in the next chapter, can thus present an opportunity to acquire an extra cache of emotional distinction. Rather than encouraging defection from meaty practices, discomfort conversely becomes part of the productive network available to those engaged in practices involving ethical meat and animals, because bravery, the emotion required to overcome it, is so universally feted. In the pursuit of eating animals ethically, the association with bravery becomes a mechanism of power that can be accessed especially via an engagement with killing. In this way, emotional comfort and discomfort work together to not only maintain, but double down on, 'better' meaty practices and animals' edibility. Hence, Foucault's nexus of power/knowledge/pleasure can be more broadly construed as a nexus of power/knowledge/(*dis*)pleasure.

4   Conclusion

Having explored how bravery shapes the negotiation of discomfort in relation to specific meats, animals, and practices, in the next chapter, it is more substantial challenges to the edibility of meat and food animals that are my focus. Hence, I look at various degrees of transgression, when meat or food animals cross over and become emotionally inedible, and when 'bad' or otherwise unethical meat or animals become edible. How these two-way transgressions are corrected, policed, or ameliorated reveal further mechanisms by which these animals are unable to make sense as anything but food.

References

ACE. (2017). Length of Adherence to Vegetarianism. *Animal Charity Evaluators*. Online. Retrieved April 2018, from https://animalcharityevaluators.org/research/dietary-impacts/vegetarian-recidivism/.

Adams, C. J. (2010). *The Sexual Politics of Meat: A Feminist-vegetarian Critical Theory*. London; New York: Continuum International Publishing Group.

Ahmed, S. (2004a). Affective Economies. *Social Text, 22*(2), 117–139.

Ahmed, S. (2004b). *The Cultural Politics of Emotion*. New York; London: Routledge.

Ahmed, S. (2010). *The Promise of Happiness*. Durham; London: Duke University Press Books.

Anderson, E. C., & Barrett, L. F. (2016). Affective Beliefs Influence the Experience of Eating Meat. *PLoS One, 11*(8), 1–16.

Appadurai, A. (1996). *Modernity at Large: Cultural Dimensions of Globalization*. Minneapolis; London: University of Minnesota Press.

Asher, K., Green, C., Gutbrod, H., Jewell, M., et al. (2014). *Study of Current and Former Vegetarians and Vegans*.

Bastian, B., & Loughnan, S. (2016). Resolving the Meat-Paradox: A Motivational Account of Morally Troublesome Behavior and Its Maintenance. *Personality and Social Psychology Review, 21*(3), 278–299.

Bastian, B., Loughnan, S., Haslam, N., & Radke, H. R. M. (2012). Don't Mind Meat? The Denial of Mind to Animals Used for Human Consumption. *Personality and Social Psychology Bulletin, 38*(2), 247–256.

Beedie, P., & Hudson, S. (2003). Emergence of Mountain-Based Adventure Tourism. *Annals of Tourism Research, 30*(3), 625–643.

Beekman, V. (2006). Feeling Food: The Rationality of Perception. *Journal of Agricultural and Environmental Ethics, 19*(3), 301–312.

Bittman, M. (2013). *VB6: Eat Vegan Before 6:00 to Lose Weight and Restore Your Health … for Good*. New York: Hachette Books.

de Boo, J. (2016, May 19). How Many Vegans? One of the Fastest Growing Lifestyle Movements. *The Huffington Post*. Online. October 2016.

Brymer, E., & Schweitzer, R. (2013). Extreme Sports Are Good for Your Health: A Phenomenological Understanding of Fear and Anxiety in Extreme Sport. *Journal of Health Psychology, 18*(4), 477–487.

Burnett, C. (2001). Whose Game Is It Anyway? Power, Play and Sport. *Agenda, 16*(49), 71–78.

Campbell, S. (1994). Being Dismissed: The Politics of Emotional Expression. *Hypatia, 9*(3), 46–65.

Carnicelli-Filho, S., Schwartz, G. M., & Tahara, A. K. (2010). Fear and Adventure Tourism in Brazil. *Tourism Management, 31*(6), 953–956.

Chan, E. Y., & Zlatevska, N. (2019). Jerkies, Tacos, and Burgers: Subjective Socioeconomic Status and Meat Preference. *Appetite, 132*(1), 257–266.

Chau, A. Y. (2008). The Sensorial Production of the Social. *Ethnos, 73*(4), 485–504.

Chiles, R. M. (2013). Intertwined Ambiguities: Meat, *in Vitro* Meat, and the Ideological Construction of the Marketplace. *Journal of Consumer Behaviour, 12*(6), 472–482.

Citrin, L. B., Roberts, T.-A., & Fredrickson, B. L. (2004). Objectification Theory and Emotions: A Feminist Psychological Perspective on Gendered Affect. In L. Z. Tiedens & C. W. Leach (Eds.), *The Social Life of Emotions* (pp. 203–226). Cambridge; New York: Cambridge University Press.

Cole, M. (2011). From "Animal Machines" to "Happy Meat"? Foucault's Ideas of Disciplinary and Pastoral Power Applied to 'Animal-Centred' Welfare Discourse. *Animals, 1*(4), 83–101.

Crawford, E. (2015, March 17). Vegan Is Going Mainstream, Trend Data Suggests. *Foodnavigator.com.* October 2016.

Dalziell, J., & Wadiwel, D. J. (2016). Live Exports, Animal Advocacy, Race and 'Animal Nationalism'. In A. Potts (Ed.), *Meat Culture* (pp. 73–89). Leiden; Boston: BRILL.

De Leersnyder, J., Boiger, M., & Mesquita, B. (2015). Cultural Differences in Emotions. In *An Interdisciplinary, Searchable, and Linkable Resource* (pp. 1–15). Hoboken, NJ: John Wiley & Sons, Inc.

Devi, M. (1997). *Rudali, from Fiction to Performance.* Kolkata: Seagull Book Pvt. Ltd.

Evans, A. B., & Miele, M. (2012). Between Food and Flesh: How Animals Are Made to Matter (and Not Matter) within Food Consumption Practices. *Environment and Planning D: Society and Space, 30*(2), 298–314.

Flower, S. (2018). *The Part-time Vegan.* Hachette Books.

Foucault, M. (1978). *The History of Sexuality.* New York: Pantheon Books.

Foucault, M. (1982). The Subject and Power. *Critical Inquiry, 8*(4), 777–795.

Foucault, M. (1989). *The Order of Things*. London; New York: Routledge.

Goud, N. H. (2005). Courage: Its Nature and Development. *The Journal of Humanistic Counseling, 44*(1), 102–116.

Hayes-Conroy, A., & Hayes-Conroy, J. (2008). Taking Back Taste: Feminism, Food and Visceral Politics. *Gender, Place & Culture, 15*(5), 461–473.

Hayes-Conroy, A., & Hayes-Conroy, J. (2010). Visceral Difference: Variations in Feeling (Slow) Food. *Environment and Planning A, 42*(12), 2956–2971.

Hollows, J. (2003). Oliver's Twist. *International Journal of Cultural Studies, 6*(2), 229–248.

Hsu, E. (2008). The Senses and the Social: An Introduction. *Ethnos, 73*(4), 433–443.

Hurlstone, L. D. (2011). *Performing Marginal Identities: Understanding the Cultural Significance of Tawa'if and Rudali Rough the Language of the Body in South Asian Cinema*. Dissertation and Thesis. Paper 154. Master of Science in Communication. Portland State University.

Illouz, E. (2009). Emotions, Imagination and Consumption: A New Research Agenda. *Journal of Consumer Culture, 9*(3), 377–413.

Kemper, T. D. (1978). *A Social Interactional Theory of Emotions*. Hoboken, NJ: John Wiley and Sons.

Kinsella, E. L., Ritchie, T. D., & Igou, E. R. (2017). On the Bravery and Courage of Heroes: Considering Gender. *Heroism Science, 2*(1), 1–14.

Kirkwood, K. (2016, May 26). Dude Food vs Superfood: We're Cultural Omnivores. *The Conversation*. Online. October 2016.

Knight, D. (1999). Why We Enjoy Condemning Sentimentality: A Meta-aesthetic Perspective. *The Journal of Aesthetics and Art Criticism, 57*(4), 411.

Kolcaba, K. (2003). *Comfort Theory and Practice: A Vision for Holistic Health Care and Research*. New York: Springer Publishing Company.

Korsmeyer, C. (2002). *Making Sense of Taste*. New York: Cornell University Press.

Levi-Strauss, C. (1991). *Totemism*. London: Merlin Press.

Lewis, P., & Simpson, R. (2007). *Gendering Emotions in Organizations*. Basingstoke; New York: Palgrave.

Lim, N. (2016). Cultural Differences in Emotion: Differences in Emotional Arousal Level between the East and the West. *Integrative Medicine Research, 5*(2), 105–109.

Locher, J. L., Yoels, W. C., Maurer, D., & van Ells, J. (2005). Comfort Foods: An Exploratory Journey into the Social and Emotional Significance of Food. *Food and Foodways, 13*(4), 273–297.

Lockwood, A. (2019, January 7). Do 84% of Vegans and Vegetarians Really Go Back to Eating Meat? *Plant Based News*. Online. April 2019.

Loughnan, S., Bastian, B., & Haslam, N. (2014). The Psychology of Eating Animals. *Current Directions in Psychological Science, 23*(2), 104–108.

Lupton, D. (1996). *Food, the Body and the Self*. London; Thousand Oaks; New Delhi: SAGE Publications Ltd.

von Massow, M., Weersink, A., & Gallant, M. (2019, March 12). Meat Consumption Is Changing But It's Not Because of Vegans. *The Conversation*. Online. April 2019.

Menely, T. (2007). Zoophilpsychosis: Why Animals Are What's Wrong with Sentimentality. *Symploke, 15*(1/2), 244–267.

Mesquita, B., & Frijda, N. H. (1992). Cultural Variations in Emotions: A Review. *Psychol Bulletin, 112*(2), 179–204.

Mesquita, B., & Walker, R. (2003). Cultural Differences in Emotions: A Context for Interpreting Emotional Experiences. *Behaviour Research and Therapy, 41*(7), 777–793.

Miller, J. (2012). In Vitro Meat: Power, Authenticity and Vegetarianism. *Journal for Critical Animal Studies, 10*(4), 41–63.

Molz, J. G. (2007). Eating Difference: The Cosmopolitan Mobilities of Culinary Tourism. *Space and Culture, 10*(1), 77–93.

Mykletun, R. J., & GyimOthy, S. (2010). Beyond the Renaissance of the Traditional Voss Sheep's-Head Meal: Tradition, Culinary Art, Scariness and Entrepreneurship. *Tourism Management, 31*(3), 434–446.

Narula, S.K. (2014, April 9). What's Wrong with Sentimentality? *The Atlantic*. Online. January 2018.

Parry, J. (2010). Gender and Slaughter in Popular Gastronomy. *Feminism & Psychology, 20*(3), 381–396.

Piqueras-Fiszman, B., & Spence, C. (2015). Sensory Expectations Based on Product-Extrinsic Food Cues: An Interdisciplinary Review of the Empirical Evidence and Theoretical Accounts. *Food Quality and Preference, 40*, 165–179.

Pluhar, E. B. (2009). Meat and Morality: Alternatives to Factory Farming. *Journal of Agricultural and Environmental Ethics, 23*(5), 455–468.

Probyn, E. (2000). *Carnal Appetites: FoodSexIdentities*. London; New York: Routledge.

Reckwitz, A. (2002). Toward a Theory of Social Practices: A Development in Culturalist Theorizing. *European Journal of Social Theory, 5*(2), 243–263.

Reckwitz, A. (2016). Practices and Their Affects. In A. Hui, T. Schatzki, & E. Shove (Eds.), *The Nexus of Practices: Connections, Constellations, Practitioners* (pp. 114–125). London; New York: Taylor & Francis.

Reed, B. (2013). *The Ethical Butcher: How Thoughtful Eating Can Change Your World*. Berkeley, CA: Soft Skull Press.

Ruby, M. B., & Heine, S. J. (2011). Meat, Morals, and Masculinity. *Appetite, 56*(2), 447–450.

Seltzer, L. F. (2015, October 21). The Complex Emotion of Courage: Do You Really Understand It? *Psychology Today*. Online. October 2016.

Shove, E. (2003). Converging Conventions of Comfort, Cleanliness and Convenience. *Journal of Consumer Policy, 26*(4), 395–418.

Singer, H. (2016). Writing the Fleischgeist. *Animal Studies Journal, 5*(2), 183–201.

Singh, J., Khanna, A., & Khanna, P. K. (2014). Rudali' as an Epitome of Caste, Class and Gender Subalternity: An Analysis of Mahasweta Devi's Rudali. *Indian Journal of Applied Research, 4*(7), 282–283.

Sobal, J. (2005). Men, Meat, and Marriage: Models of Masculinity. *Food and Foodways, 13*(102), 135–158.

Solomon, R. C. (2004). *In Defence of Sentimentality*. New York; Oxford: Oxford University Press.

Stallen, M., & Sanfey, A. G. (2015). The Neuroscience of Social Conformity: Implications for Fundamental and Applied Research. *Frontiers in Neuroscience, 9*, 337.

Stephens, N. (2010). In Vitro Meat: Zombies on the Menu? *SCRIPTed, 7*(2), 394–401.

Stone, P., & Sharpley, R. (2008). Consuming Dark Tourism: A Thanatological Perspective. *Annals of Tourism Research, 35*(2), 574–595.

Sutton, D. E. (2010). Food and the Senses. *Annual Review of Anthropology, 39*, 209–223.

Tarlow, P. E. (2007). Dark Tourism: The Appealing 'Dark' Side of Tourism and More. In M. Novelli (Ed.), *Niche Tourism: Contemporary Issues, Trends and Cases* (pp. 47–58). London; New York: Routledge.

Turner, J. H., & Stets, J. E. (2005). *The Sociology of Emotions*. Cambridge University Press.

Turner, J. H., & Stets, J. E. (2006). Sociological Theories of Human Emotions. *Annual Review of Sociology, 32*(1), 25–52.

Twine, R. (2014). Vegan Killjoys at the Table—Contesting Happiness and Negotiating Relationships with Food Practices. *Societies, 4*, 623–639.

Vannini, P., Waskul, D., & Gottschalk, S. (2013). *The Senses in Self, Society, and Culture: A Sociology of the Senses*. New York; London: Routledge.

Walker, K. (2016, June 17). Become a Part-Time Vegan and Get Healthy. *Body and Soul*.

Wansink, B., Cheney, M. M., & Chan, N. (2003). Exploring Comfort Food Preferences Across Age and Gender. *Physiology & Behavior, 79*(4–5), 739–747.

Wetherell, M. (2012). *Affect and Emotion: A New Social Science Understanding*. London; Thousand Oaks; New Delhi: SAGE Publications Ltd.

Williams, A. M., & Irurita, V. F. (2006). Emotional Comfort: The Patient's Perspective of a Therapeutic Context. *International Journal of Nursing Studies, 43*(4), 405–415.

Williams, A. M., Lester, L., Bulsara, C., Petterson, A., et al. (2017). Patient Evaluation of Emotional Comfort Experienced (PEECE): Developing and Testing a Measurement Instrument. *BMJ Open, 7*, e012999.

7

Feelings of Meat

In contrast to the relatively easily resolved discomforting associations I examined in the previous chapter, here I home in on instances where these associations are intense enough to disrupt my participants' meaty practices to greater and lesser degrees. I therefore explore, in the first half of the chapter, when meat and food animals transgress, or cross over, emotional boundaries to become inedible, how 'proper' emotions are then re-associated, and also how potentially transgressive emotional trajectories are continually policed.

The second part of the chapter focuses on the opposite transgression—circumstances where 'bad' or unethical meat and food animals become edible. This occurs when typical sources of discomfort, as discussed in Part II and thus far in Part III, give way to more comforting associations within the context of the practice as a whole. In both transgressive cases, from edible to non-edible and vice versa, the power exercised through the normalised, multi-sensory, and embodied nexus of edible orders is foregrounded, whereby animals are persistently constrained within anthroparchal cartographies of meat. Encompassing aspects of this nexus that are hitherto not accounted for, either by validating discourses or by emotions of distinction, I identify a third 'type' of mechanism that maintains food

© The Author(s) 2020
P. Arcari, *Making Sense of 'Food' Animals*,
https://doi.org/10.1007/978-981-13-9585-7_7

animals' domination. The type of mechanism is an ethico-aesthetic, and the first (of two) that I identify here is 'moral approval'—essentially an invocation to do what is constituted as 'right' and morally 'good'. The second appears in Part IV.

1   Crossing Over: From Discomfort to Emotional Inedibility

As my discussions with producers and consumers have revealed, occasionally, discomforting emotions associated with one type of knowledge or 'truth' (about meat, animals, or meaty practices) can be intense enough to momentarily arrest, and sometimes completely shatter, normalised emotional routines, or "practice-specific emotionality" (Reckwitz 2002: 254). A 'normal' and 'practice-specific emotionality' is comfortable with the use of cows, pigs, chickens, and sheep as food, and so when Grace (C) was growing up on her family's farm, the process of killing them "didn't bother" her. Except, that is, "when Dad killed my only pet chicken". Suddenly, her formerly comfortable associations were unsettled.

Highlighting how emotional associations can wrangle with orders of knowledge that designate animals more or less as pets/food, Helen (C), who was vegetarian for three to four years, described living with various 'pets' and how "all the animals interacting, made me feel more uncomfortable about eating meat". Elucidating this further, Alison (P) observes, "You just don't eat an animal that trusts you and follows you around"— that is, they become emotionally inedible. This association of trust with certain animals, and the subsequent discomfort associated with betraying that trust, is a distinct thread in my data and is especially evident in the relationship between producers and 'their' food animals. Their discomfort can be intense enough to challenge and even dismantle normalised practices, but only to a degree still limited by cartographies of meat.

Recalling his childhood on a farm, Simon (P) remembers that "any old lamb from the mob" would be killed on-farm for its flesh; however, "if it was a pet, it'd always go off to market", thus creating distance from what he suggests might have been an uncomfortable on-farm process for him and/or his family. Melanie (P) is similarly bothered by eating their own animals, who she has watched running around in the paddock. She says, "I'm ok with it going to the market, knowing that it's going to the abattoir from the market", but she "wouldn't have been happy" to know that the same animal "came back into my freezer". Melanie (P) needs the distance and anonymity that comes from selling their animals to a sale yard and then buying meat from the butcher—to "not know where it's come from", because "knowing that it was one of mine, I couldn't have done that" (i.e. eat it). Emphasising the sense of connection explicitly, Florence (P) describes how her father loves a "good steak", but for him, "don't overly connect it, which is probably why he won't eat his own". In fact, for her whole family, there is a heightened connection with 'their' animals so that "if they have to stop and think about that, they sort of go … something about knowing the animal". If these producers did think about it, perhaps the associated emotional discomfort would unsettle their practices, as it did for Alison (P). After acquiring a few pigs to raise for their meat, Alison (P) explains, "we've raised them and then gone, Oh, we can't even, we can't eat them … cos they've got a personality, probably they've become pets".

Alison's (P) comment highlights that for these producers, a different kind of connection with their animals allows a blurring and occasional transgression of normalised boundaries between orders of edibility and non-edibility, and sometimes necessitates active policing of those boundaries. The status of *their* animals as food has become less unequivocal. Naming can play a big part in shaping transgressions. Nicki (C), who runs a hobby farm, tells me they "don't name our boys [who go to slaughter], we only name the girls". Similarly, Brigid (P) refers to her named pet sheep:

He's a boy and he was made an orphan … two years ago. He's got such a personality, so yeah, there's always exceptions I suppose.

Naming implies subjectification, as against the normalised, wholesale objectification of these animals and their flesh. It risks an emotional connection that is not appropriate and often incompatible with the knowledge of these animals as edible. It is at this point, where experience is not conforming to the order of things, that Douglas sees the opportunity "to force attention into less habitual tracks" and "examine the filtering mechanism itself" (2002: 38).

Every year, there are media reports of pigs, cows, chickens, and sheep who have avoided becoming meat through dramatic acts of escape, often at the point of slaughter (Conaty 1998; Pleasance 2014; Bluestone 2014; Chasan 2015; Schelling 2015; Krause 2015; Steinbuch 2016; Embury-Dennis 2018; Petenko 2018; Demers 2019). Most of these stories are not reported in mainstream media. When they are, they depict the animals as "wily" and "devious" outlaws "on the run", "evading capture" through their "fugitive antics", and being chased by keystone cops as they "make monkeys of the police". These descriptions are often accompanied by humorous epithets such as "houdini", "save their bacon", or "walking steak frites" (the UK's Butch Cassidy and the Sundance Pig being the most famous case to date). Invariably, the "daring", "courage", and "spirit" shown by these "intrepid" individuals sets them apart from the billions of others killed each year for food. There is often a "huge public outcry" around their capture and the negotiation of their fate (if they have not been killed in the process), usually leading to a "quashing" or "reprieve" of their death sentence, which would now be "unsporting". Even if they are killed in the process, they die "a hero's death" (Bluestone 2014).[1] The lucky ones are taken in by animal sanctuaries where they can instead lead "cushy" lives (Steinbuch 2016). Of course in reality, these individuals are not any different from the billions who pass unnoticed through Noske's (1989) animal-industrial complex. However, their momentary humanisation, which occurs as a process of their individualisation from the herd, often (not always) allows them to cross boundaries of use, evading their former designation as food, to assume a new role no longer predicated on the edibility of their flesh.

My participants reveal further instances of 'crossing over'. For instance, Julie (C), who eats meat once or twice a week, makes it clear that "we don't eat our poultry, they're too cute". Diane (C) similarly relates the following story:

---

[1] Implicit in these discourses is the idea that the 'normal' manner of their death in slaughterhouses is otherwise sporting and what the majority of the non-heroes deserve.

I had two chickens, and we called them rhubarb and polenta to remind ourselves that they were to be eaten. But we just never got around to killing them and when they got sick we actually got them very expensive vet treatment and then when they died we had funerals, so we got very attached.

Alison (P) makes a more direct connection, explaining that "sometimes animals just become humanised"—these individuals, "that were our pets, we didn't eat". However, this ability of certain individuals to 'cross over' serves to emphasise the status of others as food, as Michael (C) illustrates:

We had cows—there was Dinner and Rosie, which was my dad's first cow and he had to put it down. He found that quite distressing … we definitely knew both Mary [the pig] and Rosie [the cow] to be more household pets … but yeah, Dinner was just dinner.

Crossing over appears to happen when a discomforting emotional association that would normally be inappropriate within meaty practices is provided an acceptable outlet when it is limited to just one animal. Such animals are constituted as being unique in some way from all the others on account of their more 'human' qualities and capacities, or their outstanding courage and will to live, which make them more relatable and bring them closer to those animals we would commonly regard as pets. These examples would seem to illustrate Todd May's (2014) theorisation of moral individualism (where the characteristics or capacities of the animal determine its moral status) and moral relationism (where this status is determined by the animal's relations with humans). As Coghlan points out, "capacity-possession appears to be necessary for RBRs [relation-based reasons] to arise" (2016: 1243). They also highlight the fickle or shaky ground on which these moral understandings rest, given that capacity-possession is selectively perceived and inconsistently applied.

What appears ultimately more decisive in the fate of these animals is the socially constituted narrative that accompanies, and constitutes, their individual transgressions, which Coghlan (2016) attends to in his "narrative-style philosophy". However, taking this further, I would conceive the selective moral elevation of certain animals rather as their location within a broader array of practices where occasional transgressions are *allowed*, and perhaps even necessary. They maintain the illusion of a relationship where resistance is possible, rather than the reality, which is domination. For while individual animals are being differently constituted, the prevailing, normalised constitution of their species as edible remains unchallenged. Additionally, moments of transgression become focal points for any accumulated discomfort, or discomfort associated with a momentary visibility of mechanisms of domination. Much like requisite bravery, they serve to productively channel and transmute discomfort in ways that reaffirm the normalisation and hence the persistence of meaty practices. Through their public enactment, these emotional flashpoints exercise the power to absolve billions of deaths, and thereby efface the domination that characterises the normal order. As Douglas aptly explains:

> Then suddenly we find that one of the most abominable or impossible is singled out and put into a very special kind of ritual frame that marks it off from other experience. The frame ensures that the categories which the normal avoidances sustain are not threatened or affected in any way. Within the ritual frame the abomination is then handled as a source of tremendous power. (2002: 166)

Indeed, Douglas identifies the capacity to accommodate transgressions as essential to systems of power (2002: 40).

Central to these alternate, 'abnormal' constitutions of certain animals is the existence of temporal and spatial locations where the normal order can be suspended—where what would be inappropriate emotions to associate with food animals (compassion, kindness, responsibility) become appropriate. An altered knowledge of these animals then becomes permissible, as well as a different emotional association, as long as that association "do[es] not deviate too far from the emotional script provided by culture" (Turner and Stets 2006: 26). For these transgressions are playing out with their own inherent order, and in ways that reinforce what is 'normal'.

Transgressions are therefore by definition, and also necessity, the exception. They are singular instances where dominant emotional associations with food animals are permitted to escape the normative script, allowing individual animals to cross boundaries, but only one or two animals, and crossing over only to 'othered' spaces that are legitimised within the overall system (e.g. farm sanctuaries). For all of my participants, the animals in question remain highly individualised and in some way unique, while the species as a whole remains persistently edible. To humans, a certain pig is still Pig, the edible animal that 'produces' pork, and the same with an individual cow, lamb, or chicken—the individual does not escape their species representation. As Berger says of animals in myth and history:

> each lion was Lion, each ox was Ox. This—maybe the first existential dualism—was reflected in the treatment of animals. They were subjected and worshipped, bred and sacrificed. (Berger 1992: 7)

In terms of competing emotions, I would argue that the edibility of these animals persists because these often intense emotional flashpoints of individualisation do not present a sufficient challenge to the overwhelmingly comfortable associations with meat and practices involving meat, and the discomforting associations with not eating meat. Furthermore, these transgressions are located in spaces constituted as being outside normalised practices where 'non-normal' emotions are uniquely construed as appropriate and are thereby unproblematically isolated from everyday meaty practices. Discomfort is thus corralled and subsumed within the dominant order of emotional rules. As Simon (P) emphasises, "at the end of the day they're meat", demonstrating how order shapes emotions, whereas emotions are more likely to merely unsettle order, and then usually only temporarily—"cultural categories are public matters. They cannot so easily be subject to revision" (Douglas 2002: 40). This pull of order and 'knowledge', which shapes negotiations between appropriate versus inappropriate emotions, is clearly illustrated by my participants' accounts of particularly memorable transgressions and how these were managed.

1.1   Correcting and Policing Transgressions

Often during the course of my interviews, participants would recall past events and experiences that had made a particularly strong impression on them and were associated with intense emotions—intense enough to change their practices around food, even if only temporarily. Their memories of these events were typically very vivid and visceral, highlighting the dynamic, rhizomatic interplay between the emotions, senses, knowledge, and the imagination. What I want to focus on in these accounts are the intensities of the emotional associations, how they shaped participants' practices, and how they were corrected or policed in line with how meat and food animals are understood to make sense. Ultimately, they demonstrate the dominance of a normalised nexus of knowledges/senses/emotions associated with meat and its role, along with associated mechanisms of power, in maintaining the edibility of these animals.

Maria (C) recounts a particularly graphic experience on a farm, which involved the inept shooting of a kangaroo:

> The kangaroo got shot from a distance and we kind of thought it was dead but it turns out that it had only been shot in the leg … so my friend had a knife and he cut the throat and the animal was first of all angry, or scared, and lashing out and scratching, and then it didn't die. Cutting through the throat was quite a process and … death didn't come instantaneously … I was thinking, 'shit, this animal didn't die as quickly as I expected it to, this was not a good death'. … so then my friend broke it down and I stripped off the skin and that was the first time I'd touched meat or flesh that was still blood warm, which is a very confronting thing.

Maria (C) remembers, "I felt terrible about that … very complicit". Although she and her friends "ate it later and it was delicious", she goes on to say, "I was much more uncomfortable with eating meat for a period of time". Quite spontaneously during her interview, Gillian (C) remembered that she worked in a chicken plucking farm when she was young, "that's why I became a vegetarian, oh my god, of course … I'd totally forgotten that". This experience created lasting and intensely discomforting emotional associations with animals and meat for Gillian (C) who says, "I don't think I ate chicken for 20 years and I went off all food". Alison (P) grew up on a farm and remembers that "as a child, [on-farm slaughter] was really hard. I actually found that quite traumatising". She vividly recalls that she would take the chosen animals away, "go and hide" when the slaughterman arrived at the farm:

> I'd be like, 'no, no, don't kill them', and then … that feeling of betrayal, when they talked me into handing [the animals] over. I remember going into my bedroom and just bawling my eyes out.

Alison (P) was vegetarian for six years as a teenager, which, she says, was a direct result of this sense of 'betrayal'.

The altered emotional associations with meat and animals that these participants' experiences engendered were intense enough to disrupt, to varying degrees, what had until then been reportedly 'normal' associations. Nevertheless, these 'ab-normal' associations only lasted a certain period of time (from months to years)—they were not persistent and each returned to normalised practices involving meat, and more appropriate emotional associations with food animals. Similar hiatuses, of between six months and four years, were triggered by a visit to a deer farm (Finn P), a graphic PETA (People for the Ethical Treatment of Animals) campaign (Lisa C), and coverage of strike action at Victorian abattoirs (Heather C).

The rhizomatic pathways by which practice-appropriate emotions are constituted in relation to meat and food animals are both product and productive of the historically and systemically legitimised use of animals for food. These pathways are deeply inscribed on human and nonhuman bodies. As well as being self-reinforcing, they are also magnetised, capable of drawing 'errant', outlying emotions back into line. This process is assisted by the naturalising discourses (Part II) and the plethora of normalised social practices that continually offer alternative ways to channel and thereby mollify discomforting or inappropriate emotional associations with animals and meat.

These 'socialising' practices that shape human relations with all nonhuman animals begin at an early age and are the focus of Cole and Stewart's volume on the cultural construction of human-animal relations in childhood in Western cultures (2014: 211). Using a four-quadrant dia-gram, the authors depict the positioning of categorised animal 'others' along two intersecting continuums according to the degree to which they are subjectified or objectified, and then the extent to which they are expe-rienced multi-sensorially "as living, breathing individuals" (22–24), rang-ing from sensibility to non-sensibility. For example, pets and working animals are primarily subjectified and sensible while farmed and labora-tory animals tend to be objectified and non-sensible. Critically, Cole and Stewart note how some categories constitute their associated animals as more "appropriate repositories for affect" than others (even though they may be the same species), instantiating, at childhood, a normalised "channelled affect" that pervades and shapes all future relations (24). Therefore, building on their theory regarding socialising practices, I iden-tify the socialisation of emotions as another significant part of this process.

In addition to the corrective socialisation of discomforting associations with particularly graphic experiences (e.g. of killing) and the consump-tion of 'betrayed' animals, my data also reveal instances where inappro-priate emotional associations are actively policed—that is, they are ameliorated, subdued, and properly socialised in favour of the 'appropri-ate' emotion before any transgression can occur.

Although children were not included among my research participants, my data nevertheless indicate the kind of childhood socialisation process that Cole and Stewart (2014) have examined. Henry (C) speaks of taking his young children with him to trap rabbits, which they will eat, but thinks, "it might be a bit confronting for them still", with the implication that this will eventually abate. Illustrating this is likely to be the case, Brigid (P) recalls that her father would name the animals on their family farm "Mince, Roast, Chops, Stew, Casserole" and says, "good on him, it's good, it's in your face all the time and you do become desensitised".

Honed via this socialisation process, knowledge and awareness of the appropriate emotional associations within practices that depend on kill-ing animals for meat are learned. This awareness regulates the presence of, or potential for, inappropriate discomfort and ensures the dominant "affective rut" (Wetherell 2012: 14) remains in place. Policing this 'rut' or norm involves, quite simply, not thinking about it too much. As Grace (C) very plainly attests, "I try not to think about the animals too much".

Even Geoffrey (C), a committed and enthusiastic meat eater, finds offal, and especially tongue, occasionally confronting: "It's a bit disgusting when you're peeling it". However, as he says, "you just try not to think about it". What is perhaps at stake in this instance is the potential for what Rosemary Deller refers to in relation to bio-artworks as "ontological blurring" (2016: 75). This ontological blurring ensues when the "mutability" (ibid.) of animals and humans and their proximity at the level of body parts, textures, sinews, blood and flesh are foregrounded, or, rather in this case, fore-knowledged, fore-sensed, and fore-felt.

Even without this ontological blurring, a strategy is required to deal with certain discomforting knowledges, because, as Douglas comments:

> The search for purity is … an attempt to force experience into logical categories of non-contradiction. But experience is not amenable and those who make the attempt find themselves led into contradiction. (2002: 163)

For example, considering kangaroo a more ethical meat because "it's better for the environment" and "it's lived out in the wild", Dan's (C) comfortable associations with it were contradicted when a professional culler, visiting from interstate, started to regale him with stories of his culling experiences. Dan (C) explains, "I ended up having to tell him to shut up because I was like, I don't want to know how it's done, because it's not quite sounding like the idyllic situation that I wanted it to be". Similarly, Heather (C) remains "really uncomfortable" and "anxious" about how animals are slaughtered—"I haven't gotten my head around it". Unable to reach a resolution, she says, "I kind of avoid thinking about that … I remain in denial about it". There is an implied emotional risk in thinking about 'it', whatever the 'it' might be (possibly the animal as an individual, sentient being, or their feelings, or how they are killed)—a sense of resistance to pulling back the veil on something potentially harmful. It is harmful because the associated emotions would be inappropriate and therefore transgress the affective norms of meaty practices. As Grace (C) continues, "If I actually think about the animal then I feel bad", while Brigid (P) reveals, "sometimes I'll let myself think about it too much and I freak out and I'm like fuck, what gives me the right to take another life in order to satisfy my cravings".

Emphasising the interdependent and co-constitutive nexus of knowledge/senses/emotions operating within meaty practices, emotional associations that are inappropriate to this nexus can bring about a questioning of the supporting knowledge and sensorial perceptions—or how these perceptions are made sense of. However, short-circuiting and denying potentially problematic emotional associations, for example by 'not thinking about it', more often than not successfully corrals the associated emotions and prevents inappropriate transgressions.

1.2   Keeping to the (Normalised) Emotional Path

Practices involving meat and food animals that have been deeply normalised over time also carry certain expectations regarding the experience of performing them (shaped by emotional and perceptual mimesis)—specifically, that the emotional and sensory associations experienced will be within a range that has over time become 'normal' (Castelfranchi and Miceli 2011; Passafaro et al. 2014). These practice-constituted expectations exert a kind of self-fulfilling regulation on how each repeat performance unfolds. However, if some aspect of the actual experience falls outside the expected range, the practitioner experiences a rupture in the 'normal' performance that has to be negotiated. The elements, and their usual relation to each other, have become unsettled.

For example, consider the differing perspectives on 'normal' noted by Evans and Miele between meat eaters, for whom "the act of eating an animal is an astonishingly smooth and unremarkable practice", and vegetarians and vegans, for whom "this situation is simply intolerable" (2012: 312). They add that work needs to be done to maintain this smoothness—to maintain order, keep animals in their proper place, and avoid ambiguity creeping in; the *possibility* of thinking 'that'. In her analysis of relations between human and animal bodies used in psychological experiments, Vinciane Despret (2004) shows how expectation works to create the animal subject, and the experiment, in ways that produce the expected outcome. She describes expectations as "availability to an 'affecting' that both creates events and is created by them" (125). Seeing attempts to theorise dividing lines between what belongs to the inner and outer body as misguided and missing the point, she instead draws on William James' account of emotion as belonging to a "more ambiguous sphere of being" (125):

> Ambiguous experiences, ambiguous bodies, experiences making bodies and bodies making experiences; signs that wander, hesitate to fix themselves: we produce emotion, and it produces us. (Despret 2004: 127)

An availability to ambiguity seems to describe what happens when my participants' 'normal' practices involving meat are ruptured in some way—when a different way of relating, one that runs contrary to expectations, somehow gains traction. Echoing the notion of heterotopia as where order, logic, and language are detonated (Dehaene and De Cauter 2008), according to Wetherell (2012: 9), Despret is describing:

> Moment[s] of hesitation in emotion when it is possible to launch body and mind on new alternative trajectories and choose other forms of becoming. In this moment, body/mind is unlabelled potential—unscripted and undifferentiated. The old scripts, figurations, positions and narratives are always available waiting in the wings.

For Douglas, these moments are "thresholds" that promise the "beginnings of new statuses" (2002: 115). They could thus be read as glimpses of, and potential pathways to, spaces of heterotopic abnormality. In terms of the persistence of meat, it would seem from my participants' accounts that there are many potential 'moments of hesitation in emotion' that carry the possibility of a rupture in these practices. However, the emotions associated with these moments are rarely intense enough, distinctive enough, or aligned with the 'right' kind of knowledge, to compete with those associated with normalised knowledges of meat and food animals. The 'moment' or 'ambiguity' never takes off on a new trajectory but remains grounded, steps back from the threshold. Even when the emotions are sufficiently intense and the practice as a whole is ruptured, allowing new associations to endure for a few months or even years, the magnetising effect of normalised knowledge/senses/emotions eventually reasserts, and emphasises, their overarching persistence. As Wetherell explains, some "affective practices" are "very densely knotted in with connected social practices where the degree of knitting reinforces the affect and can make it resistant and durable, sometimes unbearably so" (2012: 14).

Normalised emotional associations with meat and food animals would appear to represent (for most humans) just such a densely knotted set of affective practices that are constituted by, and constitutive of, the wider complex of social practices of which they are part. The constitution of these 'appropriate' emotional associations simultaneously constitutes those deemed inappropriate. A tacit knowledge thus emerges regarding what is considered an inappropriate emotion and how to negotiate, or police, such associations to maintain the integrity of what is considered 'normal'. Hence, both the self-imposed injunction to 'not think about it', and the distinction that comes from requisite bravery, in combination with the validating discourses, serve a regulatory function that helps to contain emotional associations within the realm of what is expected with regard to meat and food animals. Alison (P) and Brigid (P) further illustrate this regulatory effect.

Every week, Alison (P) selects the animals that are to be sent to the abattoir. However, she says, "I'm struggling with it a lot more now, I don't know why, I'm just really not feeling good about it". Reflecting for a moment, she continues, "I'm probably becoming a bit more aware of their feelings, and thinking, Oh goodness, we're really taking their lives". Given that Alison (P) later reckons that a scenario where "you can just have animals for pets, you wouldn't have to eat them" would be her preference, and based on my participants' allusions to the inherent risk of thinking too much, I suggest that Alison (P) is in fact now thinking too much, or more than she was previously. Hesitating in an unscripted space, she is becoming available to ambiguities in bodies and their 'signs', to different possibilities and alternate emotional trajectories. As suggested, unless more positive associations can outweigh these newly discomforting emotional associations, the legitimacy of Alison's meat eating as a whole may be ruptured (as it was previously).

From admonitions to not 'overthink' or 'think too much' and the concomitant danger of doing so, there is a sense of emotional associations being actively policed, to keep them within the bounds of what is 'appropriate'. For when Brigid (P) 'freaked out' after 'letting herself' think about it too much, she quickly went on to explain, "I'll rectify that". Other, more appropriate, associations that she has with practices involving food animals, such as "I believe we're meant to eat meat", enable her to correct her flight path and re-associate her emotions according to validating discourses. As she affirms, "this is what you're doing, this is why you're doing it, and then just enjoy it". Similarly, Julie (C) explains that she "can't" allow her uncomfortable feelings associated with the unknown-ness of any meat she might be offered, to "overwhelm everything". Allowing feelings free rein so they become 'overwhelming' is broadly constituted as an inappropriate in relation to practices involving food animals and meat, which instead reward fortitude and bravery. As she says, "just suck it up princess".

There is an implicit understanding being delineated here that untoward emotional associations with animals and their flesh need to be brought back into line and should continue to toe that line. This line is shaped by the emotional community as part, and product, of meaty practices (and other social practices) and represents a practice-constituted and practice-regulating pressure to rein in and "rectify" inappropriate and potentially unruly or 'overwhelming' emotional associations. Anything else becomes problematic and poses a challenge to the tacit conventions of associated practices.

Returning briefly to my introduction, behavioural approaches with their focus on the individual would take a different perspective on these participants' accounts, charging them with inconsistent beliefs or hypocrisy, variously articulated as cognitive dissonance, or the value-action/attitude-behaviour gap (Vigors 2018; Dowsett et al. 2018). However, to re-emphasise, instead of individuals being responsible (or sometimes 'to blame') for their choices and actions, a practices perspective sees people performing a range of practices with different associated meanings, competencies, teleoaffectivities, and material arrangements. Various co-located practices (in time and space) may be broadly aligned or more or less conflicting leading to competition, whereupon different practices can vie for attention and "temporary defection, multi-tasking and contamination between practices" become likely (Shove et al. 2012: 94.) Hence, to paraphrase Douglas, meaty practices and their sense of form make demands on behaviour, govern assessments of desire, permit some and over-ride others (2002: 101). Practices could thus be said to exert a "hold on conduct" (Foucault 1977: 172) through the 'sense of form' they engender in their participants. Effects of power therefore permeate the whole nexus of power/ knowledge/pleasure, reflecting and reproducing unequal relations between humans and animals not only in systems of knowledge, but also in their associated emotional rules.

However, it is to those inappropriate, excluded, or 'over-rided' emotions—the ones that, according to Hochschild's 'feeling rules', are "misfits between feelings and situations" (1979: 566)—that my attention is persistently drawn. While he acknowledges that the 'fit' (appropriateness) or 'misfit' (inappropriateness) of emotions is socialised, Hochschild, in my opinion, does not give adequate consideration to the role of 'misfits' in shaping the 'fits' for two reasons. First, inappropriate or misfit emotions are constituted through normative social practices, but the process of their stigmatisation, or exclusion, contributes in turn to the ongoing re-constitution and reinforcement of the dominant emotional rules. As with comfort/discomfort, fear/bravery, they are co-constitutive and mutually dependent. Second, the stigmatisation of these inappropriate, excluded (and typically feminised) emotions extends to and encompasses all the knowledges, emotions, objects, and practices that they are in any way associated with. These elements are not just floating around in space, but tend to become aggregated in various collectively imagined practices and the constructed abstract 'figures' that participate in them (e.g. 'the vegan').

For example, Natalie (C) explains that her "soft spot for ducks" means that she has what she considers "too much of an emotional objection to duck hunting"—why too much, too much for whom or what, and according to what criteria/rules, is not clear. Is an emotional objection inferior to other kinds of objection? Also, where is 'too much emotion' located/directed? What sort of meanings, practices, and abstract figures are associated with too much emotion? There is not the space here to explore these co-constitutive 'mis-fits' of the normalised social order. However, my participants collectively delineate certain 'alien' constructs that align with Ahmed's (2010a, b) notions of affect aliens and feminist killjoys, and also Richard Twine's (2014) conception of the vegan killjoy, and therefore warrant further analysis.

Putting these misfit aliens aside, in the next section, I turn to the reverse transgression, when what is ethically inedible, or at least ambiguous, becomes productively edible.

2 'Bad' Schm-ad: Disturbing Practices and the Maintenance of Edibility

Despite the lengths that my participants go to in order to ensure that their particular construction of ethical meat is being associated with the 'right' farm, producer, practices, animal and even parts of the animal, in certain situations, all of this order and knowledge is disturbed, and more or less dispensed with, when other practice 'knowledges' (including sensory and cultural dimensions) take precedence. Meat that in one context is understood as 'bad' suddenly becomes acceptable, and even 'good' because another knowledge is now being associated with it. I observe this especially when my participants discuss their practices in relation to eating out, seeking novelty, travelling overseas, and being polite.[2]

What these situations reveal is, on the one hand, the arbitrary and unstable basis that an 'ethical' framework provides for the idea of incremental improvement and taking 'a step in the right direction', and on the other, the knowledge and 'truth' underlying this ethical framework, which is that eating meat—that is *any* meat from *any* animals—is still prioritised over not eating it at all. In addition to validating discourses and emotions of distinction, this is due, I argue, to the persistence of two fundamental ethico-aesthetic mechanisms of power, this first of which is moral approval.

---

[2] These are, of course, not discrete bundles of practices with their own associated systems of knowledge and emotional rules. That is not how social practices operate (Shove et al. 2012) and so there are inevitable overlaps and synergies between them.

2.1   Eating Out

Eating out, as described by my participants, is one of the main practices involving food where they prioritise different knowledges such that their usual configurations of knowledge/senses/emotions in relation to 'good' and 'bad' meat no longer apply. Instead, such concerns are outweighed by the comforts and pleasures associated with 'going out', relaxing, 'letting go', eating things you do not have at home, and being with friends. This creates a sense of all bets being off. Anne (C) explains, "when I eat out, I don't worry about it … we tend to just give ourselves a bit of a break". Similarly, when Maria (C) is eating out, it is "for more hedonistic reasons and so sometimes [I] will order pork without knowing where it's come from". The alternate emotional rules shaped through the non-routine 'space, time and relations' of eating out (Zembylas 2005: 56) are charac- terised by hedonism and the opportunity to "indulge" (Dan C). These teleoaffectivities (Schatzki 1997) take emotional precedence over, and are constituted as more appropriate than, formerly stringent distinctions between 'good' and 'bad' meat. As Anne (C) further illustrates:

> When we go to Melbourne it is just an eating fest and we'd eat the things that we can't get here. So we go to yum cha,[3] we go to Malaysian, we go to Indian and that is purely a you know, a hedonistic experience.

If the meat involved in these practices is not 'ethical' according to par- ticipants' normal criteria, then eating 'non-ethical' or unknown meat is more appropriate to the sense of form of these practices than purpose- fully rejecting it. Eating meat is therefore the normalised benchmark, associated with certain comforts of eating out that are not surrendered even if the meat is not 'ethical'. Vegetables and plant-based foods are not being associated with hedonism and comfort, only meat. In this way, normalised knowledge/senses/emotions associated with eating out can be seen as contributing to the persistence of meat.

---

[3] Yum cha is known in Australia as a traditional Chinese brunch involving varied dishes of small steamed or fried savoury dumplings. In Europe and North America, people more commonly speak of going for 'dim sum' which refers to the range of small dishes served at a yum cha.

Sensory pleasure when eating out, or the perceptual and emotional mimesis involved in the anticipation of it (Illouz 2009), can even outweigh associations with what is known to be factory farmed meat, typically regarded as the worst of the 'bad'. Despite it being "pink pig", Sally (C) cannot resist the pork crackling at one of her favourite restaurants, "it was amazing … It was soooo good", and Maria (C) acknowledges that a particular meat that she buys is "still an industrial product so I feel a little bit uncomfortable about [it]"; however, "it does taste pretty good". The corollary of these accounts is that to not engage with these otherwise 'bad' meats, which are now associated with hedonism and indulgence, would be to deny access to the rewards they implicitly promise as part of a comforting eating out nexus of knowledge/senses/emotions. Such denial becomes constructed as missing out on a 'good' experience, or, as Damien (C) describes it, "living in a vacuum". Anne (C) similarly maintains that she "can't cut [her kids] out of that experience", because you have to "let them have a good time". Referencing proscriptions around 'thinking too much', which might dampen or erase the pleasures promised by eating out, Anthony (C) says he tries not to "overthink the meat too much when I'm out".

Associations with certain dishes (pork crackling, steak and kidney pie, chicken parma) and also styles of food can similarly render meat edible where it might normally be regarded as 'bad'. With regard to 'fast food', Lisa (C) comments, "I'm getting fast food anyway, so I'll just eat whatever it is". Oliver (P) says that he is "guilty at once in a while grabbing fast food … we're not purists", while Bella (P) explains:

When it comes to buying pizza I probably didn't give a rat's arse, I was tired, I made a takeaway decision because of other reasons, and the same probably goes for Asian.

'Asian'-style food offers a popular sanctuary from normal distinctions. Damien (C) and his partner "eat a lot of Asian food"; however, he says they lack the right tools—"the big woks that they have", and skills—"I don't have the ability", which is why they eat out, even though "I know they wouldn't have ethically sourced meat". Sally (C) maintains, "it's no good going into a Japanese restaurant and trying to find out where the meat comes from". Similarly, for Maria (C), whose local Vietnamese restaurant "does these amazing little rice paper rolls with this kind of beef jerky", she says she has "no idea where that beef comes from and I'm not even going to bother asking because I know that it's not".

Putting aside the racist tone of these comments (another topic worthy of further research), 'fast' and 'cheap' food or take-out tends to exist in a slightly separate category of eating out where knowledge/senses/emotions associated with convenience, comfort, tastiness, and also of cooking skills are prioritised over those associated with what is ethical or 'good'. In contrast, eating out in moderate and higher-end restaurants is less about convenience and comfort, and more about experiencing the 'hedonistic', sensory, and also the commensal pleasures associated with the 'occasion' of eating out—"you go to socialise … so the food is not the actual priority" (Oliver P). However, common to both forms of eating out is that these associations are persistently constituted by my participants in relation to meat, not vegetables. Meat thus becomes synonymous with indulging, 'taking a break', enjoying hedonistic pleasure, commensality, comfort, and tastiness.

Two dimensions to the persistence of meat and the use of animals for food can be identified here: one being the reflection and reinforcement of meat as normal and pleasurable, and the second being that the same associations are not being made with vegetables and plant-based foods. Effectively, the domination of animals is inscribed in the knowledge/senses/emotions associated with practices of eating out. However, as my participants illustrated, sometimes eating out *is* also as much about the food itself as the experience, especially when it is something not generally prepared and eaten at home, which brings me to my second emotion of distinction.

2.2   Seeking Novelty: Cultural Omnivorousness as an Emotion of Distinction

The pursuit of sensory experiences regarded as novel and exciting, and indeed the constitution of novelty and excitement as things worth pursuing, is another way in which the usual knowledge/senses/emotions associated with 'ethical', 'bad' or at least doubtful meat are pushed aside.

More stable and persistent 'truths' relating to how meat and food animals make overall sense are then foregrounded, for once again, novelty and excitement are not being associated with vegetables and plant-based foods. The high regard afforded to novelty for its own sake constructs the edibility of some foods (usually from animals) as a challenge of the unknown, or, more accurately, a challenge of the never or not usually eaten, which is to be surmounted. As Maria (C) recalls, "I ate testicles. That challenged me because I don't like the look of balls and the idea of eating them is just challenging". Geoffrey (C) similarly, "saw it on the menu" and thought "that's different, I've never had bull's testicles before ... I'm going to try it".

This sort of emotional association with novelty and excitement is typically theorised from a psychological perspective where it is framed as a genetically based behaviour (Hirschman and Stern 2001), a 'motivational value' (Pepper et al. 2009: 127), and an outcome of body-mind mechanics that produces a transient psychobiological condition or a more stable behavioural trait (Gallagher 2011). 'Neophilia', as it is called, being "a love or enthusiasm for what is new or novel" (Merriam-Webster.com 2018), is therefore conceived as a personality trait that 'belongs' to individuals (Mitchell and Hall 2003). Nowhere in the emerging literature on food neophilia (e.g. Capiola and Raudenbush 2012; Ji et al. 2016) are novelty and excitement conceived as being socially constituted, this constitution dependent on the collective interplay of knowledge/senses/emotions by which different objects and practices are variously designated novel and exciting. Mitchell and Hall do allude to the "new significance and meaning" that food can take on when it is part of 'travel experiences' (2003: 60). However, this is the only occasion they suggest that the social world may play a role in understandings of novelty.

In contrast, Schatzki refers to novelty as something external to the individual—a known quantity, like innovation, that can "burst forth" (2016: 17), while Shove observes novel technologies as 'created' or arising in social practices (Shove 2003: 203). However, these less individualised conceptions do not adequately account for the discourse around novelty that I observe in my data. This is because they do not problematise the construction of novelty, and so 'novelty' is depicted as a reified, identifiable, and fixed entity that just appears or motivates. Instead, like the variable assignations surrounding 'ethical' meat, I see the knowledge and emotional value associated with novelty as being constituted primarily (if not entirely) in social practices. Novelty becomes variously associated with certain objects and practices, which thereby become exciting and 'good'. What individualistic or market-driven accounts of novelty have missed seems therefore to be better encapsulated by the idea of cultural 'omnivorousness', a term coined by sociologist Richard Peterson to signify an "openness to experiencing everything" that is "antithetical to snobbishness" (Peterson and Kern 1996: 904).

In their analysis of inconspicuous consumption, Shove and Warde remark on the social elevation of cultural omnivorousness, acknowledging it as a trend whereby being 'truly' cultural means:

> To have been everywhere, *eaten* everything, and heard as many types of music as possible in order to obtain the veneer of knowledge (and preferably hands-on experience) of all potentially discussable cultural items. (2002: 234, emphasis added)

Cultural omnivorousness, marked by "some distinctive status orientation" (Warde et al. 2007: 145), can therefore reasonably be associated with my participants' notions of being adventurous and seeking novelty through eating foods that lie outside normalised categories of what is considered socially and culturally edible. Furthermore, these scholars' accounts suggest that engaging with objects and practices understood in some way to embody omnivorousness enhances one's stockpile of 'discussable cultural items' and thereby imparts *social* distinction. Linking this to the idea of affective practices that "confer 'distinction'" (Wetherell 2012: 16), I argue that there is also an *emotional* distinction that accompanies such practices, and attaches to certain objects, so that omnivorous cultural practices are as much constituted, and marked, by a particular orientation of knowledge/senses/emotions as by social status. The palpable sense of excitement, anticipation, and pleasure that my participants exhibit in relation to what are perceived as novel meats intensifies the positive orientation of knowledge/senses/emotions associated with these edible objects, and thereby my participants' attraction to them and their associated practices. The senses (real or mimetic) add a vividness that further enlivens these associations. They become viscerally, emotionally, *and* intellectually significant.

For example, Oliver (P), for whom English is a second language, eloquently describes how he and his partner "sometimes intrigue ourself and indulge ourself with some like alternative meat". Anthony (C) and Ellis (C) similarly allude to a sense of risk and pioneer-like exploration in their approach to eating meat. If he's going to eat meat, Anthony says, "I might as well eat something weird, 'weird-wonderful', make something strange out of it", while Ellis describes her and her partner as "pretty adventurous eaters … certainly wouldn't be afraid to try things … we certainly wouldn't turn our nose up at things without trying them". Eating something constituted as uncertain, intriguing, and weird attaches an additional frisson to consuming animal flesh, on top of existing sensory and emotional associations that may already be present. As James (P) has also observed, "a huge part of the attractiveness of [our] meat is the fact that it's come from animals that people think they haven't eaten before". This attraction is accompanied by a certain underlying tacit knowledge (like that related to witnessing or participating in animals being killed) that to be 'adventurous' is a commendable quality that one *ought* to have.

Illustrating how emotional associations, both comforting and discomforting, can support and enhance each other, these associations with novelty also engage the distinction associated with requisite bravery, which is made accessible through the discomfort or fear that novel foods can engender. As Michael (C) explains, "when I go out to a restaurant, I like to eat things that I haven't, because it's new and I *should* be trying those things". If there is something different on a restaurant menu, Dan (C) similarly says he is "always keen to give them a go, just for the novelty of trying it". However, as indicated in Part II with respect to 'freaky' and typically inedible animals (such as dogs), there are understood to be limits to this adventurousness, and the perceived rewards of bravery, as James (P) explains:

> I'd look at the menu and see what's unusual, but having said that if it was unusual and it was something like, you know, endangered, I wouldn't eat it. … that's probably why I like eating offal when I go out. It depends on your mood but generally that's what I do, scan the menu and go, 'that's different, sounds nice, I know what part it is', let's see how they've cooked it.

Accompanying this attraction and pressure towards 'adventure', novelty, and omnivorousness, which earn social and emotional status across many practices involving, for example, food, music, art, travel, and sports, is a simultaneous push away from the emotional stigmatisation associated with *not* being these things. To *not* 'give things a go' is constituted respectively as 'being strict', in other words, *not* 'taking a break', *not* being adventurous, *not* being indulgent and hedonistic, and instead being regimented and potentially offensive, as Anthony (C) suggests, "I'm not too militant about anything in life".

This is where I perceive the consumption of food to be slightly different from the consumption of other cultural items in the cultural omnivore's palette such as places and music. Earlier, I described how an animal's flesh can become materially inscribed with either violence or happiness, which affects how the nexus of knowledge/senses/emotions become orientated towards the idea of its literal embodiment and therefore its edibility. In the same way, I argue that certain practices involving food, with their associated knowledges and emotions, can inscribe food animals or meat with 'novelty' so that they represent something 'good', exciting, maybe a little scary and thereby also potentially life-affirming. To then incorporate them into your own body is "in both real and imaginary terms, to incorporate all or some of its properties: we become what we eat" (Fischler 1988: 278). Hence, like the intrepid traveller, the eater is embodying novelty, risk, and excitement along with the meat, and affirming their own aliveness in the process. In being constructed as a novelty, animals or meat that might normally be considered 'bad', or associated with discomfort, become edible because of the possibilities of accessing the emotional (and social) distinction associated with them and their consumption, and of absorbing their life-affirming qualities. Moreover, as bell hooks said, and emphasising Foucault's nexus of power/knowledge/pleasure that underpins this book, it is also "by eating the Other … that one asserts power and privilege" (1998: 197).

In this section, I have discussed emotions associated with just two kinds of 'disturbing' practices—eating out and seeking novelty—that my participants frequently describe as occasions when normal 'ethical' distinctions are suspended. However, they also describe other, interrelated practices, and particularly aspects of their location (in time and space) that introduce additionally disturbing emotional associations. Notable among these are practices relating to travelling overseas, naturalised 'cravings' and 'urges', and practices that carry an expectation of 'being polite'. It is not possible to explore these more fully here. I highlight them only to emphasise that every practice and every performance of a practice bring together many different emotional associations, including those associated with overlapping practices, that variously mingle, merge, enhance, conflict, dominate, shrink, shift, or change entirely.

Thus, in certain circumstances, associations of meat with a release from 'normal' everyday constraints (e.g. eating out, travelling overseas, and/or cravings), and with novelty, excitement, and polite commensality, emerge as more important than whether it is ethical or not, and even more decisively 'bad'. The normalised nexuses of power/knowledge/pleasure associated with these practices support meat consumption and the use of animals for food regardless of how 'ethically' they are done. It is this fundamental aspect of adherences to meaty practices that I address in the next and final section, which is based on my identification of emotional associations with eating meat and using food animals that do not appear to be attached to particular practices (though they are constituted by them). Rather, they appear attached solely to a constituted anthroparchal ideology surrounding meat and food animals, and what they have come to mean, at least in Westernised cultural contexts. In this way, they are more accurately conceived as ethico-aesthetic mechanisms of power.

3   Moral Approval: An Ethico-Aesthetic Mechanism of Power

An invocation to do what is 'appropriate', 'right', and 'good' permeates the practices described in the previous sections, and also shapes their emotional associations. I propose that emotions associated particularly with "doing the right thing" (Geoffrey C) engender a particular 'ethico-aesthetic' of knowledge/senses/emotions that constitutes eating meat, in and of itself, as not just universally natural, right and good, but morally superior. Chloe Taylor traces the concept of ethico-aesthetics through Foucault's writings on diet, sexuality, and 'care for the self', where, she explains:

Foucault describes 'techniques of the self' both as an ethical relation to the self and as an aesthetics of one's own life ... To approach one's own life ethically through such techniques of the self is, for Foucault, to see one's existence as an aesthetic project or a work of art. (2010: 72–73)

In this reading, and in common with a non-dualistic, rhizomatic view of mind/body, an ethico-aesthetic is an embodied ethics that is less about meanings, causes, and effects and more about bodily connections, affects, and social formations (Malins 2004: 95). Ethico-aesthetics as a specific concept was formulated by Guattari in his 1995 volume *Chaosmosis, an Ethico-aesthetic Paradigm.* Subsequent scholars of the arts, humanities, and social sciences have drawn from and expanded on his thesis in different ways (Malins 2004, 2011; Alliez and Massumi 2014; Hynes 2013). In her work on assemblages of bodies and urban space, Peta Malins draws on Guattari (also with Deleuze 1987) to describe ethico-aesthetics as a project that aims:

To increase ethical, life-enhancing assemblages ... that increase a body's power to form creative, productive relations and which increase its capacity for life. (2004: 98)

While post-Foucauldian accounts of ethico-aesthetics can appear anthropocentric, focused on increasing the human capacity for pleasure and life, both Taylor (2010) and Springgay (2011) recognise the potential for an ethico-aesthetic to foster a care for the life and pleasure of the 'other', something Foucault was also attentive to:

Are we able to have an ethics of acts and their pleasures which would be able to take into account the pleasure of the other? Is the pleasure of the other something which can be integrated in[to] our pleasure. (Foucault 1984: 346)

Echoing Despret's (2004) moments of ambiguity, there is, therefore, the more radical potential for an ethico-aesthetic to "unsettle dominant features of reason and standardization" (Springgay 2011: 78), so that 'things' might become something else. Using eating as an example, Stephanie Springgay, quoting O'Sullivan (2007), explains:

> The ethical dimensions of eating would mean that the body no longer passively accepts what goes into it; rather the body opens itself up to deterritorializations, a multitude of surfaces, "a call to creativity, a call to become actively involved in various strategies and practices that will allow us to produce/transform, and perhaps even go beyond, our habitual selves". (2011: 78)

However, while a certain ethico-aesthetic of sensations, affects, emotions, and bodily pleasures is certainly operating in relation to ethical and sustainable meat, it is one that necessarily only pays instrumental attention to the pleasure of animal others. Former vegan Gillian (C) describes ethical meat as mind-expanding: "ethical eating has broadened my brain … and made me look at grain production and … water use and biodynamics and so much more than just animal production". For Dan (C), it is indicative of a higher consciousness: "anything that expands our empathy and understanding is going to benefit humanity long term". He continues, drawing in additional 'good' things (diversity) to emphasise the 'goodness' of ethical meat, "I think having a consciousness that respects the diversity on the planet and respects the other animals we share the planet with is a good thing". Finn (C) describes it as "being in tune with your body or your conscience". Alison (P) leverages on the idea of family to further reinforce the goodness of the association: "When you become socially conscious, you want to actually feed really good quality food to families". For Gillian (C), as well having broadened her brain, she asserts "as humans who eat ethical and sustainable meat, I think we're healthier in mind and body and spirit". "It's good for the soul", says Andrew (P), while Brigid (P) similarly affirms that her work with animals "get[s] some actual soul into you", which she feels she has "in buckets".

In his analysis of the 'happy meat' discourse, Matthew Cole notes: "Consumers of 'happy meat' … receive a gustatory reward for their vicarious pastorship, as well as moral approbation for fulfilling their *responsibility* towards animals" (2011: 94, emphasis in original). Contrary to Charlotte's (C) view that "you can't buy ethics and you can't buy morality", I argue that is exactly what my participants are doing. As both producers and consumers, they are helping to shape a market for ethics and morality as edible components of animal flesh, thereby maintaining an overall ethico-aesthetic of animal edibility as care for self and others. As Probyn observed, and as I have demonstrated in Part III, "food and eating is as much marked by pleasure as it is by power" (2000: 18).

Of course, everybody would like to think, sense, and feel emotionally that they are doing the right thing to some extent—that they are engaging in practices that correspond more or less with their knowledge of cultural codes of 'good' and 'bad', 'right' and 'wrong'. However, the fact that my participants seek 'better' emotions through increasingly refined and often variable definitions of 'good' meat (rather than no meat) suggests two things: first, that comforting emotional associations with objects and practices involving 'better' meat, and meat generally, are sufficiently normalised, and prioritised, to outweigh any discomforting associations; and second, that the threat posed by discomforting associations with practices of not eating meat (and their associated heterotopic spaces) far outweighs any discomforting feelings that may linger around objects and practices associated with eating animals. As Joyce (C) succinctly explains, "people feel bad about eating meat, and they want to feel better [about eating it]". In other words, they do not want to *not* eat meat.

The inherent moral approval associated with the entire ideology pertaining to eating meat and using food animals, and not just certain related practices, constitutes a significant limitation to generalising an acceptance that there are other ways of relating to food animals beyond the grid that reflects their existing cartographies. In whatever way associated practices are challenged—environmentally, health-wise, or in relation to animal ethics—the resilience of the ethico-aesthetic mechanism ensures the primary focus remains on improving *how* these practices are undertaken while leaving their ontological basis intact. For this reason, I identify it as a fundamental mechanism of power that maintains the edibility of food animals and therefore their state of domination.

3.1 Conclusion

Based on the three validating discourses, two emotions of distinction, and one ethico-aesthetic I have thus far identified as mechanisms of power, I suggest that a universal knowledge or 'truth' underpins all the determinations, distinctions, and associations relating to food animals and meat. That universal knowledge quite simply asserts humans' entitlement to use animals as if they were a natural resource. I argue that this entitlement is the most persistent, and naturalised, construct within human-animal relations, and nowhere is its exercise and power more apparent than in the way the increased visibility of meat production processes has been easily incorporated into its legitimising construct.

Focusing on this 'new' ethical aesthetic of visibility in meat production and consumption, Part IV further demonstrates how "deployments of power are directly connected to the body" (Foucault 1978: 151), extending Foucault's nexus of power/knowledge/pleasure to include his, and others', notions of the 'arrogance', 'ownership', privilege, and also pleasure that infuses 'the gaze'. In short, I demonstrate how these animals are not only *thought* of, and *sensed*, but also *seen* as edible. This lends further support to Probyn's (2000), and Hayes-Conroy and Hayes-Conroy's (2008), articulations of the visceral politics of food by adding sight to Probyn's textures, flavours, and taste of power (2000: 7) and illustrating how "the power of taste" (Hayes-Conroy and Hayes-Conroy 2008: 463) is also accompanied by the power of the gaze.

References

Ahmed, S. (2010a). Feminist Killjoys (And Other Willful Subjects). *The Scholar and Feminist Online, 8*(3), 1–10.

Ahmed, S. (2010b). Killing Joy: Feminism and the History of Happiness. *Signs: Journal of Women in Culture and Society, 35*(3), 571–594.

Alliez, E., & Massumi, B. (2014). Performing the Ethico-aesthetic Paradigm. *Performance Research, 19*(3), 15–26.

Berger, J. (1992). Why Look At Animals? In *About Looking* (pp. 1–14). London: Bloomsbury.

Bluestone, G. (2014, December 15). Cow Dies Hero's Death After Escaping Slaughterhouse. *Gawker*. Online. October 2016.

Capiola, A., & Raudenbush, B. (2012). The Effects of Food Neophobia and Food Neophilia on Diet and Metabolic Processing. *Food and Nutrition Sciences, 3*, 1397–1403.

Castelfranchi, C., & Miceli, M. (2011). Anticipation and Emotion. In P. Petta, C. Pelachaud, & R. Cowie (Eds.), *Emotion-Oriented Systems. Cognitive Technologies* (pp. 483–500). Heidelberg: Springer Berlin.

Chasan, A. (2015, June 1). Cow Escapes Slaughterhouse, Later Shot and Killed. *New York Daily*. Online. October 2016.

Coghlan, S. (2016). Moral Individualism and Relationalism: A Narrative-Style Philosophical Challenge. *Ethical Theory and Moral Practice, 19*(5), 1241–1257.

Cole, M., & Stewart, K. (2014). *Our Children and Other Animals*. London; New York: Routledge.

Coles. (2011, April). RSPCA Approved Chicken to Hit Shelves in Victoria, South Australia and Tasmania for the First Time. *Coles News*. Online. June 2015.

Conaty, D. (1998, January 17). Flying Pigs Saved from the Abattoir. *The Irish Times*, Online. October 2016.

Dehaene, M., & De Cauter, L. (2008). *Heterotopia and the City*. London; New York: Routledge.

Deleuze, G., and Guattari, F. (1987). *A Thousand Plateaus: Capitalism and Schizophrenia*. London; New York: Continuum.

Deller, R. (2016). The Animated Aesthetics of Cultured Steak. In K. Sellberg & L. Wanggren (Eds.), *Corporeality and Culture: Bodies in Movement* (pp. 67–79). London; New York: Routledge.

Demers, C. (2019, January 9). Three Lucky Calves Recovering at Farm Sanctuary. *WENY News*.

Despret, V. (2004). The Body We Care For: Figures of Anthropo-Zoo-Genesis. *Body & Society, 10*(2–3), 111–134.

Douglas, M. (2002). *Purity and Danger: An Analysis of Concepts of Pollution and Taboo*. London; New York: Routledge Classics.

Dowsett, E., Semmler, C., Bray, H., Ankeny, R. A., & Chur-Hansen, A. (2018). Neutralising the Meat Paradox: Cognitive Dissonance, Gender, and Eating Animals. *Appetite, 123*, 280–288.

Embury-Dennis, T. (2018, February 16). Cow Escapes on Way to Slaughterhouse, Smashes through Metal Fence, Breaks Arm of Man Trying to Catch Her Then Swims to Safety on Island in Lake. *The Independent*. Online. February 2018.

Evans, A. B., & Miele, M. (2012). Between Food and Flesh: How Animals Are Made to Matter (and Not Matter) within Food Consumption Practices. *Environment and Planning D: Society and Space, 30*(2), 298–314.

Fischler, C. (1988). Food, Self and Identity. *Social Science Information, 27*(2), 275–292.

Foucault, M. (1977). *Discipline and Punish: The Birth of the Prison.* New York: Vintage Books.

Foucault, M. (1978). *The History of Sexuality.* New York: Pantheon Books.

Foucault, M. (1984). *Foucault Reader* (P. Rabinow, Ed.). New York: Pantheon Books.

Gallagher, W. (2011). *New: Understanding Our Need for Novelty and Change.* New York: Penguin Books.

Hayes-Conroy, A., & Hayes-Conroy, J. (2008). Taking Back Taste: Feminism, Food and Visceral Politics. *Gender, Place & Culture, 15*(5), 461–473.

Hirschman, E. C., & Stern, B. B. (2001). Do Consumers' Genes Influence Their Behavior? Findings on Novelty Seeking and Compulsive Consumption. *Advances in Consumer Research, 28*, 403–410.

Hochschild, A. R. (1979). Emotion Work, Feeling Rules, and Social Structure. *American Journal of Sociology, 85*(3), 551–575.

hooks, b. (1998). Eating the Other: Desire and Resistance. In R. Scapp & B. Seitz (Eds.), *Eating Culture* (pp. 181–200). New York: State University of New York (SUNY) Press.

Hynes, M. (2013). The Ethico-aesthetics of Life: Guattari and the Problem of Bioethics. *Environment and Planning A, 45*(8), 1929–1943.

Illouz, E. (2009). Emotions, Imagination and Consumption: A New Research Agenda. *Journal of Consumer Culture, 9*(3), 377–413.

Ji, M., Wong, I. A., Eves, A., & Scarles, C. (2016). Food-Related Personality Traits and the Moderating Role of Novelty-Seeking in Food Satisfaction and Travel Outcomes. *Tourism Management, 57*, 387–396.

Krause, J. (2015, January 6). Runaway Chicken Escapes Deadly Yom Kippur Ceremony to Live Out Her Dream Life. *Animal Scoop.* Online. October 2016.

Malins, P. (2004). Machinic Assemblages: Deleuze, Guattari and an Ethico-aesthetics of Drug Use. *Janus Head, 7*(1), 84–104.

Malins, P. (2011). An Ethico-aesthetics of Heroin Chic: Art, Cliché and Capitalism. In L. Guillaume & J. Hughes (Eds.), *Deleuze and the Body* (pp. 165–187). Edinburgh: Edinburgh University Press.

May, T. (2014). Moral Individualism, Moral Relationalism, and Obligations to Non-human Animals. *Journal of Applied Philosophy, 31*(2), 155–168.

Mitchell, R., & Hall, M. C. (2003). Consuming Tourists: Food Tourism Consumer Behaviour. In M. C. Hall, L. Sharples, R. Mitchell, N. Macionis, et al. (Eds.), *Food Tourism Around the World* (pp. 60–80). London; New York: Routledge.

Noske, B. (1989). *Humans and Other Animals: Beyond the Boundaries of Anthropology*. London: Pluto Press.

O'Sullivan, S. D. (2007). Academy: The Production of Subjectivity. In I. Rogoff (Ed.), *Academy* (pp. 238–244). Revolver: Frankfurt.

O'Sullivan, S. (2011). *Animals, Equality and Democracy*. New York: Palgrave Macmillan.

Passafaro, P., Rimano, A., Piccini, M. P., Metastasio, R., et al. (2014). The Bicycle and the City: Desires and Emotions Versus Attitudes, Habits and Norms. *Journal of Environmental Psychology, 38*, 76–83.

Pepper, M., Jackson, T., & Uzzell, D. (2009). An Examination of the Values that Motivate Socially Conscious and Frugal Consumer Behaviours. *International Journal of Cultural Studies, 33*(1), 126–136.

Petenko, E. (2018, December 31). The Cow that Escaped the Slaughterhouse Gave Birth, and Her New Baby Is Udderly Adorable. *NJ Advance Media*.

Peterson, R. A., & Kern, R. M. (1996). Changing Highbrow Taste: From Snob to Omnivore. *American Sociological Review, 61*(5), 900–907.

Pleasance, C. (2014, June 3). This Little Piggy's Not Going to Market: "Babe" the Porker Escapes Slaughterhouse Van by Leaping 16ft to Freedom ... and Avoids the Chop after Being Adopted. *The Daily Mail Australia*. Online. October 2016.

Probyn, E. (2000). *Carnal Appetites: FoodSexIdentities*. London; New York: Routledge.

Reckwitz, A. (2002). Toward a Theory of Social Practices: A Development in Culturalist Theorizing. *European Journal of Social Theory, 5*(2), 243–263.

Schatzki, T. R. (1997). Practices and Actions: A Wittgensteinian Critique of Bourdieu and Giddens. *Philosophy of the Social Sciences, 27*(3), 283–308.

Schatzki, T. R. (2016). Practices, Governance, and Sustainability. In Y. Strengers & C. Maller (Eds.), *Social Practices, Intervention and Sustainability: Beyond Behaviour Change* (pp. 15–30). London; New York: Routledge.

Schelling, A. (2015, July 1). Desperate Cow Does the Unthinkable to Escape Slaughter. *The Dodo*. Online. October 2016.

Shove, E. (2003). Converging Conventions of Comfort, Cleanliness and Convenience. *Journal of Consumer Policy, 26*(4), 395–418.

Shove, E., & Warde, A. (2002). Inconspicuous Consumption: The Sociology of Consumption, Lifestyles, and the Environment. In R. E. Dunlap (Ed.), *Sociological Theory and the Environment: Classical Foundations, Contemporary* (pp. 230–251). Lanham, MD: Rowman and Littlefield Publishers.

Shove, E., Pantzar, M., & Watson, M. (2012). *The Dynamics of Social Practice: Everyday Life and How It Changes*. London; Thousand Oaks; New Delhi: SAGE Publications Ltd.

Springgay, S. (2011). The Ethico-aesthetics of Affect and a Sensational Pedagogy. *Journal of the Canadian Association for Curriculum Studies, 9*(1), 66–82.

Steinbuch, Y. (2016, January 22). Cow that Ran from Slaughterhouse Gets Cushy New Life. *New York Post*. Online. October 2016.

Taylor, C. (2010). Foucault and the Ethics of Eating. *Foucault Studies*, (9), 71–88.

Turner, J. H., & Stets, J. E. (2006). Sociological Theories of Human Emotions. *Annual Review of Sociology, 32*(1), 25–52.

Twine, R. (2014). Vegan Killjoys at the Table—Contesting Happiness and Negotiating Relationships with Food Practices. *Societies, 4*, 623–639.

Vigors, B. (2018). Reducing the Consumer Attitude–Behaviour Gap in Animal Welfare: The Potential Role of 'Nudges'. *Animal, 8*(2), 232.

Warde, A., Wright, D., & Gayo-Cal, M. (2007). Understanding Cultural Omnivorousness: Or, the Myth of the Cultural Omnivore. *Cultural Sociology, 1*(2), 143–164.

Wetherell, M. (2012). *Affect and Emotion: A New Social Science Understanding*. London; Thousand Oaks; New Delhi: SAGE Publications Ltd.

Zembylas, M. (2005). *Teaching with Emotion: A Postmodern Enactment*. Connecticut: Information Age Publishing.

Part IV

The Power of Transparency

Introduction to Part IV

Part IV constitutes the final leg of my trek through the terrain of cartographied meat, in which I gather together the preceding themes and arguments to demonstrate how power, knowledge, and pleasure are brought to bear on the subject of the gaze. My contention is that in order to see through and beyond, to "the far side of this grid" (Foucault 1989: 175), where there is the possibility of thinking differently about animals, it is not more visibility that is required, but a radically alternate regime of power/knowledge/pleasure, and knowledge/senses/emotions. This is the only way to 'un-code' the human-centric, or anthroparchal, eye and alter the gaze from one of entitlement to one that is open to the fully independent being-ness of 'others'. Over two chapters, Part IV therefore addresses my fourth objective by examining **what effect increased visibility of 'food' animals and increased transparency of meat production processes have on the edibility of animals, and what this says about how animals are 'made sense' of.**

Foucault's thesis on the power of transparency, conceived as part of the regime of power/knowledge/pleasure relating to food animals and meat, provides the analytical framework for this third dimension of my investigation. With reference to Bentham's prison Panopticon, a "technology of power" which achieves "subjection by illumination" (Foucault 1980: 148; 154), the visibility of practices relating to the use and edibility of food animals similarly becomes a means of achieving absolute access and control over "their bodies, their gestures, and all their daily actions" (1980: 151–152). Under the systemically normalised regime of power/knowledge/pleasure, food animals are seen, just as much as they are 'made sense of', as objects of consumption—mapped bodies of meat. Therefore, visibility cannot avoid functioning first and foremost under an expansion of this logic of power (Pachirat 2011: 289). Only when the logic is subverted, when the "already encoded eye" (Foucault 1989: xxii) is un-coded, can they become visible as something else.

Of course, it could be argued that the discomfort of visibility could itself generate a subversion of this logic, which is a theory on which many animal advocacy organisations base their campaigns. However, as I have demonstrated in the previous sections, and will further illustrate in Part IV, there are many ways that discomfort associated with visibility can be re-interpreted and/or ameliorated to reinforce the normalised regime of power/knowledge/pleasure. For mechanisms of power, as Foucault explains, are "caught up in the very pleasure of their exercise" (1980: 186), and sight is one of the pleasures of the senses:

> There is something in surveillance, or more accurately in the gaze of those involved in the act of surveillance, which is no stranger to the pleasure of surveillance, the pleasure of the surveillance of pleasure, and so on. (Foucault 1980: 186; also 1985)

Weaving together the power of knowledge, the pleasure of knowing, and the knowing of pleasure, with Foucault's further theorisations on the gaze, the power of transparency and the pleasure of surveillance, the full extent of the regime of power/knowledge/pleasure that bears upon animals becomes clear. This culminates in a "gaze of power" (Foucault 1979: 81). Drawing on complimentary theorisations of the gaze and the pleasures of looking as articulated by Mulvey (1999), Berger (1992), Pick (2015), and other scholars of visual media, I argue that the gaze of power is one that is infused, through social practices and the regime of power/knowledge/pleasure that permeates them, with entitlement.

References

Berger, J. (1992). Why Look At Animals? In *About Looking* (pp. 1–14). London: Bloomsbury.

Foucault, M. (1979) *Power, Truth, Strategy* (M. Morris & P. Patton, Eds.). Feral Publications Sydney.

Foucault, M. (1980). *Power/Knowledge: Selected Interviews and Other Writings, 1972–1977*. New York: Pantheon Books.

Foucault, M. (1985). *The Use of Pleasure*. New York: Random House.

Foucault, M. (1989). *The Order of Things*. London; New York: Routledge.

Mulvey, L. (1999). Visual Pleasure and Narrative Cinema. In L. Braudy & M. Cohen (Eds.), *Film Theory and Criticism: Introductory Readings* (pp. 833–844). New York; Oxford: Oxford University Press.

Pachirat, T. (2011). *Every Twelve Seconds: Industrialized Slaughter and the Politics of Sight*. New Haven; London: Yale Agrarian Studies Series.

Pick, A. (2015). Why not Look at Animals? *Necsus European Journal of Media Studies*. Spring, Online. November 2015. 17 pp.

8

Entitlement

Thus far in this book, I have demonstrated that what Probyn describes as "a politics of feeding" (2000: 18) is a decidedly speciesist politics where a fundamental entitlement to animals' bodies and lives is always unquestioningly assumed and reproduced at every turn. Drawing on Acampora (2016: 18), I argue that this entitlement, or 'ideology of human privilege', underpins the persistence of meat consumption and the use of animals for food, and the moral approval that associated practices attract. For not only does it "short-circuit" (ibid.) attempts for animals to be *known* in any other way, it also prevents them being *seen* as anything but edible. Acampora insists these " ideologies of human privilege must be exposed and analyzed for progress to be made in overcoming animal oppression" (2016: 1).

Therefore, noting that discursive and material transparency/visibility is a strategy employed by both animal advocates and the meat industry, particularly over the past decade, I argue that a Foucauldian 'power of transparency' is in operation whereby this visibility is always parsed under an ideology of human privilege or entitlement, that is, the conviction that humans are entitled to use the bodies of animals

© The Author(s) 2020
P. Arcari, *Making Sense of 'Food' Animals*,
https://doi.org/10.1007/978-981-13-9585-7_8

for their own purposes. I conceive entitlement as an effect of power—a productive way of legitimating and interpreting relations of oppression. Like bravery/fear, and comfort/discomfort, entitlement and oppression are mutually constitutive and dependent. The production and exercise of entitlement through social practices shape systems of oppression, which in turn reinforce conditions of entitlement. Both are required to create a state of domination. Entitlement to animals' lives and bodies, as an effect of the systemic power relation, permeates all its mechanisms—the validating (naturalising) discourses, emotions of distinction, and ethico-aesthetics. However, it is the increased visibility of food animals and meat, as it is constituted by my participants and in the broader literature on ethical meat, that especially foregrounds this entitlement and how it works to maintain the persistent edibility of food animals.

I begin Chap. 8 with an outline of my approach to entitlement, drawing on theorisations of intersectionality to link privilege (conceived as entitlement) with the domination of animals. I then focus on how human entitlement manifests in my data, highlighting how a popularised notion of 'respect' functions, especially within the ethical meat discourse, to cloak this entitlement, and presumed ownership of animals, in a benign sense of mutuality and the 'giving' up/over of life (as opposed to taking). Following this, I illustrate how this idea of food animals 'giving' their life is woven through my participants' accounts of various phases of the meat production process. While echoing my discussion of the 'natural contract' in Part II, here I foreground the entitlement that suffuses my participants' reflections. Animals are attributed with thoughts and emotions, sometimes also a voice through which these are expressed, and the capacity to voluntarily enter into contracts, all of which attest to the benefits *to* animals of an ethical approach to eating their flesh. In this way, food animals are cast as willing, or at least preordained, participants in a natural process of self-mortification (Cole 2011).

1   Entitlement, Privilege, and Oppression

In his imagining of the postcolonial animal, Armstrong (2002: 413) observes:

> In identifying the costs borne by non-European 'others' in the pursuit of Western culture's privileged entitlement, post-colonialists have concentrated upon 'other' humans, cultures and territories but seldom upon animals.

Entitlement can thus be thought of in terms of privilege as both terms connote a special right that is granted or enjoyed by some person, group, or entity with more advantage. Black and Stone (2005: 244) identify the following five components of privilege:

1. Privilege is a special advantage; it is neither common nor universal
2. It is granted, not earned or brought into being by one's individual effort or talent [i.e. a birthright]
3. Privilege is a right or entitlement that is related to a preferred status or rank
4. Privilege is exercised for the benefit of the recipient and to the exclusion or detriment of others
5. A privileged status is often outside of the awareness of the person possessing it

According to this conception, privilege, or entitlement, is not a commodity that is located in any person or entity but is, like Foucault's power, understood as a dispersed phenomenon of social practices. I have chosen the word 'entitlement' over 'privilege' because it offers a more active way of conceiving privilege. Entitlement immediately conjures the preposition 'to' while privilege does not. It therefore inherently implies an act to which the entitlement relates (to use, to kill, to eat), and a material object (human or nonhuman) the act is applied to.

While Black and Stone acknowledge that 'race'/ethnicity and gender have traditionally been the primary focus of literature dealing with privilege, they seek to expand this ambit to include "the five socially constructed categories of sexual orientation, SES [socio-economic status], age, differing degrees of ableness and religious affiliation" (2005: 244). Privilege, hereafter referred to as entitlement, is embedded in social practices via these (and many other) distinctions that set apart those who, knowingly or unknowingly, are on the 'right' side, and therefore enjoy its benefits, from those who are not. The latter 'others' may as a result experience different and limited access to practices, and be regarded differently within those practices compared with those who are entitled. Those regarded as 'other' can do little to access the 'special advantage', the innate birthright, and the 'preferred status' that Black and Stone's schema suggests would alter their position as the 'excluded'. They are therefore being oppressed by systemic mechanisms of power that are suffused with entitlement and reflected in social practices—oppression and entitlement go hand in hand.

Emphasising this point, Iris Young's (1990) account of oppression correlates closely with Black and Stone's account of privilege. Young draws on differences in gender, age, socio-economic status, sexual identity, and 'racial', ethnic, and religious groups to illustrate how oppression operates, and also prefigures the theory of intersectionality by highlighting that there are "similarities and overlaps in the oppressions of different groups" (64). Young furthermore highlights the systemic nature of oppression, and underscores the fact that entitlement and oppression are not attributes possessed or experienced by individuals, but the product of everyday practices (39). While acknowledging that within a system of oppression, individuals may certainly cause intentional harm to those who are oppressed, and that entitled individuals and groups may benefit from the continuation of this system, it is, like power, ultimately the result of structural phenomena. The causes of oppression, Young explains, are "embedded in unquestioned norms, habits and symbols, in the assumptions underlying institutional rules" (1990: 39). They are exercised in "ordinary interactions, media and cultural stereotypes … bureaucratic hierarchies and market mechanisms … systematically reproduced in major economic, political, and cultural institutions" (ibid.).

There are therefore multiple and distinct expressions of entitlement and oppression and more than one form can be associated with one body so that people are "shaped by the interaction of different social locations (e.g. 'race'/ethnicity, indigeneity, gender, class, sexuality, geography, age, disability/ability, migration status, religion" (Hankivsky 2014: 2). 'Intersectionality' is a term originally coined by critical race scholar Kimberle Crenshaw in 1989, but it designates a strand of critical thought that has a much longer history. In her account of intersectionality, Hankivsky explains that it is the interaction between these 'social locations' in "connected systems and structures of power (e.g., laws, policies, state governments and other political and economic unions, religious institutions, media)" that gives rise to "interdependent forms of privilege and oppression shaped by colonialism, imperialism, racism, homophobia, ableism and patriarchy" (ibid.). This mirrors Young's (1990) depiction of overlapping oppressions and, in a nod to Armstrong's (2002) search for the postcolonial animal, I would add speciesism to these conceptualisations of systemic, structural, and intersectional entitlements and oppressions.

While being well established in the fields of feminist theory, social theory, and critical race studies, the extension of intersectionality to include animals through the operation of speciesism (prejudice or discrimination based on species) or more broadly anthroparchy (the social system of human domination) is for many scholars a logical and essential move due to similarities and overlaps with other forms of oppression.

A recognition of how species difference and our treatment of animals are entangled in and shape discourses and 'knowledge' of 'race', ethnicity, gender, sexuality, and so on acknowledges that these discourses are co-constitutive (Cudworth 2011: 174). The mechanisms of power through which all forms of oppression, domination, inequality, and social difference gain expression are relational and "continue to fundamentally shape questions of in/justice across human and nonhuman cultural terrains" (Deckha 2008: 267).

Taking this argument further, Sorenson suggests animal oppression is even more central and that speciesism "constitutes a basic form of oppression that provides a structure for the oppression of other humans" (Sorenson 2014: xv). Wolfe (2003: 8) is more explicit and underlines the importance of recognising the intersectional dimensions of our use of animals:

as long as this humanist and speciesist structure of subjectivization remains intact, and as long as it is institutionally taken for granted that it is all right to systematically exploit and kill nonhuman animals simply because of their species, then the humanist discourse of species will always be available for use by some humans against other humans as well, to countenance violence against the social other of whatever species or gender, or race, or class, or sexual difference.

Offering empirical evidence of these connections, Caviola et al. (2018) found speciesism to be positively associated with other prejudicial attitudes such as racism, sexism, and homophobia. Furthermore, compared to these human-human prejudices, speciesism is the "dominant and explicitly accepted social norm and ideology in current Western societies" (16). I propose that the speciesism to which Sorensen and Wolfe refer manifests in my data as an entitlement to animals' lives and bodies. From this perspective, it is the ideology by which humans 'receive' an entitlement to objectify and de-individualise animals while using them as a natural resource, which permeates and is normalised through social practices, that provides the enduring baseline model for a similarly entitled approach to the objectification and oppression of human 'others'. These 'others' being those who are also objectified and used as part of social practices organised according to an entitled group, whether male, white, heterosexual, cis-gendered, able-bodied, affluent, or otherwise socially advantaged. Cole similarly identifies privilege as symptomatic of fundamental relations of power between humans and animals, filtered through the benign "architecture" of pastoral care (2011: 97). As Gillian (C) says, "you can love animals and eat them, you know it is possible".

Here I would suggest that calls for an applied ethics or politics of care, empathy, compassion, and love that have appeared increasingly over the past one to two decades[1] (Held 2006; Bekoff 2007; Gruen 2014; McKibben 2019) are incomplete because, as Diana Tietjens Meyers (2017: 422) notes in relation to Gruen's *Entangled Empathy*, they do not sufficiently engage with the "social norms that specify who owes what to

---

[1] And previously with Carole Gilligan's *In a Different Voice: Psychological Theory and Women's Development* (1982) and Joan Tronto's (1993) *Moral Boundaries: A Political Argument for and Ethics of Care*.

whom", the histories that have shaped them, and the current practices that reinforce them. As a result, a *rhetoric* of love or care can be layered largely untroublingly over persistent inequities and become another mechanism by which the exercise of power is concealed. This phenomenon is not new and can be traced back to missionaries and slave owners who professed to care deeply for their subjects and treat them with kindness and love (Genovese and Fox-Genovese 2011; McLisky 2015).

Echoing the care and love most commonly directed at food animals, these forms of 'pastoral paternalism' (Cruickshank 2014) betray an interest in the products of life—for example, labour, souls, meat—as part of a larger project of self-care, rather than an interest in the life in and of itself. It is a specific kind of love that is allocated to living others over which an advantaged group assumes dominion—by right of judicial, religious, or 'natural' law. It is the kind of love that accepts control, coercion, mistreatment, and even death as part of its remit. This rhetoric of love, which is commonplace in relation to animals that are used in various ways, needs to be acknowledged and addressed in discussions of care ethics. At the very least, its currency highlights the persistence of normalised orders and associated entitlements, and emphasises that emotions (like love) are not uniformly understood or expressed. They are shaped by the practices of which they are part and can thus have vastly different implications for the objects with which they are associated.

Returning to Cole's point regarding privilege and the architecture of pastoral power, I consider entitlement part of this power's productive network, while oppression is its negative counterpart. Recall here Foucault's assertion that it is the productive network of power that "induces pleasure, forms knowledge, produces discourse" and, in so doing, more than negative instances of power (e.g. oppression) "makes power hold good ... makes it accepted" (1980: 119). Consequently, while the nature and operation of oppression are commonly the focus of intersectional studies, I focus instead on its counterpart—the nature and operation of the entitlement that facilitates oppression and greases its wheels. With reference to my empirical data, I show how entitlement manifests in producers' and consumers' accounts of their practices involving ethical and sustainable meat and how it describes a particular, apparently justified, regard for the 'other' who is oppressed.

1.1   Entitlement: It's Your Right

One of the basic cornerstones of the use of animals for food is breeding, which Zipporah Weisberg (2009: 30) describes as:

> A practice built directly out of humans' entitlement to the bodies and lives of other animals and to the latter's reduction to the mere stuff of control.

Weisberg suggests two related aspects of entitlement here—an entitlement to animals' bodies and lives, and an entitlement to reduce them to 'mere stuff' besides meat (i.e. money) as part of broader (human) systems of value exchange. Both aspects are especially evident in how producers of ethical and sustainable meat refer to food animals as a means to deliver their lifestyle goals. Many highlight that it is the 'country lifestyle' that they were primarily seeking, and one of the most obvious ways to make this economically viable is to raise animals and sell their flesh. Oliver (P) relates how he and his partner "fell in love with the environment … wanted to become part of it … wanted to call it home". Will (P) was similarly "always very interested in moving to the country". "It's a lifestyle choice" as James (P) describes it, but one that provides limited options for making money. Bella (P) and her partner were trying to move to the country but were unsure how they would make a living. Joel Salatin became an inspiration, as he is for many of my participants, and showed them "how you can make a living growing meat as a small farm" (Bella P). The stuff of meat, an outcome of the naturalised and normalised edibility of food animals, therefore provides a means for these producers to define and control their lives. However, economically speaking, not every animal is 'created' equal. As Bella (P) explains, "you get very little meat off a sheep, so you have to have lots more of them". Agreeing, Melanie (P) says, "you can survive on pigs, you can't survive on beef", because, as Simon continues, pigs "create cash flow". Alternately, for Andrew (P), it is "doing beef" that has enabled him to create an income and stay on the farm.

The two aspects of entitlement foregrounded here—the entitlement to use animals as food, and the right to financially leverage this entitlement to deliver a desired lifestyle—neatly illustrate what happens "when anthropocentrism sh[akes] hands with capitalism" (Wadiwel 2017). The assumption that raising animals to sell their flesh is the only way to make an income in the country also emphasises the extent to which entitlement defines the colonisation of animals as well as land in what Azeezah Kanji calls "a matrix of Eurocentric-anthropocentric-androcentric power" (2017: 65; see also Armstrong 2002; Nibert 2013). Both land and animals are subject to the same "civilizing mission" (Armstrong 2002: 414), conceived respectively as terra or res nullius[2] until mapped by anthroparchal cartographies of use. The understandings and practices that support the colonisation of the 'other' (human and nonhuman) as part of a civilising and beneficial project mean the authority of the implicit entitlement is not questioned. Indeed, it is rarely recognised or acknowledged that a fundamental, a priori entitlement is even being exercised. For example, Julie (C) considers herself "privileged to the able to eat ethical meat, it's a privilege", but the origin or foundation of that privilege is not questioned; it is taken for granted that eating animals is an advantage, benefit, or birthright. The power of knowledge in the constitution of entitlement, and the social order, is evident, as Palmer explains:

> Power, truth, and *right* are intricately related because power produces truth; it produces knowledge; it constructs particular kinds of people; and it creates particular kinds of societies. (Palmer 2001: 344, emphasis added)

Colonialism objectifies as it also commodifies—seeing value in, and, more fundamentally, enacting an implicitly understood entitlement to 'blank slates' that can be "mastered and exploited" (Kanji 2017: 51). This is the 'mission' that underpins practices of breeding, genetically altering, raising, and killing food animals in order to consume their flesh, however benign these practices are constituted as being.

Every producer describes having some level of emotional attachment to their animals, because, as Graham (P) says, "that's why you do it, because you like to be around these animals, you've just got to love animals". But the bottom line is always commodification, as Florence (P)

---

[2] Res nullius is "a thing owned by no person" which can "be reduced to private ownership by capture" (Epstein 2002: 3).

illustrates, "we love our sheep, they're grass eaters, aside from one a year [for eating]". Betraying the same underlying instrumentalism, one of Blake's (P) newly acquired pigs turned out to be barren "so I got rid of [her]"; Andrew (P) "kill[s] and butcher[s] the old layers [hens]"; and when Alison's (P) cows reach a certain age, "we dry them off, let them fatten up ... and turn them into sausages and mince". James (P) explained to me how one of his cows simply has the "wrong attitude" and so "unfortunately she's going to god". Clearly, she is not conforming to the proper order wherein "'docility' is ... an explicitly sought after and marketed characteristic of 'farmed', especially female, animals, in the discourses of breeders, especially as it implies ease of handling (i.e., control)" (Cole 2011: 86, referencing Cudworth 2008). A docile body is the desired effect of Foucault's disciplinary power. This is one of the more visible mechanisms of power exerted over animals that maintains the lie of there being the possibility of resistance, for, as James' cow discovered, resistance is a mere trick that can only end one way for her as she inevitably falls off the spectrum of power.

This casts Alison's (P) benign conception of multiple enterprises where "the animals [are] doing what's natural to them and have a quality of life" in a different light. For 'loving animals' while also deciding the terms and duration of that life suggests that they are being loved only as far as they adhere to their function, which is to provide a cash crop of meat. Farmer and author Catherine Friend makes this 'loving' instrumentalism explicit in her account of her quest to become *The Compassionate Carnivore*. She explains:

> I continue to farm because I love animals ... It's why we do what we do. But unless a landowner can afford to keep animals around just to look at, the rest of us animal lovers must find a way for the animals to earn their keep and contribute to the economic health of the farm. (2009: 34)

In most Westernised countries, food animals, or 'livestock', are legally accorded certain minimum rights under welfare standards and guidelines[3]

---

[3] For examples, see www.animalwelfarestandards.net.au/, www.gov.uk/guidance/animal-welfare, and www.nal.usda.gov/awic/standards-and-guidelines. Emphasising how systemically normalised orders of animals have become, the standards and guidelines are different (and much less protective) for livestock compared with cats and dogs, aquatic animals, animals used in research and

because, as Melanie explains, "all animals deserve to have shelter, shade, water and the right food and be able to move around freely". However, as Blake clarifies:

> As ruthless as it is, it's a business you're running. Ultimately if you kept it as a pet, sheep get overweight and kind of sit, and at seven years old they die anyway you know as an old, unproductive ewe, cos you're not breeding it. So, it's just better just to have it go when it's in its prime.[4]

Finn (P) says, "I just love animals", as do most of my participants, but also, "they're *given* one and a half to two years before they're killed, which is a good amount of time" (emphasis added). The way my non-producer participants describe the relationship between a producer and their animals is perhaps more accurate. For Julie (C), an ethical producer is "somebody who is *giving* their animals" a 'good' life (emphasis added). Maria (C) similarly describes "the life that you *give* the animal" (emphasis added), while for Ellis (C), an ethical producer "*allows* them to have a good life" (emphasis added). After all, "we've domesticated these animals", as Ellis (C) explains—implying a sense of human ownership over the lives that are 'given' to animals and are not theirs to live. These accounts complement the validating discourse relating to the value of contingent life, reinforcing the narrative whereby these animals literally owe their lives to humans, and that life is, in and of itself, good. The implicit appropriation of animals' lives and bodies manifests discursively, though subtly in my discussions with participants.

Consumers of ethical and sustainable meat repeatedly refer to *their* meat, and *their* animals. I am excluding producers here to avoid conflating their knowledge of legal ownership with consumers' 'knowledge' of their entitlement. Gillian (C) refers to "*my* lamb farmers"; Sally (C) describes visiting a local producer where she can "give *my* bacon a scratch behind the ear" (note also the rendering of the living animals as already

---

teaching, and exhibited animals (e.g. zoos). There are also different standards and guidelines for pigs, cattle, and sheep, and then depending on whether they are at saleyards, in "processing establishments" (i.e. slaughterhouses), or in transit.
[4] As indicated in the story of Baa in Chap. 4, sheep may in fact live to 17 years.

meat)[5]; David (C) has been "slaughtering *my own* chickens" for a number of years; Anne (C) knows the producers of "*my* pork and *my* beef"; while Geoffrey (C) investigates "how a lot of *my* cows are slaughtered", and purchases "*my* pork products" from "*my* other butcher" (all emphases are added).

This possessive language may seem innocuous; however, Adams (2010 [1990]), Dunayer (1995), Stibbe (2001), and others have demonstrated the contribution of everyday, normalised language to the reproduction of speciesism; language being just one of the ways that Butler's "Imperializing gestures of dialectical appropriation" gain their foothold (2000: 310). Furthermore, I construe this 'dialectical appropriation' differently to the use of similar possessive terms in the context of more mutual and loving forms of relationality. The instrumentality that overshadows relations between humans and food animals, and the constant presence of the power spectrum that they will all inevitably 'fall off' at some point, cast the use of these terms in a less genuinely, or necessarily differently loving, light. Instead, I suggest they are indicative of an enhanced sense of entitlement associated with meat and food animals that are local, 'known', and therefore better. The fact that what these participants are commonly referring to is the meat, that is, bacon, pork, or beef, rather than the animals, lends support to this view.

These animals, and their flesh, are therefore wholly constructed as the property of humans, much like natural resources are described as 'ours' when they are in fact appropriated. However, Armstrong explains that unlike 'natural' resources, animals somewhat resist representations of nature "as a passive object or blank slate ready for mapping" (2002: 415). This is because their sentient aliveness creates an additional requirement for their use as a resource and that is death. While this aspect of eating meat remains problematic to some degree for most of my participants, as I have shown, it is mostly assuaged by the dominant regimes of power/ knowledge/pleasure and knowledge/senses/emotions that constitute these animals as food, and also the entitlement therein that sanctions their use. In this sense, an awareness of entitlement sits comfortably alongside a knowledge of the 'proper' order. As Bella (P) illustrates, "even

---

[5] This, and the reverse, where meat is referred to as a living animal, are common features in my data.

though I think of them very much as a living creature that deserves its own decent life, I know that they're for meat". Dan (C) similarly comments, "I always try to consciously see it as still as an animal … a sentient being. It had interests, it had needs and now it's becoming part of me".

In the context of this deeply felt entitlement, the 'realities' of eating animals, however confronting in their content and delivery, can at best be discomforting for a period, but often merely thought-provoking and 'interesting', as Diane (C) illustrates:

> but pork is still a pig, poultry is really still a chicken. So, you're eating their dead bodies and putting it in your body, which is why that vegan meme is kind of cool. So I was like 'Ooooh, yeah, they die, and their grave is my body', it's kind of dark right?

Sentience, aliveness, being-ness, and even relatedness can seamlessly and unproblematically segue into edibility, or even be perceived simultaneously in a strange kind of co-existence. Jennifer (C) intriguingly comments, "if you've got a relationship with something, you're less likely to just be frivolous with it and leave it half uneaten in the fridge". Relateability and edibility are more easy bedfellows than perhaps many animal advocacy organisations are aware. For those participants more indifferent about, and/or reducing, their meat consumption, and even the two consumers (one vegetarian and one pescetarian) who only buy and cook meat for their family or partner, there remains an acceptance of the basic human entitlement to meat—if that is what 'they' want. Julie (C) believes, "we need to accept that people enjoy meat and use and eat and want to eat meat".

However, the increasingly difficult-to-ignore exercise of lethal power over animal 'resources', coupled with the social imperative to move away from the principles and conditions of factory farming, can trouble the knowledge, and practices, of innate entitlement, as the preceding accounts of discomfort illustrated. This introduces the need for an additional narrative and modified regimes that cast humans in a different role than that of animals' oppressor and maintain the normalised order that denies this state of domination. A similarly productive mechanism is required to reinterpret colonialism, where what Bentley terms "self-validating narratives" re-articulate the aim of colonialism from one of resource appropriation to "a selfless project bestowing European gifts of progress" (2015: 65). The validating discourses, distinctive emotions, and ethico-aesthetic I have identified in this book are all constitutive of self (or practice)-validating narratives that re-articulate meaty practices in a more pleasant light. But there is one further ethico-aesthetic mechanism that features strongly in my data, and across broader discourses of ethical and sustainable meat.

To use these animals that humans are 'unalterably and unquestionably' entitled to use (ideologically as well as legally) and whose lives are therefore in every sense owned, but use them in a 'better' way, requires that, (1) the manner (rather than the fact) of their death is given consideration (as discussed in Parts II and III), and, (2) their former life is somehow acknowledged in the subsequent use of their flesh. The latter consideration often translates to not wasting meat, "so it doesn't die in vain" (Lisa C), meaning to no purpose—that is, human purpose. Because, as David (C) emphasises, even though "you exploit an animal for a particular human need … you should get the best [human] use out of it". However, more commonly, the 'unavoidable' killing of animals is incorporated into an ethical aesthetic of meat eating through the notion of respect for life, as Diane (C) explains, "the key point, and the first point, should be about the quality of life of those that we eat, those sentient beings that we eat". This is a discourse of respect that is predicated on the knowledge of entitlement and ownership and which therefore generates a necessarily human-centric consideration for the life and pleasure of the other, as I will demonstrate.

2   Respect: An Ethico-aesthetic Mechanism of Power

'Respect' is a fairly ambiguous term and can convey subtly different meanings. In the *Oxford* and *Merriam Webster* dictionaries, respect includes the notions of having 'due regard'; treating someone in an 'appropriate way'; admiring what is 'valuable' or 'a sense of the worth' of a person; and deference to 'certain rights'. However, when used in reference to a desire or 'need' for a non-renewable 'natural resource', such as a mineral, a plant, or a life, respect is highly problematic and even disingenuous.

The word 'respect' carries different connotations depending how, and towards whom, it is directed. It is this ambiguity that I suggest is exploited in relation to animals, because the word can in one sense sound benevolent and caring, and yet, as I will show, animals are being 'respected' only to the extent that they are due, worth, or valued (i.e. as food animals), and given 'certain rights' that are 'appropriate' based on how they are thought about or looked at (i.e. as food animals). It is therefore a respect already infused with entitlement and violation. A recent documentary series by food writer–turned-gourmet farmer Matthew Evans, which aired on Australia's public television network SBS in 2016, exemplifies this perspective. As a reviewer from one of Australia's leading national newspapers commented, Matthew urges people to become "more knowledgeable, thoughtful and respectful about the meat we consume", and have "a sense of respect for the life that has been taken for our enjoyment" (Quinn 2016; see also Arcari 2018).

This is a call that has certainly been taken up so that respect has become an integral part of the discourse of 'better' meat practices (Gutjahr 2013), one with a particular ethico-aesthetic of knowledge/senses/emotions. However, as with the ethico-aesthetic of moral approval, the regard being accorded to the life or 'value' of animal others is purely, or 'appropriately', instrumental. The discourse relating to respect is therefore, as yet, uncritical of the fact that normalised and systemic relations of power, oppression, and domination permeate the ethico-aesthetics surrounding a body that is shaped by a predominantly white, male, and, ultimately, human sensibility. This observation draws me back to my focus on entitlement and what it indicates about the apparent intransigence of the 'natural' order that decrees animals as edible resources for human use. For as long as this order remains in place, along with the entitlement it engenders, sensations, emotions, and pleasures become similarly ordered so that a prevailing ethico-aesthetic is incapable of moving 'beyond' to allow the decolonisation of food animals. I suggest that the ethico-aesthetic constitution of 'respect' for meat and food animals plays a significant role in reinforcing this order.

Helen (C) asserts animals should be "treated with respect the whole way through … with respect when they're alive, when they're killed, and how the meat is then cut up and put together and distributed … because it was a living creature". The two meanings of respect (as in admiration or esteem, and also 'due' or 'appropriate' regard) are more clearly conflated by Ellis (C) who explains "respecting another life is sort of that respect for mother nature … you're treating things with respect … reflects how I feel about nature … respecting nature". Nature can at once be admired for its qualities but also its value (i.e. an 'other's' flesh), which is where an understanding of entitlement shifts the nature of that respect (for life) towards an appreciation of use. Gillian (C) demonstrates this shift when she says, "I think the main thing is [their use is] respectful, it's respectful to *what they are*, and to their lives" (emphasis added). Similarly, Helen (C) re-emphasises the respect due all animals, "even if they are brought up to be killed". Respect is allocated only according to what is appropriate to a food animal, as Damien (C) clarifies, "animals need to be respected … they're put on this earth for us, so we need to respect them … respect for the product". As Damien's comment indicates, respect is directed at the meat that an animal 'provides', alluding to the 'knowledge' of this as a 'natural' process where it is the provision of this 'natural' gift (of meat) as an outcome of a natural contract that is being respected, rather than the animal's life.

My participants cast themselves as humble and passive receivers of nature's unsolicited rewards. For example, it is important to Grace (C) to "recognise that you're eating an animal, and be, you know, aware of it, and appreciative". For Julie (C), it's more about "respecting the animals and knowing you've *given* them the best life that you can before slaughtering them for human consumption, and using as many parts of the animal as possible" (emphasis added). The knowledge that they are exercising the appropriate respect adds an important dimension to my participants' use of animals for food. It allows this use to be construed benignly and even reverently, where animals are "treated with respect rather than like a piece of meat *which they are*" (Helen C, emphasis added).

Drawing on the three validating discourses discussed in Part II, and especially the benevolent 'natural contract', the ethico-aesthetic of respect for food animals and their flesh interprets their death as a naturalised gift that they offer up in return for the opportunity firstly to live, and secondly to live a 'good' life. This is an opportunity that humans are understood to have provided and have a responsibility to keep providing to contingent lives. Much like benevolent discourses of colonialism, the benefits to animals are emphasised such that the inherent entitlement and oppression are productively erased. The harsh, individual realities of civilising missions against humans and animals are effectively transmuted and absorbed into a larger story of universal 'good', characterised by a reciprocal and respectful, balance between giving and taking life—as if life were simply a renewable resource.

The apparently simple notion of 'respect' is thus supportive of a multi-storeyed, interwoven narrative that constructs a benevolent, mutually beneficial relationship between humans and food animals. This narrative also pervades popular discourses on ethical and sustainable meat. In *Animal, Vegetable, Miracle*, Barbara Kingsolver's part-memoir, part investigation of her journey to rural self-sufficiency, she emphasises, "It's not without thought and *gratitude* that I slaughter my animals" (2008: 224).

The contractual discourse reinforces the ethico-aesthetic of 'respect' and legitimises it as being 'appropriate' to the 'worth' of these renewable food animals. But, perhaps more than the terms 'ethical' or 'natural', respect is a demonstrably ambiguous term. There is no measure of respect and participants largely hope and assume that animals are 'being respected'. This seems primarily to equate to the *way* they are killed, or even less significantly for animals, simply the attitude of the person carrying out the killing. Tracey (C) reasons that ethical producers who care for their animals "are not going to kill them in a way that is disrespectful", and Finn (P) relies on abattoir employees "having some level of respect for the animals".

Much like the ritual surrounding Douglas' (2002) sacrificial tribe member, respect does little to alter the circumstances or fate of food animals' lives, but its association with ethical practices allows consumers to feel comfortable with this, and modify or mediate their knowledge and experience of it accordingly. For Helen (C), the ritual of breaking fast at

Easter allows her to view the consumption of offal as unproblematic, and "part of that respecting the animal", and David (C) describes the positive affects he and others derive from his 'ethical' connection with his chickens before he kills them:

> Being with them … their entire lives and bringing about the end of their lives. And particularly … sharing it with others … that was very much something [that] I guess helped to really solidify the point of trying to eat, particularly meat, in an ethical way.

As with 'ethical' killing, which I discussed in Part II, 'respect' is constituted and pursued for the productive associations it offers to humans— how it enables them to feel differently and 'better' about eating animals, how it leverages their experience of community and knowledge of what makes a 'good', 'ethical', and 'moral' human. This is a clearly anthropocentric rendering of ethico-aesthetics that does not take into account the 'pleasure of the [animal] other', except via another veiled mechanism of power known as animal welfare. As long as the terminal or non-renewable use of some body, human or nonhuman, is being uncritically construed as respectful and in accordance with their 'social location', then under the normalised and systemic order of entitlement, it is difficult for this use to be regarded as problematic.

Consequently, under the ethico-aesthetic of respect for these naturalised contractual gifts, it can be seen how the increased transparency of meaty practices and increased visibility (and subjectification) of food animals do not pose a significant challenge to the normalised 'order of things' that keeps their state of domination substantially hidden. For this subjectification is capable of being narrativised with human voices (like the voice of the colonist speaking for the colonised) that normalise the secular ritual surrounding eating animals. Any other kind of 'true' subjectification, in which the subaltern animal might speak and be heard (Spivak 1988), still 'glimmers' beyond the grid. Hence, as Armstrong emphasises:

> Encountering the *post*colonial animal means learning to listen to the voices of all kinds of "other" without either ventriloquizing them or assigning to them accents so foreign that they never can be understood. (2002: 417, emphasis added)

2.1   Conclusion

Entitlement permeates all the mechanisms of power discussed thus far that contribute to maintaining the persistent edibility of animals. These include three validating discourses (the value of contingent life, invoking nature, and the benevolent natural contract), two distinctive emotions (requisite bravery and cultural omnivorousness), and two ethico-aesthetics (moral approval and respect). Through the normalised and rhizomatic nexus of power/knowledge/pleasure, entitlement encodes everything that is known, felt, and sensed, emotionally and viscerally, about meat and food animals. It characterises the gaze of power and infuses practices relating to the visibility of food animals. It is also, I argue, why support for increasing the material visibility of food animals, as a means to challenge not only their use as food but also their treatment, is misguided and even counter-productive to any radical change in the status quo.

I take up this argument in the next chapter with reference to the heightened visibility of meat, food animals, and practices of meat production that shapes my participants' accounts of their meaty practices. I describe the forms this 'new' visibility takes in the ethical foodsphere and identify two 'types' of visibility in effect. Drawing on Foucault's accounts of the look or 'gaze' and panoptic surveillance, and also perspectives from visual culture and cinema studies, particularly Mulvey's (1999) apt perspective on the 'pleasure of looking', I articulate how I arrive at, and what I mean by, my notion of the entitled gaze.

References

Acampora, R. (2016). [Provocations from the Field] Epistemology of Ignorance and Human Privilege. *Animal Studies Journal, 5*(2), 1–20.

Adams, C. J. (2010). *The Sexual Politics of Meat: A Feminist-vegetarian Critical Theory*. London; New York: Continuum International Publishing Group.

Arcari, P. (2018). The Ethical Masquerade: (Un)Masking Mechanisms of Power Behind 'Ethical' Meat. In M. Phillipov & K. Kirkwood (Eds.), *Alternative Food Politics: From the Margins to the Mainstream* (pp. 169–189). Abingdon, UK: Routledge.

Armstrong, P. (2002). The Postcolonial Animal. *Society and Animals, 10*(4), 413–419.

Bekoff, M. (2007). *Animals Matter: A Biologist Explains Why We Should Treat Animals with Compassion and Respect.* Boston, MA: Shambhala Publications.

Bentley, T. (2015). *Empires of Remorse: Narrative, Postcolonialism and Apologies for Colonial Atrocity.* London; New York: Routledge.

Black, L. L., & Stone, D. (2005). Expanding the Definition of Privilege: The Concept of Social Privilege. *Journal of Multicultural Counseling and Development, 33*(4), 243–255.

Butler, J. (2000). Subjects of Sex/Gender/Desire. In S. Saguaro (Ed.), *Psychoanalysis and Woman: A Reader* (pp. 309–322). New York: NYU Press.

Caviola, L., Everett, J. A. C., & Faber, N. S. (2018). The Moral Standing of Animals: Towards a Psychology of Speciesism. *Journal of Personality and Social Psychology, 116*(6), 1011–1029.

Cole, M. (2011). From "Animal Machines" to "Happy Meat"? Foucault's Ideas of Disciplinary and Pastoral Power Applied to 'Animal-Centred' Welfare Discourse. *Animals, 1*(4), 83–101.

Cruickshank, B. (2014). Review: Disciplining the Poor: Neoliberal Paternalism and the Persistent Power of Race. *Contemporary Political Theory, 13*(1), e1–e3.

Cudworth, E. (2008). Most Farmers Prefer Blondes: "The Dynamics of Anthroparchy in Animals" Becoming Meat. *Journal for Critical Animal Studies, 6*(1), 32–45.

Cudworth, E. (2011). *Social Lives with Other Animals.* Basingstoke; New York: Palgrave Macmillan.

Deckha, M. (2008). Intersectionality and Posthumanist Visions of Equality. *Wisconsin Journal of Law, Gender and Society, 23*(1, Spring), 249–267.

Douglas, M. (2002). *Purity and Danger: An Analysis of Concepts of Pollution and Taboo.* London; New York: Routledge Classics.

Dunayer, J. (1995). Sexist Words, Speciesist Roots. In C. J. Adams & J. Donovan (Eds.), *Animals and Women: Feminist Theoretical Explorations* (pp. 11–31). Durham; London: Duke University Press Books.

Epstein, R. A. (2002). *Animals as Objects, or Subjects, of Rights.* John M. Olin Law and Economics Working Paper No. 171. The Law School: The University of Chicago.

Foucault, M. (1980). *Power/Knowledge: Selected Interviews and Other Writings, 1972–1977.* New York: Pantheon Books.

Friend, C. (2009). *The Compassionate Carnivore*. Philadelphia: Da Capo Lifelong Books.

Genovese, E. D., & Fox-Genovese, E. (2011). *Fatal Self-Deception: Slaveholding Paternalism in the Old South*. New York: Cambridge University Press.

Gilligan, C. (1982). *In a Different Voice*. Cambridge, MA; London: Harvard University Press.

Gruen, L. (2014). *Entangled Empathy: An Alternative Ethic for Our Relationships with Animals*. New York: Lantern Books.

Gutjahr, J. (2013). The Reintegration of Animals and Slaughter into Discourses of Meat Eating. In H. Röcklinsberg & P. Sandin (Eds.), *The Ethics of Consumption: The Citizen, the Market and the Law* (pp. 379–385). Wageningen, Netherlands: Wageningen Academic Publishers.

Hankivsky, O. (2014). *Intersectionality 101*. Institute for Intersectionality Research and Policy, Canada, 34 pp.

Held, V. (2006). *The Ethics of Care: Personal, Political, and Global*. Oxford: Oxford University Press.

Kanji, A. (2017). Colonial Animality: Constituting Canadian Settler Colonialism through the Human-Animal Relationship. In M. Woons & S. Weier (Eds.), *Critical Epistemologies of Global Politics* (pp. 63–78). Bristol: E-International Relations Publishing.

Kingsolver, B. (2008). *Animal, Vegetable, Miracle: A Year of Food Life*. New York: HarperCollins.

McKibben, P. (2019). *Love Notes: For a Politics of Love*. New York: Lantern Books.

McLisky, C. (2015). "And They'll Know We Are Christians by Our Love": Exploring the Role of Christian Love on Maloga Mission, 1874–1888. *Journal of Religious History, 39*(3), 333–351.

Meyers, D. T. (2017). Commentary on *Entangled Empathy* by Lori Gruen. *Hypatia, 32*(2), 415–427.

Mulvey, L. (1999). Visual Pleasure and Narrative Cinema. In L. Braudy & M. Cohen (Eds.), *Film Theory and Criticism: Introductory Readings* (pp. 833–844). New York; Oxford: Oxford University Press.

Nibert, D. A. (2013). *Animal Oppression and Human Violence: Domesecration, Capitalism, and Global Conflict*. New York; Chichester: Columbia University Press.

Palmer, C. (2001). 'Taming the Wild Profusion of Existing Things'? A Study of Foucault, Power, and Human/Animal Relationships. *Environmental Communication: A Journal of Nature and Culture, 23*, 339–358.

Probyn, E. (2000). *Carnal Appetites: FoodSexIdentities.* London; New York: Routledge.

Quinn, S. (2016, April 18). Number of Vegans in Britain Rises by 360% in 10 Years. *The Telegraph.* Online. February 2017.

Sorenson, J. (2014). Thinking the Unthinkable. In J. Sorenson (Ed.), *Critical Animal Studies: Thinking the Unthinkable* (pp. xi–xxxiv). Toronto: Canadian Scholars' Press Inc.

Spivak, G. C. (1988). Can the Subaltern Speak? In Marxism and the Interpretation of Culture. In C. Nelson & L. Grossberg (Eds.), *Marxism and the Interpretation of Culture* (pp. 271–313). Basingstoke: Macmillan Education.

Stibbe, A. (2001). Language, Power and the Social Construction of Animals. *Society and Animals, 9*(2), 145–161.

Tronto, J. (1993). *Moral Boundaries: A Political Argument for an Ethic of Care.* New York: Routledge.

Wadiwel, D. (2017). *The Werewolf in the Room: Animals and Capitalism.* Conference Keynote Lecture. Seventh Australian Animal Studies Association Conference: Animal Intersection. 3–5 July, Adelaide, Australia.

Weisberg, Z. (2009). The Broken Promises of Monsters: Haraway, Animals and the Humanist Legacy. *Journal for Critical Animal Studies, 7*(2), 22–62.

Wolfe, C. (2003). *Animal Rites: American Culture, the Discourse of Species, and Posthumanist Theory.* Chicago and London: University of Chicago Press.

Young, I. (1990). *Justice and the Politics of Difference.* Princeton; Oxford: Princeton University Press.

9

Visibility: Inviting an Untroubled Gaze

This is the second chapter on the power of transparency, and the final empirical chapter of this book. Here, I attend to my argument that, regardless of attempts by animal advocates to make visible the (mis)treatment and abuse of food animals, of how transparent producers are about production practices, and how familiar consumers are willing to become with all these practices, a fundamental sense of entitlement, as described in the previous chapter, prevents food animals from being seen any differently.

Whether in life or in death, the visibility of food animals is saturated with entitlement, with the knowledge and 'truth' of the purpose they were born to, and with the entwined sensory and emotional associations that contribute to masking their actual state of domination. With the understanding of how entitlement works through the mechanisms of power, sight only adds to an entirely normalised ethico-aesthetic, or embodied ethics, of eating animals to constitute, and (re)articulate, a more stable, ethically oriented order. From Foucault's conceptions of the nexus of power/knowledge/pleasure—via power/knowledge and the (dis)pleasure of knowing—to Probyn's (2000) and Hayes-Conroy and Hayes-Conroy's (2008, 2010) insights regarding the power of taste, I now arrive

© The Author(s) 2020
P. Arcari, *Making Sense of 'Food' Animals*,
https://doi.org/10.1007/978-981-13-9585-7_9

at film theorist Laura Mulvey's seminal ideas on the pleasure of looking. This brings my tour of Foucault full circle by highlighting the "power through transparency" or "subjection by illumination" that is exercised by the gaze of power in the market for ethical meat (Foucault 1980: 154).

1   Perfect Vision?

In 1975, Peter Singer said those who "require" animals to be killed for food "do not deserve to be shielded from this or any other aspect of the production of the meat they buy", and a few years later Linda McCartney's made her now-famous assertion that "if slaughterhouses had glass walls the whole world would be vegetarian".[1] Over 40 years later, consumers are increasingly 'choosing'[2] *not* to be shielded, as my data have indicated, while per capita meat consumption continues to outpace population growth. It would seem, therefore, that visibility does not have the anticipated transformational effect on social practices, and yet, visibility and 'exposure' is still one of the primary methods by which campaigners for dietary change and/or animal rights seek to jolt consumers out of their habitual routines. For example, The *Animal Activists' Handbook* claims:

> if the realities of factory farms and slaughterhouses were as visible as the meat they produce, all thoughtful compassionate individuals would be vegetarian advocates. (Ball and Friedrich 2009: 17)

During the 40 years since Singer's book, issues associated with the intensive, industrial-scale production of food in general have attracted increasing public and media attention and this has translated to the market place with more specialised product labelling (e.g. organic, fair trade, sustainable) and big retail partnerships with conservation groups, animal

---

[1] Although commonly attributed to Sir Paul McCartney, it was Linda McCartney who first said this. As far as I can tell, it first appears in her 1995 cookbook, *Linda's Kitchen: Simple and Inspiring Recipes for Meals without Meat*.

[2] It is recognised that consumer 'choice' is not an accurate representation of the many factors that contribute to this shift in social practices involving meat. However, it depicts the nature of my participants' engagement with a more transparent model of meat production, which they characterise as voluntary and even sought after, rather than imposed and unwelcome.

welfare organisations, and leading animal advocates, including Singer (Lewis and Huber 2015; Satya 2006). In addition, there are frequent boycott campaigns designed to draw attention to, and commercially penalise, companies for a range of (alleged) nefarious practices. Reflecting the colonisation of the foodsphere with multiple agendas around corporate dominance, social justice, health, environmental sustainability, and animal rights, this gradual politicisation of food has been well documented (e.g. Cook and Crang 1996; Goodman et al. 2010, 2012; Goodman 2004; Micheletti and Stolle 2010; Clarke et al. 2008; Barnett et al. 2005, among others).

Common to the politics of food across all these agendas is the goal of "making transparent" (Goodman 2004: 902) according to a broader thesis of commodity de-fetishisation (Gunderson 2013). Under this thesis, previously hidden conditions of production and provision are exposed for public scrutiny and assessment, and in turn become part of the constitution of new orders of distinction—distinctions that determine not only whether a product is organic, environmentally friendly, socially just, or ethical, but whether one is more/differently organic or ethical than the next. What has been de-fetishised thus becomes re-fetishised as consumers become reconnected, and immersed, in the practices of production (Pottinger 2013). However, this immersion is often so highly mediated and managed that many authors refer to it rather as a double fetishism whereby a "diverse range of commodity signs" contribute to their continuing aestheticisation (Cook and Crang 1996: 136; see also Goodman 2004; Lewis 2011).

Transparency is therefore a key feature of what are understood as alternate, and 'better', practices involving food, and most especially meaty practices. However, it manifests in many forms, through different media and spaces, in ways that reflect different kinds and different degrees of mediation.[3] For example, improved surveillance techniques have opened up the largely invisible world of industrial meat production to the public gaze. Depending on how this transparency is organised and mediated,

---

[3] Acknowledging that sight, like the other senses of perception, is always mediated and never independent of its co-constitutive environment of social practices which shape how what is sensed is made sense of.

and by whom, it can be used to create distance from popularised accounts of conditions in factory farms and other industrial spaces of animal production, thereby providing comforting reassurance of more benign treatment of food animals. The trend in 'chicken cams', 'cow cams', and 'pig cams' showing real-time footage of these food animals going about their daily routines are illustrative of this benign visibility. Alternatively, it can be used to expose poor conditions and mistreatment of food animals, commonly through the use of 'raw footage', with the aim of inciting discomfort in viewers and contributing to changes in how these animals are used. Examples of this raw visibility include the Australian ABC's Four Corners report on the live export trade titled *A Bloody Business*, Animals Australia's Make It Possible campaign, and various campaigns targeting the use of animals for food by organisations such as People for the Ethical Treatment of Animals (PETA), Mercy for Animals, and others. This capacity of visibility to be variously benign or raw is explored further in the next section.

First, however, I need to address the obvious relevance here of the large body of literature drawing on Foucault's theorisations of biopower and biopolitics (e.g. Rabinow and Rose 2006; Esposito 2008). This includes the application of these theories to the lives of, and human relations with, animals (e.g. Wolfe 2012; Wadiwel 2008; Chrulew 2012), and also how these intersect with Foucault's technologies of the self (Agamben 1998; Mayes 2015; Genel 2006).

In Foucault's biopower, the subject or 'biosubject' (Wolfe 2012: 22)— here the food animal—is conceived as the object of a power to "make live and let die" in contrast to the sovereign power to "let live and make die" (Foucault 1997: 239). However, the foregrounding in theorisations of biopower of "primary material" (Esposito 2008: 29), "individual bodies" (Foucault 1997: 242), or "the molecularization of vitality" (Rose 2009: 13) places the focus on how this kind of power acts on and shapes bodies, where what I am primarily interested in is how power (in relation to food animals) is constituted in the first place. In this sense, my concern is with what Genel (2006: 48) describes as "technologies of power which are no longer presented exclusively internal to a code of legality or sovereignty, codes which in fact mask the new modes of the exercise of power". Or as Wadiwel observes, in contrast to Agamben's (1998) understanding of human-animal relations as defining of biopolitics, biopower is rather, "before anything else, a question of determining the distinction between human and animal" (Wadiwel 2008: 18).

It is with the aim of understanding how this distinction between human and animal is constituted, how it is maintained, and why it remains so persistent that I reserve my focus for now on Foucault's theorisations on power/knowledge/pleasure and also transparency. Such an understanding will provide added insight into how all forms of power in relation to animals are constituted, whether conceived as sovereign, pastoral, disciplinary, or biopower. A productive avenue for future analysis would be the question of bio- versus sovereign power in relation to food animals, given that they represent the power of humans to both make live *and* make die.

2   Two Views of Visibility

The majority of transparency currently circulating in relation to food animals and meat lies somewhere between a purely benign, anodyne portrayal of animal farming, and the rawness of activist footage. Again, where on that scale usually depends on the intention of the visibility and how it is being mediated. To illustrate this point, there are slaughterhouses in Denmark and the US that have opened their doors to the public so that, through glass walls, they can see "100% of what happens here, not 99%" (Wiper 2014, online). However, this access to the rawness of killing is done with the aim of normalising slaughter, making it seem benign and even honourable, as the Vermont Packinghouse in the US indicates: "What if you could see and know just how each animal was slaughtered, and know that it was done with respect and dignity" (company website). Similarly, the American Meat Institute produced video tours, narrated by Temple Grandin, of slaughterhouses for lambs, turkeys, pigs, and cows, in the "meat industry's sincere hope that these videos will give consumers who want to know more about meat processing the information they seek" (Animal Handling Website).[4] Then there are

---

[4] An organisation sponsored by the American Meat Institute (AMI) and used to promote the project and the accompanying videos (http://animalhandling.org/ht/d/sp/i/80622/pid/80622).

the animal advocacy/welfare organisations, such as the RSPCA and Compassion in World Farming whose campaigns are oriented towards improving conditions for food animals, for example by abolishing factory farming, farrowing crates for pigs, or promoting cage-free eggs. The practice being targeted will be presented in a raw manner intended to elicit discomfort, while animal farming in general—done the 'right' way—is presented more benignly.

Transparency also manifests discursively and the swathe of books published over the past 10 to 20 years attests to the growth in appetite for behind-the-scenes accounts of practices of meat production (Table 2.1). Again, these accounts range from the benign—for example, Simon Fairlie's (2010) *Meat: A Benign Extravagance*, to the raw—for example, Jonathan Safran-Foer's (2009) *Eating Animals*. However, it is in the more popular media sphere, such as film and TV, that transparency as actual visibility, in relation to food animals and meat production, has really hit its stride in the past decade, challenging the emancipatory thesis of McCartney's glass walls. In Table 2.1, I list at least 11 documentary films and three TV shows that include scenes of food animals being managed, confined, and killed, and variably graphic images of associated mistreatment and abuse. In her exploration of the depiction of food animals in documentaries, Smaill (2014: 34) identifies a further 14 that highlight problems with agricultural practices and 'agribusiness'. New forms of digital media that have simultaneously sprung up over this time enable the dissemination and sharing of this and other kinds of visibility more rapidly and to a wider audience than ever before. As a result, increasing numbers of consumers are becoming aware of ethical (and not just environmental and health) issues associated with meat consumption, even if not the principle of using animals as food. Depending on how the visibility of these issues is framed, the intent is to encourage a shift to more benign practices of meat production and consumption, or encourage an outright rejection of any use of food animals.

This is, however, a mere speck on the tip of the visibility iceberg. Digital media such as YouTube, Facebook, Twitter and Instagram are not only outlets for the secondary promotion of the kinds of material described above, but have also become platforms for original and instructional material, often linked to personal blogs or other websites. A few minutes spent online will unleash a deluge of visual material relating to meat production, including practices of breeding and raising food animals, and especially slaughter and butchery, much of it produced by meat companies.[5] Indeed, the killing of food animals, as the most widely discomforting and concerning aspect of meaty practices, features prominently in such material. This is especially so in light of the increasing popularity of keeping backyard chickens and even hobby farms as sources of more ethical, DIY meat, provided you have the 'bravery' and skills to carry out your own killing and butchery.

However, not all of this material aims to be instructional. Reflecting prescriptions around being able to witness slaughter, as discussed in Part III, detailed accounts of killing food animals appeal to broader understandings of ethical meat and the practices associated with it. Newspaper and magazine articles, opinion pieces, and blogs illustrate a discursive side of this second-hand transparency whereby the reader sees through the author's eyes as they recount their experience of watching animals being slaughtered. Despite including fairly raw accounts, such stories are typically cast in overall benign and life-affirming terms, urging their readers to access the same experience.[6] While the volume of this online material relating to slaughter and butchery is quite staggering, televised depictions also capture wide audiences and garner the kind of media

---

[5] For example:

- Chico Locker and Sausage Co. Inc. "A look inside the glass walls of a slaughterhouse." 27 August 2012. Web. 20 October 2015.
- Smith Meadows. "Behind the scenes at a local butcher shop." 15 April 2013. Web. 20 October 2015.

[6] Just a fraction of such accounts include:

- Eisendrath, Ben. "Loving a good slaughterhouse." *The Atlantic* 30 March 2012. Web. 21 October 2015.
- Grover, Sami Grover. "A look inside a humane slaughterhouse." *Treehugger*. 2 May 2011. Web. 20 October 2015.
- McEvedy, Allegra. "My visit to the slaughterhouse: crossing the line between life and meat." *The Guardian* 29 August 2014. Web. 23 October 2015.
- Rivera, Lizzie. "Can Eating Meat Ever Be Ethical?" *The Independent*. 8 April 2016. Web. 21 January 2019.
- Bell, Ryan. "The Smell of the Slaughterhouse." *National Geographic*. 29 February 2016. Web. 21 January 2019.
- Timms, Katie. "Inside an Abattoir". Plymouth Herald. 10 June 2018. Web. 21 January 2019.

attention that suggests a certain mainstreaming of associated practices (Mesirow 2011; Fearnley-Whittingstall 2008; Blundell 2016).

TV celebrities imbue the visibility trend with added social status, and lend an air of authority and expertise to the increasingly 'known' social prescription to be more aware of, and embrace, the 'realities' of meat consumption. In 2011, Oprah aired a tour of a Cargill 'beef processing plant' (i.e. slaughterhouse). Cargill agreed because "[consumers] want to know and we want to show them". A steady stream of celebrities have likewise been keen to educate consumers about the harsher, raw realities of slaughter. In addition to performing a routine cull of underweight or injured chickens in his 2008 series *Hugh's Chicken Run*, Hugh Fearnley-Whittingstall had earlier (in 1995) filmed the shooting, stringing up, and butchery of Genghis the pig as part of the BBC series *Food File*. In 2005, Jamie Oliver and Gordon Ramsay both slaughtered lambs as part of their respective TV shows, *Jamie's Great Escape* and *The F Word*. Ramsay's footage was later endorsed by PETA, who said it "will turn many compassionate people into vegetarians" (Adams 2006). More recently (2016), Matthew Evans, the 'Gourmet Farmer', co-wrote and narrated *For the Love of Meat*, where he follows three food animals—chicken, pig, and cow, from the farms where they are raised, to the slaughterhouses where they are killed and processed, and finally their consumption. Matthew proclaims, "Consumers deserve more transparency from the meat industry so we can see how *our* animals are farmed in our name" (Broadcast November 3, 2016; see also Arcari 2018).

While there is generally some degree of public outcry over these televised displays of killing, they are predominantly considered to have been beneficial to the extent that they disclose exactly what is required to produce meat. As Julie (C) comments:

> If we're going to eat it we need to understand and know it. I reckon even ridiculous stunts like that Jamie Oliver one where he slaughtered an animal on television, was a bit of a catalyst towards some of this kind of thing, and I think that's where those, you know, those hipster things where you go and do a day of nose to tail eating with celebrity chefs, is kind of starting to help that.

They therefore illustrate the resolution of raw visibility with benign visibility—how they can work together so that "sight and sequestration exist symbiotically" (Pachirat 2011: 252). For under the normalised regime of power/knowledge/pleasure relating to food animals and meat, visibility tends to be commodified by "engendering commerce for witnessing or participating in killing" (Acampora 2016: 10) more often than it elicits any fundamental shift in how food animals are seen and made sense of.

Again, the different forms and spaces of visibility I have described in this short space represent just a fraction of the ways in which people can gain visual access to the lives and deaths of food animals. Besides these, first-hand experiences can be purchased from farmers and permaculture hubs offering 'artisanal' workshops where attendees learn how to kill and 'dress' chickens, humanely slaughter lambs, and butcher cows. Local events may offer the novelty of meeting your meat before dining on it nose to tail (Fig. 9.1), a foodie predilection that received satirical treatment in an Australian stage play by Eddie Perfect simply titled *Beast* (2013).

Less confronting, though still a major part of an industry-mediated visibility strategy, many local farms now invite the public to tour their operations, meet the animals, and, of course, usually purchase some 'product' before they leave. This is a purposefully benign use of visibility as part of an embodied ethics of eating animals, as compared with the more raw approach to slaughter. However, both strategies are emblematic of the New Carnivore movement as theorised by Parry (2010) and the "celebrated carnivorism" noted by Cudworth whereby "killing 'food' is naturalised" at the same time as compassion for animals is resisted (2011: 94). As Parry also explains, the New Carnivore movement "strive[s] to present animals' becoming meat as a humane, benevolent, and wholly 'natural' process" (2010: 4). This ethical reconfiguration of humans' inherent entitlement to kill animals lends it additional legitimation, and through these rose-tinted lenses, visibility poses little threat to the normalised order of meaty practices.

In the following sections, I draw on my data to show how both benign and raw visibilities operate in my participants' accounts of their ethical meaty practices. In so doing, I further demonstrate how visibility simply reflects, and in turn reinforces, the entitlement born of the normalised

Fig. 9.1 Advertisement for a local nose-to-tail dinner featuring Kaiser the pig (Author's own image)

regime of power/knowledge/pleasure. The result is a persistently mediated visibility whose horizons are still constrained by the proper order of things.

2.1 Benign Visibility

It is primarily the *lives* of food animals that are visually constituted in a benign way. On the websites of the producers that participated in my study, images are widely used, and in them, cows, pigs, chickens, and sheep/lambs are made to appear relaxed, amiable, and content with their lives; their relationships with humans appear affectionate, warm, and gentle; and their environments evoke the bucolic idyll and romantic 'folksy charm' discussed in Part II.

Portraying a sense of trust between the farmer and 'their' animals seems to be a key ingredient in emphasising their pastoral role in caring for the wellbeing of food animals, rather than their ultimate role as 'growers' of meat, via the slaughterhouse, for money. Of the 15 producer websites, three include images of people and animals in close proximity, the animals appearing interested or following closely behind in open fields or along tracks. A further seven show more intimate scenes of companionship: two piglets snuggling on the farmer's lap; a lamb being gently cradled and caressed by the farmer in an open field; a fluffy, yellow chick being carefully held and kissed on the head by a small child; pigs being gently petted by the farmer as they feed; the farmer lying splayed in an open field in the centre of a large group of cows, all gathered around him peering curiously, just a few inches away; the farmer reclining in a field against the body of resting pig, gazing out over the countryside; and a large cow lying peacefully in a straw-laden area with a person resting prone along its back. The assurance that farmers not only love their animals, but that the animals in turn love and trust the farmer, is the implicit message here. Family scenes are also common, further conflating this open and transparent model of farming with all the moral 'goods' associated with family, as well as with nature and the Australian rural tradition.

These promotional images, along with the textual components of these websites, which, as demonstrated in Part II (Table 4.1), tap into the discursive field of authentic, natural, and sustainable idylls, are speaking to a viewer concerned to 'know' that the meat they eat is ethical, but also to substantiate that knowing with seeing. Producers are very aware of the power of the visual medium in this era of transparency and politicised food practices, as Andrew (P) illustrates:

Words are just words, the more powerful thing is actually the media, like visual things, photos, videos. Those are the kinds of things that I try to do because they really, they're powerful in what they can convey to people, much more than a word.

Phillipov and Goodman (2017) describe this as "affective capitalism" (347), noting that farmers are increasingly selling themselves using presentational forms of media more typically associated with celebrity culture, such as family life stories and photos. In this way, they invite "affective investments" (347), and, based on my consumer participants, these promotional images are receiving just the affective investments these farmers desire—but that are also demanded and shaped by the new embodied ethics of meat. For example, Heather (C) describes a nearby 'ethical' farm as "where they pretty much go all out in responding to the needs of the animal during its lifetime … to make it natural, healthy, happy for the animal". Julie (C) follows a particular farmer on Facebook and is assured his animals "have a better life, they're more comfortable, they're happier, they can socialise". She explains:

> He's very transparent, he does tours at least once a month kind of thing.
> Hopefully he's opening a shop because … I'd be able to take the kids out
> there to get his meat. The reason why people follow him and support that
> farming practice is because he's transparent about it.

Images are also used at farmers markets where meat producers often display photographs of their farm and their animals. These are 'addressed' and 'received' (Sturken and Cartwright 2009) in a similar way, as Natalie (C) indicates, "often they'll bring photographs kind of as proof, which is very sweet". Validating producer Andrew's assertion that images are more powerful than words, Natalie says, "I don't need that certifying body to tell me that their animals are well treated because they bring photos".

Whether used on websites or at farmers markets, these photographic images reflect a very familiar but highly ordered and constructed understanding of a food animal, one that is built on the knowledge and exercise of entitlement, thus illustrating Sturken and Cartwright's point that "the photographic gaze … helps to establish relationships of power" (2009: 100). This benign visibility of animals who are unerringly food focuses on promoting the happiness of the lives they live, in which they are loved and cared for in natural surroundings, and avoids polluting this scene with images of their death.

Scenes of death and slaughter may pollute a visibility constructed to be purposefully benign, where associations with eating animals remain positive and unchallenged/ing. However, as discussed previously, the heightened visibility of industrialised and intensive meat production practices, especially slaughter, has challenged the stability of these associations for many concerned consumers. For them, a new regime of knowledge/ senses/emotions that incorporates the more raw aspects of visibility is required to re-stabilise their practices. Under an ethical configuration, visibility becomes a route by which to attain social distinction by facing the reality of slaughter with the requisite bravery and respect.

2.2   Raw Visibility

Visibility (or its interpretation) is shaped by, and can in turn reinforce, particular systems of knowledge (and power and pleasure), as with the ordering of 'good' and 'bad' meat by its visual appearance.[7] In *The Order of Things*, Foucault refers to this as the "already encoded eye" (1989: xxii). Normalised systems of knowledge need to be unsettled in some way before this eye can be de- (or re-)coded. There are many ways this can happen in relation to meat and food animals, but the most likely way, based on my data, is in relation to visible practices of killing and their potential association with alternate, contrary systems of knowledge.

First-hand experiences of animal death, as recounted by my participants, being not just more visually but multi-sensorially embodied, might be expected to present more of a challenge to normalised systems of knowledge compared with second-hand media representations. However, in every case the animals being killed are known definitively as *food* animals, thereby appearing as already 'encoded' at a fundamental level. Confronting the realities of killing is thus associated primarily with bravery and, in turn, a more moral and "clear-eyed" (Joyce C) approach to their meaty practices. As Finn (P) says, "I feel like [going to an abattoir] educated me". It appears the dominant regime of power/knowledge/

---

[7] In the same way that knowledge contributes to how other sensory (taste, smell, touch, sound) experiences, real or mimetic, are interpreted.

pleasure in relation to meaty practices asserts itself through the encoded eye even when the potential for instability seems greatest.

However, the value constituted around raw visibility not only includes the social distinction and the comforting sensory and emotional associations that this distinction then inscribes on meat, the practices that it is part of, and the individuals who eat it. It also includes its function as a regulatory measure to ensure that food animals are always being treated in the 'appropriate' and 'proper' way—that is, in ways that belie any notions of domination. From this perspective, consumers seek more, and not less, visibility. For example, Grace (C) raises the issue of the proposed Ag-Gag laws[8] in Australia, "to stop Animals Australia". She continues, "Why go to that trouble when you can just show that you're doing the right thing?"

Supposedly reliable distinctions between 'good' and 'bad' slaughter practices could therefore be largely based simply on how visible they are. As Anne (C) remarks, "there are abattoirs closer than Kyneton here but they're not transparent, and so the people who care don't use them". The implication is that mechanisms of transparency in themselves generate the 'right' sort of practices, acting as a form of visive discipline—not of the animal 'inmates' of the system, like Foucault's prisoners under the surveillance of the panopticon (1980: 146–165), but of the 'prison guards'—that is, the slaughterhouse owners and workers. This reversal of the panoptic mechanism to become the surveillance of the few by the many is more accurately termed synoptic by Lyon (2006). Synoptic forms of surveillance are a characteristic feature of the kind of visibility that modern communication and information technologies have enabled, and which Lyon notes Foucault largely neglected. As Joyce (C) illustrates:

If there is a legislative requirement for more humane practices, if there was a legislative requirement that all slaughterhouses be scrutinised with CCTV that would automatically change practices.

---

[8] Based on US-style legislation, these laws are designed to criminalise the (*unmanaged and unmediated*) monitoring and investigative activities undertaken by various animal advocacy groups and organisations in relation to any industry that uses animals for profit, but particularly those that raise 'food' animals and produce meat.

However, this transparency is being applied to practices that are already 'unalterably and unquestionably' (and legally) justified. Attention is therefore directed at the monitoring and control of the techniques of those practices, passing over any question of the actual work of killing. This is the same process by which Pachirat notes that a focus on food safety within industrial slaughterhouses deflects attention away from the routinised killing of animals and onto elaborate performances of hygiene (2011: 206). Additionally, regulation inherently carries an a priori and authoritative approval of what is being regulated, which once again bypasses any critical questions regarding the actual practice being regulated. Knowing that a certain practice is being regulated and monitored, especially visually, becomes sufficient in itself to instil a sense of comfort in the knowledge that it is a 'normal' practice being carried out in the 'right' way.

This principle extends also to local farmers as alternate 'prison guards', whose practices are similarly regulated, although the mechanisms of transparency here, in addition to online promotional images and photographs at farmers markets, include the farm tour. Of the 15 producers who participated in my research, six have an open farm-gate policy, meaning the public is welcome to visit anytime or on specially arranged tour days. A further seven are happy to accommodate visitors by arrangement. Indeed, these seven were all interviewed on-farm and were happy to show me around their everyday operations. Of the remaining two, one has since gone out of business, and one is a retail operator. It is again the possibility of this raw visibility, as much as its actual experience, that exerts a normalising and legitimising effect whereby all is perceived to be operating as it 'should'. As Anthony (C) explains:

> There needs to be access, so if you claim to be something people should be able to come and check it out if they want to … I think that would go a big way in changing my perception of a producer if I could go there … even if I didn't go there myself, I trust that if they were doing something terrible then people would be speaking out about it.

It is according to more connected (Schor and Fitzmaurice 2015)[9] and experiential (Holbrook and Hirschman 2015)[10] understandings, or constitutions, of consumption that raw visibility exercises and extends a power that is already systemic. As Pachirat notes, "The politics of sight feeds off the very mechanisms of distance and concealment it seeks to overcome", they are, "modes of power capable of acting in concert to reinforce relations of domination" (2011: 252). The power of this politics of sight lies not only in the 'already encoded eye' it directs at food animals through a focus on ensuring 'appropriate' methods of meat production, but also in the fact that it is discreet, functioning "permanently and largely in silence" (Foucault 1977: 177). Foucault highlights that the principle of power, according to Bentham, is that it should be visible and unverifiable:

> Visible: the inmate will constantly have before his eyes the tall outline of the central tower from which he is spied upon. Unverifiable: the inmate must never know whether he is being looked at at any one moment; but he must be sure that he may always be so. (1977: 201)

Hence, the glass walls, cameras, farm-visits, websites, and photographs become synoptic 'technologies of power' that increase the efficiency of the entire 'food animal to meat' system by "increasing its *own* points of contact" (Foucault 1977: 206, emphasis added). For these technologies are essentially productive rather than repressive—they are "an internal part of the production machinery" whereby the goal is "to strengthen the social forces – to increase production, to develop the economy … to increase and multiply" (Foucault 1977: 175, 208). And the consumer is very much part of this optic machinery, being:

> neither in the amphitheatre, nor on the stage, but in the panoptic [synoptic] machine, invested by its effect of power, which we bring to ourselves since we are part of its mechanisms. (Foucault 1977: 217)

---

[9] Characterised by the reallocation of wealth "away from middlemen and towards small producers and consumers" (Schor and Fitzmaurice 2015: 410).

[10] Characterised by an orientation towards consumers as "hedonic feelings-centered fun-oriented flesh-and-blood living creature[s]" (Holbrook and Hirschman 2015: 2).

The commodification of visibility, as part of the capitalist project, becomes the ultimate testament of the rhizomatic co-production of power, knowledge, and pleasure and the success of this normalised regime as it relates to food animals. The increase in public access to local farm operations and the emergence of glass-walled slaughterhouses charging people admission to witness or participate in killing food animals—a consequence of the original 'glass walls' theory that Pachirat anticipated (2011: 253)—are exemplars of this commodification. Furthermore, as with other sensory and emotional associations (like violence and happiness), visibility becomes inscribed and thereby secondarily commodified on the meat itself through labelling and promotion. In this way, knowledge and signification of 'transparent' production practices carry sufficient authority alone to increase the ethical, and also monetary, value of ethical meat. Continuing this logic, the more visibility, the better for my participants, as Gillian (C) illustrates:

> I would like to see it lead towards transparent labelling and transparent practices. So that a label on the piece of meat would have say, a photo of the way that animal was raised, the kind of life it led, verified, and the kind of death it had.

Producers are also keen to further increase opportunities for visual contact. In addition to his current farm tours, Andrew would also like to include "a photo of the cow in the box the people were getting … just so they understand it was a living animal, and now it's meat [that is] going to nourish them". Just as Pachirat says, "the act of making the hidden visible may be equally likely to generate other, more effective ways of confining it" (2011: 253). But, as indicated, visibility would not be such an effective economic technique if it were not also productive of a certain amount of pleasure, as Laura Mulvey (1999) observes in her account of the pleasure of looking. It is this pleasure associated with visibility—an essential part of the power of transparency—that I now explore in more detail.

3   The Pleasure of Looking

How can the role of emotions and pleasure in relation to visuality be understood? Here I draw on Mulvey's (1999 [1975]) seminal account of the male gaze and Columpar's (2002) extension of her politicised rendering of visuality to encompass also the ethnographic and colonial gazes. All three gazes, Columpar argues, "project their own fantasies" onto an objectified other as a "site/sight of difference" (2002: 34, 40). The bearer of the gaze is thus situated in a position of mastery from which they derive pleasure, but which may also be associated with other emotions, including anxiety and fear (Columpar 2002).

Mulvey's theory concerns the pleasure derived by the specifically male gaze in narrative cinema, a focus that has over time been critiqued for failing to account for more fluid gender identities in general, and the female gaze in particular (Patterson and Elliott 2010; Glenn 2017). These are critiques Mulvey herself responded to (Mulvey 1989, 2001; Sassatelli 2011), and which have been taken up by other scholars to offer ways to think beyond the male gaze or reaffirm its enduring relevance (Oliver 2017). Notwithstanding these dissections of the intra-human gaze, Mulvey's construction of the relationship between a spectator and the object of the gaze (and the pleasures thereof) is in many ways analogous to that between humans and the visible food animal. Moreover, I contend that the gender fluidity underpinning critiques of Mulvey's male gaze does not translate to inter-species relations where no analogous fluidity can be argued, where unequal power relations are economically, legally, and socially sanctioned, and where, therefore, the gaze, in all its practical implications, is always primarily unilateral.

Mulvey draws on Freud's (1955) notion of scopophilia, broadly defined as the pleasure derived from looking. Though more often applied in reference to sexual pleasure, Mulvey highlights that Freud associated it "with taking other people as objects, subjecting them to a controlling and curious gaze" (1999: 835). She describes the woman in popular narrative cinema as "(passive) raw material for the (active) gaze of man ... [as] demanded by the ideology of the patriarchal order" (843). Paraphrasing Sturken and Cartwright on Mulvey, one could therefore propose that 'conventions of 'food animal visibility' are structured by a normative speciesism, positioning animals represented in photographs, film, slaughterhouse tours, farm tours, and so on as objects of an 'entitled' human gaze. Images are geared towards human viewing pleasure'.[11] Further demonstrating the parallels, Mulvey explains how:

> The power to subject another person to the will sadistically or to the gaze voyeuristically is turned on to the woman as the object of both. Power is backed by a certainty of legal right. (1999: 841)

Voyeurism is routinely associated with pleasure in media and cultural studies (van Zoonen 1994; Padva and Buchweitz 2014; Rodosthenous 2015). It is also implicated in a long history of using animals as objects of visual pleasure in circuses, zoos, and other forms of visual entertainment. According to Acampora, this voyeurism is indicative of a "structure of possessive consciousness" (2005: 76), or what I would call normalised entitlement, that is backed, as with Mulvey's female object, by legal rights. Emphasising the similarities with Mulvey's theory of looking, Acampora even goes so far as to describe zoos as "a form of pornography defined as visive violence", in the sense that they "engage a destructive desire in relation to the object of inspection" (2005: 71). The wildness of these 'wild' animals is cauterised in the service of a look that demands a docile body. In this sense, "the zoo to which people go to meet animals, to observe them, to see them, is, in fact, a monument to the impossibility of such encounters" (Berger 1977, in Acampora 2005: 71)—the impossibility of seeing them as anything but objects to which humans are entitled, mere "biotic entertainment" (Acampora 2005: 74).

Taking this idea of animals as 'biotic entertainment' further, I combine it with the demonstrated trend in dark tourism as touched on in Part III, and the elevation of requisite bravery and also cultural omnivorousness as associated emotions of distinction, to explain why the visibility of food animals is similarly productive of pleasure.

---

[11] Sturken and Cartwright's original phrase: "the conventions of popular narrative cinema are structured by a patriarchal unconscious, positioning women represented in films as objects of a 'male gaze'. In other words, Mulvey argued that Hollywood cinema offered images geared toward male viewing pleasure" (Sturken and Cartwright 2009: 76).

In her work on tourism and voyeurism, Debbie Lisle (2004) demonstrates how the pleasure derived from looking may also incorporate confronting and difficult subject matter as part of the viewer's quest for "authenticity and reality" (17). Drawing on Bauman's idea of the world as a giant theme park "commodified and packaged for the enjoyment of tourists" (ibid.), Lisle refers to the combination of "titillation and shame" (19), "terror and beauty" (17), desire, and repulsion that has characterised "the consumption of catastrophe" (14) since the time of the Gladiators in the Roman Colosseum, to public executions and now the trend in visiting sites of tragedy. She even notes how Ground Zero has slotted into "a normalized circuit of tourist consumption", being just one of many 'attractions' on a list to be ticked off in a "[b]een there done that" manner (9).

I argue that this same 'theme park' sensibility is at work with regard to animals, most obviously in the numerous spectacles designed specifically to amuse and excite, including racing, rodeos, bullfighting, the running of the bulls, and many others. In the same way, these have become experiences to be 'ticked off' the adventure list of many cultural omnivores. Less obviously, I include here the renewed interest in not just knowing, but seeing, how meat is produced and exactly how food animals are killed. Indeed Andrew (P) explicitly references the notion of tourism:

> People also want an experience, like there's a lot more agro-tourism[12] happening because people really want an experience, there is a calling to actually kind of be more in touch.

As my participants indicate, this has become a 'must see' spectacle for the ethical consumer, one that they "should" tick off at some point, essentially because it increases the pleasure associated with eating meat, or ensures that the association with pleasure remains in the face of increased

---

[12] Agritourism has been a popular topic in the tourism and agricultural literature for more than a decade, where it is examined from the perspective of economic development and sustainability—that is, its respective benefits and impacts (e.g. McGehee and Kim 2004; Carpio et al. 2008; Phillip et al. 2010). As far as I know, there has as yet been no sociological or more critical exploration of its constitution as a social practice as there has been for other forms of tourism—notably dark or thanatourism.

knowledge and raw visibility of production practices. The pleasure derives from the social distinction of being brave enough to face such 'reality', demonstrate an 'authentic' and 'respectful' approach to meat, and, more simply, being able to say, "I've done that" and gaining the associated social and moral approval. In this way, the emotional and sensory pleasures of eating food animals are heightened.

Interestingly, Lisle remarks that it is the public nature of the consumption of catastrophe that legitimises these experiences by moving them from the private to the public sphere (2004: 17), a view that Lyon agrees with (2006). When filtered through appropriate, codified, and almost ritualised kinds of public practices—visiting memorials, attending scenes of death (human or nonhuman), or visiting zoos, any problematic and discomforting feelings are more easily socialised and, I would argue, somewhat sanitised. What essentially emerges, usually by design, is a collectively authorised, socially approved narrative by which these 'spectacles' are made sense of. Hence, just as it is impossible to see the wild animal behind the zoo animal, as Berger remarked, it becomes impossible within these 'spectacles' to see the passive, docile animal behind the 'fighting' bull, or to see the pig, cow, sheep, or chicken behind the food animal. These animals are unable to be anything else, the highly constructed nature of their visibility (whether raw or benign) effectively "disempowering those before its gaze" (Sturken and Cartwright 2009: 76).

Transparency is, then, productive not only of power, as Foucault articulates, but also of pleasure. Transparency is equated with knowledge, knowledge with pleasure, and looking with pleasure to weave a rhizomatic nexus of power/knowledge/pleasure that translates visually to affirm and assert an entitlement that is normalised, unquestioned, and whose associated mechanisms of power are substantially hidden. As Foucault says, it is its capacity to induce pleasure and forms of knowledge, and to also "hide its own mechanisms", that "makes power hold good" (1978: 86, 1980: 119). Thus, to repeat Pachirat, "sight and sequestration exist symbiotically" (2011: 252) and it is for this reason that he questions the transformational potential of making the hidden visible (248): "This politics of sight ... must acknowledge the possibility that sequestration will continue even under conditions of total visibility" (255).

In sum, while Foucault describes the panopticon as a "technology of power designed to solve the problems of surveillance" (1980: 148), I suggest that in the case of food animals, the various synoptic technologies I have described are being used by producers and consumers of meat to manage and mediate a slightly different *perceived* 'problem' of surveillance. This being one where these same technologies, in the hands of animal advocates, promote an un- or differently encoded eye that identifies mechanisms of domination instead of benign relations of pastoral care. Under the "arrogant" (Frye 1983: 66) or entitled eye,[13] these differently encoded views gain little traction, but they do resist and challenge the dominant, normalised order and are therefore countered by a meat-friendly ethico-aesthetic of visibility to ensure animals' state of domination continues to remain hidden. However, I say *perceived* problem because further challenging the common assumption, and expectation, that visibility can achieve the work of uncoding is the fact that what is often referred to as a 'new visibility' in meat production and consumption (Parry 2010; Linne 2014) is more accurately a repackaged return to visibility. From this perspective, the period of 'sequestration' of production processes at the onset of industrialisation could be regarded as a circumstantial departure from visibility that is in the process of being reversed and repackaged.

4   Old as New: Visibility Repackaged with an Ethical Twist

The ethically charged 'new' visibility of food animals and meat production practices stands in contrast to earlier historical arrangements of, and encounters with, the visibility of animal slaughter and death (Arcari 2017b). Although largely a consequence of the organisation of community living and food provisioning, these arrangements and encounters

---

[13] In her volume *The Politics of Reality*, Marilyn Frye describes the arrogant eye as one that organises everything seen with reference to itself and its own interests: "Everything is either 'for me' or 'against me'" (1983, 67). Also: "The arrogant perceiver ... coerces the objects of his perception into satisfying the conditions his perception imposes" (ibid.).

were not only in connection to routine practices of meat production. They also characterised spectacles of more acceptable public entertainment and became, for many, an experience that was highly sought after—another form of manufactured visibility but with a different association with pleasure. Although this association is still in evidence today—for example in rodeos, and bullfighting arenas, and practices of animal baiting and fighting—these practices are regarded more problematically under the revised terms of today's repackaged visibility.

The primary function of the 'dark' experience, whether involving humans or animals, is thought to be the opportunity to confront the reality of one's own mortality within an everyday context where this is more routinely denied (Stone 2009: 24). This recalls the discussion in Part III in relation to requisite bravery where I noted how experiences that incite heightened emotions are sought after, especially those that confront death and dying (Stone and Sharpley 2008: 576). Drawing on Berger, Stone explains that death is associated with "those 'fateful moments' and 'marginal situations' whereby individuals have to confront problems which society has attempted to conceal from public consciousness" (28). He continues:

> Hence, if death and mortality are not dealt with by adequate confrontation mechanisms, not only will the individual have to face up to challenges of personal meaninglessness and a significant loss of ontological security, but the social framework as a whole becomes vulnerable to collapse into chaos. (2009: 29)

Again, concealment is being linked to potential chaos and existential crisis, while confrontation is linked to order and stability—recalling Foucault's "subjection by illumination". The power of transparency thus lies in its capacity to contain death and dying within a normalised, and normalising, system of knowledge. This system keeps the power grid (and a sense of ontological security) in place and prevents any glimpse of a beyond. Within the narrative of this knowledge system, death is "abstracted, intellectualized and depersonalized"—"socially neutralized" (Stone 2009: 31–32). It has also become increasingly mundane, familiar, and unproblematic (O'Brien 2012).

Since its sequestration during industrialisation, the repackaged return to the visibility of animal death has been normalised as well as neutralised. Conceived as a response to a demand for visual knowledge of death, the re-visibility of practices of meat production and especially animal slaughter can be seen as a continuation of the largely humanist project which seeks to reassure, enrich, and nourish a sense of the human at the centre, and in control, of everything. What has changed, however, is the added ethical twist in animals' visibility, shaped by shifts in understandings of consumption in response to the de-fetishisation of relations of production and provision, the foregrounding of animal rights, and the broader politicisation of food.

In *Why Look at Animals*, Berger (1992) describes animals as "raw materials" to be visually consumed in zoos or literally consumed as food. The burgeoning of new ways and opportunities to look at animals ethically—whether at zoo animals with a conservation eye, or food animals with a welfare eye—using a range of synoptic technologies—emphasises how persistently unproblematic these kinds of pre-coded visibility really are. What this then foregrounds is the ultimate synergy of power, knowledge, and pleasure in relation to food animals, whereby they can be looked at *ethically* before being eaten.

Integrating the preceding discussions of entitlement, respect, ethical visibility—both benign and raw, and the pleasure of looking—I now arrive at my notion of the entitled gaze. This clearly references Marilyn Frye's account of the arrogant eye and broader theorisations of the gaze, but it is an effort to more fully articulate the enduring essence, or persistent core at the heart of meat consumption and the use of animals as food that makes the perception of any other "that" beyond Foucault's grid of naturalised domination so seemingly impossible. In using the word 'entitlement', I am purposefully drawing on its positive connotations with privilege, right, proper due, and being deserving, in contrast with arrogance, which carries more clearly negative connotations. For the gaze that is most evident in my data is one that is constituted in decidedly positive, and importantly for Foucault's account of power, productive terms. My contention is that throughout evolutions in meat production practices, techniques, sensibilities (as both sense and emotion), ethics, and levels of visibility over the centuries and decades, what has remained constant is this entitled gaze directed at food animals.

I acknowledge that critiques of human-centrism shaped by the impacts of climate change and theorisations of the Anthropocene are certainly offering alternate systems of knowledge on the production of meat compared to pre-industrialisation. However, food animals remain at the centre of what are widely understood to be sustainable, equitable, nutritious, and efficient food systems, now and for the future (Arcari 2017a). The ethics being incorporated into each of these agendas are therefore persistently anthropocentric and they play a key role in shaping the repackaged constitutions of visibility that surround ethical meat. In other words, the visibility of food animals has been repackaged, but the entitlement of the gaze remains the same.

5   The Entitled Gaze

My definition of the entitled gaze is one that is suffused with a fundamental knowledge and sense of entitlement to use food animals. This systemic entitlement, as explained previously, is constituted through social practices permeated with the normalised regime of power/knowledge/pleasure in relation to animals and meat. The validating discourses, emotions of distinction, and ethico-aesthetics identified thus far are mechanisms of power that contribute to the maintenance and persistence of this regime. The gaze directed at food animals made ethically visible (in ways both benign and raw) therefore incorporates all the power, knowledge, and pleasure associated with looking at an objectified 'other' whose body and life one is ideologically and legally entitled to control. In this section, I explore this gaze in more detail. In particular, I show how it interacts with ethically repackaged constitutions of animal visibility and transparency in ways that are mutually reinforcing and also reinforcing of the associated meaty practices.

Although the exact definition of the gaze has not been problematised (to my knowledge), its usage, especially in feminist and colonial literature (Columpar 2002; Yancy 2008; Sturken and Cartwright 2009), suggests a meaning closer to that associated with 'regard' (consider; think of with a particular feeling or *in a specified way*; relate to; evaluate; judge; gaze steadily at *in a particular way*) than the more passive 'gaze' (to look steadily and intently, especially in admiration; or with great curiosity, interest, pleasure, or wonder; to fix the eyes in a steady intent look).[14] Beyond simply looking, the regard embodies and enacts certain implications for what or who is gazed at because of how they are considered, evaluated, judged, felt about, related to, and so on. As Victor Burgin says, "There can never be any question of 'just looking'" (2009: 118–119). The gaze, as it is used in theory, therefore denotes a broader, more suggestive and portentous 'sensing' than simply sight. Foucault's 1963 work *The Birth of the Clinic* is widely held to have introduced the theoretical concept of the gaze, which Lacan also took up in his 1973 volume *The Four Fundamental Concepts of Psychoanalysis*. Given that the translator of both of these works notes in a 1973 translation of Birth of the Clinic, "I have used the unusual 'gaze' for the common '*regard*'" (1973: vii), I suggest that it is the more nuanced meaning and intention of the regard that ought to be brought to an understanding of the gaze.

As well as describing the liminal point where a living being is sensed/known as belonging to a particular classification and not another, the gaze, as I conceive it, simultaneously draws from and re-enacts historical relations—it "always-already includes a history of the subject" (Burgin 2009: 119). These historical relations prescribe, and proscribe, the sort of life that is considered normal and acceptable for that being to have. Essentially, then, the gaze attaches a particular ontological pathway to the human or animal in question, which they then experience as a set of more or less direct limitations in what they are and are not allowed to do while they are alive. It is in this ontological sense that "animals only exist in the ways humans imagine them" (Cudworth, commenting on Tester, 2003: 163), but with very real-life, tangible implications, because, as Foucault says, "the gaze that sees is a gaze that dominates" (1973: 39). While non-living objects are, of course, looked at and sensed, the primary significance of the gaze lies in its capacity to objectify what is living and construe it/them as a commodity with a use and exchange value—whether people, animals, or the living environment (Urry 2005). Going one step further, the most imposing sense of the gaze, and indeed the sense in which it is used most often in theory, is as a look that can be returned. It is this

---

[14] Merriam-Webster; Oxford Dictionary; Dictionary.com.

capacity of the gaze, "armed … with its privileges and qualifications" (Foucault 1973: 51) that introduces the potential for some lives to be valued more than others, feel entitled where others are not, and to unproblematically assume the power of life and/or death over some lives but not others.

The gaze is therefore never neutral or, as Burgin said, "indifferent" (2009: 118). There is a sense in which the look or gaze is a further appropriation or devouring of a body over which the subject already claims ownership—an ownership defined by "the eye that knows and decides, the eye that governs" (Foucault 1973: 89). As Pick says, looking is "an act of mastery over the other" (2015a) and, in the case of food animals, the mastery demonstrated by the devouring gaze foreshadows the literal devouring of the animal's body. What Pick describes as "vibrant assemblages"[15] of animal, human, technology, bodily fluids, breath, and nature that occur during a slaughter, and which are increasingly being experienced first-hand, captured visually or described in graphic prose, convert the animal's death into a kind of "hypnotic aesthetic" (2015a). The act of killing, poetically re-cast or re-fetishised by this aesthetic, "presupposes unlimited access to animals" and highlights the "similarity between violence and looking that simply takes for granted the fact that animals are there for the taking" (2015a).

Looking at the violence of animal death, in all its raw vibrancy, and finding it fits comfortably with an accepted knowledge and sense of entitlement, is, I suggest, the ultimate expression of the systematised regime of power/knowledge/pleasure surrounding animal edibility. Practices involving ethical meat where visibility is understood as a further mark of distinction are therefore where this entitlement can manifest most clearly.

Both Maria (C) and Michael (C) spoke of becoming more intimately connected with food animals through participating in their death (see Part II). This intimacy is not achieved by reading about or seeing images/footage of slaughter practices. It is achieved through practices

---

[15] Elsewhere (Pick 2012), Pick takes issue with certain strands of posthumanist theory, specifically the work of Jane Bennett and more especially Donna Haraway. For this reason, I suspect her use of the term 'vibrant assemblages' is intended to be slightly disparaging.

that involve sharing the animal's physical space—hearing them, smelling them, seeing them, and even touching them. Participating in such raw visibility is pivotal to Maria's and Michael's understanding of a 'good' killing. As Pick observes, the intimacy of looking while killing "redeems instrumental relations between humans and animals" (Pick 2015a). Yet these intimate encounters, or vibrant assemblages, are always pursued to nourish and fortify the human observer, to confirm his or her own identity (Pick 2015b). Thus, the realities of human entitlement, unlimited access, and unquestioned control that are presupposed in the killing of animals for food are concealed behind an appealing ethical narrative where humans participate in the normalised act of killing, just as other animals do, in acknowledgement of their natural role in the food chain and with respect for those animals whose lives 'must' be taken, after already being 'given'.

Naturally, not everyone can raise and kill their own animals for food, or has the capacity to be as engaged in the process as they might like.[16] The bucolic narrative of the 'natural' is therefore not accessible to the majority of consumers who are reliant on the large-scale, mostly invisible, intensive meat industry. However, the synoptic technologies of websites, social media, film, labels, and CCTV, and merely the knowledge of such technologies, allow all consumers to participate in the increased visibility of production processes in whatever way they can and thereby appropriate a part of the ethical narrative while also contributing to its re-fetishisation. Visibility of, and participation in, the production, slaughter, and butchery of food animals merely expand the scope of a commodification where one no longer simply buys the meat, but may also own some part of the life, and his/her story, that provided it. Rather than turning everyone vegetarian, visibility is being used to enhance, expand, and refine the ethical remit of meat consumption in a direct and 'productive' response to the issues being associated with 'sequestered', industrialised production processes.

---

[16] I acknowledge that practices of ethical consumption, including those involving meat, have become reflective, and therefore reproductive, of systematised relations of class, gender, and 'race'. However, as noted in the Exclusions and Limitations section of Chap. 1, the parameters of the research process and the priorities of my research precluded me from including aspects of social location in my data collection activities.

Throughout his work, Foucault variously articulates the medical or clinical gaze, the disciplinary gaze, the institutional gaze, the sovereign gaze, and the normalising gaze, describing the latter as one "that makes it possible to qualify, to classify, and to punish" (1977: 184, also 1973, 1978). It is this already and always ordered aspect of the gaze, inflected (or infected) by "a permanent corpus of knowledge" (Foucault 1977: 190), that makes the quest for 'truth' via visibility doomed to fail. For visibility is already enrolled in technologies of 'truth'. It already plays an integral role in shaping, affirming, and maintaining understandings of what is real, normal, proper, and right. In her analysis of the gaze, Corinne Columpar notes the capacity for ethnographic cinema to close a geographical gap while at the same time

> manufactur[ing] the theoretical gap on which the power of the gaze rests and by which people from other cultures were rendered reassuringly distant, yet utterly knowable. (2002: 37)

Through such films, she says, viewers are taught "how to 'read bodies'" (ibid.). I argue that the visibility of animals used as food works in a similar manner by bridging a spatial gap—the much maligned disconnect between consumers and the origin of their food—while leaving the theoretical/ontological gap between humans and 'their' food animals firmly in place.

It is for this reason that I suggest Freeman and Tulloch's (2013) (and many others') faith in the power of animal liberation documentaries to deconstruct "barriers to seeing" (3) is overstated. These authors' focus on "a *violent* human/animal hierarchy" that is "covered in *dark recesses*" (4) and the "*violent* practices" of "industrial animal *suffering*" (5) neglects to consider that, using Foucault's terms, this form of resistance to normalised power relations can serve to productively redefine *the terms* by which food animals are dominated. This is because, "to a gaze forearmed by linguistics, things attain to existence only in so far as they are able to form the elements of a signifying system" (Foucault, 1989: 416). Consequently, a 'violent human/animal hierarchy covered in dark recesses' can become a benign human-animal relation that is transparent and perpetually visible, and similarly 'violent practices of industrial animal suffering' can become humane practices of animal care in a pastoral context. The manner in which these animals are used may become knowable, and they may become physically proximate and visible, but they are still, as food animals, a (cognitively) distant 'other'. As Pick (2012: 73) astutely observes:

> violence does not solely depend on mechanisms that distance and estrange, that frame lives in such a way as to render them ungrievable, but also and simultaneously on mechanisms that presuppose kinship and precariousness: violence is always also domestic violence.

While accepting that these documentaries have not necessarily brought about any radical change in the situation of animals, the authors maintain that undercover footage is a powerful means to establish the preconditions for such change. I remain sceptical, as do others (e.g. Aaltola 2015), especially given that many of my participants are familiar with such footage, and also with certain documentaries, and yet their so-called 'barriers' remain in place. Recent research supports this point, finding that meat consumers generally ignore online animal activist material, questioning its credibility and association with a vegan agenda that is "ignorant about 'real' farming conditions" (Buddle et al. 2018: 8). To constitute a more radical challenge to the use of animals as food, it is my contention that it is precisely the transparent, visible, supposedly benign, humane, and ethical models of this use that need to be more squarely critiqued. For it is here that entitlement reveals its staunchest resistance.

Through a regime of power/knowledge/pleasure, the bodies of food animals are 'read' in a particular way that has been socially, culturally, and economically normalised over centuries. Given this 'unalterable', 'unquestionable', and unlimited power over animals' lives and deaths, which is contained in, and exercised through, the entitled gaze, being able to observe what was previously invisible only emphasises their absolute accessibility, confirms their status as usable, and highlights the systemic relations of privilege that entitle us to that use. Repackaged ethical visibility is therefore highly problematic and not some more evolved, "clear-eyed" (Joyce C) way of seeing our world. To insist that visibility and transparency is the route to revealing the domination of animals is to fail to appreciate the full meaning and implication of 'the regard', and the extent to which:

> Power has its principle not so much in a person as in a certain concerted distribution of bodies, surfaces, lights, gazes; in an arrangement whose internal mechanisms produce the relation in which individuals are caught up. (Foucault 1977: 202)

The visibility of food animals is, as Foucault implies, part of a rhizomatic nexus of entitlement. It is therefore a visibility that predominantly invites voyeurism and observation (Gillespie 2016: 573) as opposed to Kelly Oliver's 'loving eye' (2001) and is therefore resistant to the sort of politicised witnessing described by Emel and Wolch (1998), Oliver (2004), and others since (Dave 2014; Gillespie 2016). For while Oliver (2004) describes witnessing and 'bearing witness to' as a way to challenge the objectifying gaze and overcome associated "dynamics of hierarchies, privilege, and domination" (79), the key to that lies in being open to "the infinite response-ability" of the subject (81). The problem I perceive with this is twofold. First, animals used as food are so tightly tethered to their designations, in multi-sensorial ways, that the response-able individual behind the gaze is not readily accessible. To be open to that individual requires that the normalised order also be challenged in some way; otherwise their visibility continues to be observed without being 'witnessed'. Second, the sort of emotional engagement or 'entangled empathy' that Oliver (2001) and also Gillespie (2016) describe as a necessary part of witnessing is not what is normally associated with these animals. Gillespie notes that it is 'unintelligible' in the space of the auction yard, and I have demonstrated how equally unintelligible it is in the ethical meatscape where a concern or 'love' for these animals and their treatment sits comfortably alongside the visibility of, and participation in, their slaughter. Vasile Stănescu (2013) mounts the same challenge to the rhetoric of love for commodified animals found in the locavore and humane meat movements.

In short, as indicated in the previous chapter, love, care, and other emotions are neither homogenous nor fixed states. They are understood differently and can have very different implications for their 'object' of focus depending on how those objects are already constituted, and the practices they are part of. This means that the love expressed for food animals is of an entirely different order to that commonly expressed for companion animals or significant other humans. The 'loving eye' that is critical and invites a loving response (Oliver 2001) cannot, therefore, be directed at what is regarded on any level as a food animal. Similarly, witnessing food animals, in the politically engaged sense, is largely inappropriate and unintelligible and is therefore excluded from, or policed from within, associated practices. As long as living others are ordered and interpreted from a purposeful point of view, then the gaze they are subject to is infused with entitlement and all its consequences. As bell hooks states (in relation to feminism), "there can be no love when there is domination" (2000: 103).

5.1   Conclusion

Practices involving ethical and sustainable meat show how discomfiting associations with the industrial treatment of animals have been translated, to varying degrees, into a carefully constructed kind of transparency, or "graspable materiality" (Pottinger 2013: 662), that neatly "inoculates" against these concerns (Pachirat 2015). A fundamental reason for this is the entitled gaze which is always encoded with, and in turn reinforces, the 'proper' order of things. Under the conditions of this gaze, increasing the visibility (both discursively and literally) of food animals merely emphasises and extends the unlimited access humans assume, as their natural right, over their lives and death. Moreover, perceived in ethico-aesthetic terms, visibility contributes to a more substantial masking of food animals' domination by augmenting 'ethical' mechanisms of power (that respect and care for the 'other') at the same time as the industrial mechanisms, which have become difficult to deny, are illuminated. When concealment, rather than the domination, of animals is identified as the problem, whether by animal advocates or by the meat industry, the use of visibility can work perversely to quell discomfort and resettle consumers in new and 'improved' practices of meat consumption. This emphasises the power that can be exercised through transparency, or how illumination can work as a mechanism of power that keeps animals subjected to a persistent state of domination (Foucault 1980).

In his analysis of human privilege in relation to animals, Acampora posits: "the oppressive society is that of human privilege and the dominant view of its injustice is anthropocentrism" (2016: 3). On this basis, I speculate whether it might be more accurate to refer to what I have dubbed the entitled gaze as simply 'the human gaze', pulling back and extending theorisations of the specifically male, colonial, and ethnographic gazes. This avoids what I suspect is the dubious question of whether a human gaze can ever be unentitled. It is from this viewpoint that I approach my conclusion. Drawing together the evidence, arguments, and discussions I have presented over these three main parts, I clarify what constitutes the grid that "by a glance, an examination, a language" (Foucault 1989: xxi) keeps animals persistently edible. After Foucault, I then ask what sort of animals might exist beyond this grid and how we might see them. Can we exit the panoptic/synoptic machine? Is there a territory where there are no cartographies of meat? Is "the purity of the unprejudiced gaze" (Foucault 1973: 195), and all that this would imply in terms of power/knowledge/pleasure, even possible?

References

Aaltola, E. (2015). Animal Suffering: Representations and the Act of Looking. *Anthrozoos, 27*(1), 19–31.

Acampora, R. (2005). Zoos and Eyes: Contesting Captivity and Seeking Successor Practices. *Society and Animals, 13*(1), 69–88.

Acampora, R. (2016). [Provocations from the Field] Epistemology of Ignorance and Human Privilege. *Animal Studies Journal, 5*(2), 1–20.

Adams, G. (2006, August 9). Ramsay Reduced to Tears as Pigs Go Underknife. *The Independent.*

Agamben, G. (1998). *Homo Sacer: Sovereign Power and Bare Life.* Palo Alto: Stanford University Press.

Arcari, P. (2017a). Normalised, Human-Centric Discourses of Meat and Animals in Climate Change, Sustainability and Food Security Literature. *Agriculture and Human Values, 34*(1), 69–86.

Arcari, P. (2017b). Perverse Visibilities? Foregrounding Non-Human Animals in 'Ethical' and 'Sustainable' Meat Consumption. *The Brock Review, 13*(1), 1–30.

Arcari, P. (2018). The Ethical Masquerade: (Un)Masking Mechanisms of Power Behind 'Ethical' Meat. In M. Phillipov & K. Kirkwood (Eds.), *Alternative Food Politics: From the Margins to the Mainstream* (pp. 169–189). Abingdon, UK: Routledge.

Ball, M., & Friedrich, B. (2009). *The Animal Activists' Handbook: Maximizing Our Positive Impact in Today's World*. New York: Lantern Books.

Barnett, C., Cloke, P., Clarke, N., & Malpass, A. (2005). Consuming Ethics: Articulating the Subjects and Spaces of Ethical Consumption. *Antipode, 37*(1), 23–45.

Berger, J. (1992). Why Look At Animals? In *About Looking* (pp. 1–14). London: Bloomsbury.

Blundell, G. (2016, October 15). Matthew Evans's for the Love of Meat Draws in Viewers. *The Australian*. Online. June 2017.

Buddle, E. A., Bray, H. J., & Ankeny, R. A. (2018). Why Would We Believe Them? Meat Consumers' Reactions to Online Farm Animal Welfare Activism in Australia. *Communication Research and Practice, 19*(2), 1–15.

Burgin, V. (2009). *Situational Aesthetics: Selected Writings by Victor Burgin* (A. Streitberger, Ed.). Leuven: Leuven University Press.

Carpio, C. E., Wohlgenant, M. K., & Boonsaeng, T. (2008). The Demand for Agritourism in the United States. *Journal of Agricultural and Resource Economics, 33*(2), 254–269.

Chrulew, M. (2012). Animals in Biopolitical Theory: Between Agamben and Negri. *New Formations, 76*, 53–68.

Clarke, N., Cloke, P., Barnett, C., & Malpass, A. (2008). The Spaces and Ethics of Organic Food. *Journal of Rural Studies, 24*(3), 219–230.

Columpar, C. (2002). The Gaze as Theoretical Touchstone. *Women's Studies Quarterly, 30*(1/2), 25–44.

Cook, I., & Crang, P. (1996). The World on a Plate: Culinary Culture, Displacement and Geographical Knowledges. *Journal of Material Culture, 1*(2), 131–153.

Cudworth, E. (2003). *Environment and Society*. London: Routledge.

Cudworth, E. (2011). *Social Lives with Other Animals*. Basingstoke; New York: Palgrave Macmillan.

Dave, N. (2014). Witness: Humans, Animals, and the Politics of Becoming. *Cultural Anthropology, 29*(3), 433–456.

Emel, J., & Wolch, J. (1998). Witnessing the Animal Moment. In J. Wolch & J. Emel (Eds.), *Animal Geographies: Place, Politics, and Identity in the Nature-Culture Borderlands* (pp. 507–531). New York: Verso Books.

Esposito, R. (2008). *Bíos: Biopolitics and Philosophy*. Minneapolis; London: University of Minnesota Press.

Evans, M. (2016). *For the Love of Meat*. Narrator and Co-author: Matthew Evans. Producer, Director and Co-author: Stephen Oliver. Distributor/Broadcaster: SBS One.

Fairlie, S. (2010). *Meat: A Benign Extravagance*. Hampshire: Chelsea Green Publishing.

Fearnley-Whittingstall, H. (2008, January 20). Poultry Is Not a Class Issue. *The Guardian*.

Foer, J. S. (2009). *Eating Animals*. New York; Boston; London: Little, Brown and Company.

Foucault, M. (1973). *The Birth of the Clinic: An Archaeology of Medical Perception* (R. D. Laing, Ed.). London and New York: Routledge.

Foucault, M. (1977). *Discipline and Punish: The Birth of the Prison*. New York: Vintage Books.

Foucault, M. (1978). *The History of Sexuality*. New York: Pantheon Books.

Foucault, M. (1980). *Power/Knowledge: Selected Interviews and Other Writings, 1972–1977*. New York: Pantheon Books.

Foucault, M. (1989). *The Order of Things*. London; New York: Routledge.

Foucault, M. (1997). *Society Must Be Defended* (M. Bertani & A. Fontana, Eds.). New York: Picador.

Freeman, C. P., & Tulloch, S. (2013). Was Blind but Now I See: Animal Liberation Documentaries' Deconstruction of Barriers to Witnessing Injustice. In A. Pick & G. Narraway (Eds.), *Screening Nature: Cinema Beyond the Human* (pp. 110–126). New York; Oxford: Berghahn Books.

Freud, S. (1955). *The Standard Edition of the Complete Psychological Works of Sigmund Freud*. London: The Hogarth Press.

Frye, M. (1983). In and Out of Harm's Way: Arrogance and Love. In *The Politics of Reality: Essays in Feminist Theory* (pp. 52–83). Santa Cruz: Crossing Press.

Genel, K. (2006). The Question of Biopower: Foucault and Agamben. *Rethinking Marxism, 18*(1), 43–62.

Gillespie, K. (2016). Witnessing Animal Others: Bearing Witness, Grief, and the Political Function of Emotion. *Hypatia, 31*(3), 572–588.

Glenn, C. (2017). Complicating the Theory of the Male Gaze: Hitchcock's Leading Men. *New Review of Film and Television Studies, 15*(4), 496–510.

Goodman, M. K. (2004). Reading Fair Trade: Political Ecological Imaginary and the Moral Economy of Fair Trade Foods. *Political Geography, 23*(7), 891–915.

Goodman, M. K., Maye, D., & Holloway, L. (2010). Ethical Foodscapes?: Premises, Promises, and Possibilities. *Environment and Planning A, 42*(8), 1782–1796.

Goodman, D., DuPuis, E. M., & Goodman, M. K. (2012). *Alternative Food Networks: Knowledge, Practice, and Politics*. Routledge.

Gunderson, R. (2013). Problems with the Defetishization Thesis: Ethical Consumerism, Alternative Food Systems, and Commodity Fetishism. *Agriculture and Human Values, 31*(1), 109–117.

Hayes-Conroy, A., & Hayes-Conroy, J. (2008). Taking Back Taste: Feminism, Food and Visceral Politics. *Gender, Place & Culture, 15*(5), 461–473.

Hayes-Conroy, A., & Hayes-Conroy, J. (2010). Visceral Difference: Variations in Feeling (Slow) Food. *Environment and Planning A, 42*(12), 2956–2971.

Holbrook, M. B., & Hirschman, E. C. (2015). Experiential Consumption. In D. T. Cook & J. M. Ryan (Eds.), *The Wiley Blackwell Encyclopedia of Consumption and Consumer Studies* (pp. 1–3). Oxford: John Wiley & Sons, Ltd.

hooks, b. (2000). *Feminism Is for Everybody: Passionate Politics*. London: Pluto Press.

Lewis, T. (2011). The Ethical Turn in Commodity Culture: Consumption, Care and the Other. *sic: Journal of Literature, Culture and Literary Translation*, Vol. 2. Online. March 2013.

Lewis, T., & Huber, A. (2015). A Revolution in an Eggcup? Supermarket Wars, Celebrity Chefs, and Ethical Consumption. *Food, Culture and Society: An International Journal of Multidisciplinary Research, 18*(2), 289–307.

Linne, T. (2014). Grazing the Green Fields of Social Media. In E. A. Cederholm, A. Bjorck, K. Jennbert, & A.-S. Lonngren (Eds.), *Exploring the Animal Turn: Human-Animal Relations in Science, Society and Culture* (pp. 19–32). Lund: The Pufendorf Institute for Advanced Studies.

Lisle, D. (2004). Gazing at Ground Zero: Tourism, Voyeurism and Spectacle. *Journal for Cultural Research, 8*(1), 3–21.

Lyon, D. (2006). 9/11. Synopticon, and Scopophilia: Watching and Being Watched. In R. V. Ericson & K. D. Haggerty (Eds.), *The New Politics of*

Surveillance and Visibility (pp. 35–54). Toronto: University of Toronto Press Incorporated.

Mayes, C. (2015). *The Biopolitics of Lifestyle: Foucault, Ethics and Healthy Choices.* London; New York: Routledge.

McGehee, N. G., & Kim, K. (2004). Motivation for Agri-tourism Entrepreneurship. *Journal of Travel Research, 43*(2), 161–170.

Mesirow, B. (2011, January 6). Kill it, Cook It, Eat It, Slaughtering in Your Living Room in HD. *LA Weekly.* Online. April 2018.

Micheletti, M., & Stolle, D. (2010). Vegetarianism - A Lifestyle Politics? In M. Micheletti & A. S. McFarland (Eds.), *Creative Participation: Responsibility-Taking in the Political World* (pp. 125–145). Boulder, CO; London: Paradigm Publishers.

Mulvey, L. (1989). *Afterthoughts on 'Visual Pleasure and Narrative Cinema' Inspired by King Vidor's Duel in the Sun (1946)* (pp. 29–38). London: Palgrave Macmillan.

Mulvey, L. (1999). Visual Pleasure and Narrative Cinema. In L. Braudy & M. Cohen (Eds.), *Film Theory and Criticism: Introductory Readings* (pp. 833–844). New York; Oxford: Oxford University Press.

Mulvey, L. (2001). Unmasking the Gaze: Some Thoughts on New Feminist Film Theory and History.

O'Brien, S. J. (2012). *Unnerving Images: Cinematic Representations of Animal Slaughter and the Ethics of Shock.* Doctoral Thesis, The Centre for Comparative Literature, University of Toronto. 274 pp.

Oliver, K. (2001). *Witnessing: Beyond Recognition.* Minneapolis, MN: University of Minnesota Press.

Oliver, K. (2004). Witnessing and Testimony. *Parallax, 10*(1), 78–87.

Oliver, K. (2017). The Male Gaze Is More Relevant, and More Dangerous, than Ever. *New Review of Film and Television Studies, 15*(4), 451–455.

Pachirat, T. (2011). *Every Twelve Seconds: Industrialized Slaughter and the Politics of Sight.* New Haven; London: Yale Agrarian Studies Series.

Pachirat, T. (2015). The Glass Walls Fallacy: Reflections from an Industrialized Kill Floor on the Promises and Pitfalls of Transparency. Keynote: Animal Publics: Emotions, Empathy, Activism. Australasian Animal Studies Association Conference, July 12–15, Melbourne, Australia.

Padva, G., & Buchweitz, N. (Eds.). (2014). *Sensational Pleasures in Cinema, Literature and Visual Culture: The Phallic Eye.* New York: Palgrave Macmillan.

Parry, J. (2010). *The New Visibility of Slaughter in Popular Gastronomy.* Masters Thesis, Cultural Studies: University of Canterbury.

Patterson, M., & Elliott, R. (2010). Negotiating Masculinities: Advertising and the Inversion of the Male Gaze. *Consumption, Markets & Culture, 5*(3), 231–249.

Phillip, S., Hunter, C., & Blackstock, K. (2010). A Typology for Defining Agritourism. *Tourism Management, 31*(6), 754–758.

Phillipov, M., & Goodman, M. K. (2017). The Celebrification of Farmers: Celebrity and the New Politics of Farming. *Celebrity Studies, 8*(2), 346–350.

Pick, A. (2012). Turning to Animals Between Love and Law. *New Formations, 76*, 68–86.

Pick, A. (2015a). *Vegan Cinema: Looking, Eating and Letting Be.* Conference Keynote Lecture. Sixth Australasian Animal Studies Association Conference: Animal Publics: Emotions, Empathy, Activism. July 12–15, Melbourne, Australia.

Pick, A. (2015b). Why not Look at Animals? *Necsus European Journal of Media Studies.* Spring, Online. November 2015. 17 pp.

Pottinger, L. (2013). Ethical Food Consumption and the City. *Geography Compass, 7*(9), 659–668.

Probyn, E. (2000). *Carnal Appetites: FoodSexIdentities.* London; New York: Routledge.

Rabinow, P., & Rose, N. (2006). Biopower Today. *BioSocieties, 1*(2), 195–217.

Rodosthenous, G. (2015). *Theatre as Voyeurism: The Pleasures of Watching.* New York: Springer.

Rose, N. (2009). *The Politics of Life Itself: Biomedicine, Power, and Subjectivity in the Twenty-First Century.* Princeton; Oxford: Princeton University Press.

Sassatelli, R. (2011). Interview with Laura Mulvey. *Theory, Culture and Society, 28*(5), 123–143.

Satya. (2006, October). The Satya Interview with Peter Singer. *Satya.* Online. July 2016.

Schor, J. B., & Fitzmaurice, C. J. (2015). Collaborating and Connecting: The Emergence of the Sharing Economy. In L. A. Reisch & J. Thogersen (Eds.), *Handbook of Research on Sustainable Consumption* (pp. 410–425). Cheltenham; Northampton, MA: Edward Elgar Publishing.

Smaill, B. (2014). New Food Documentary: Animals, Identification, and the Citizen Consumer. *Film Criticism, 39*(2), 79–102.

Stănescu, V. (2013). Why "Loving" Animals Is Not Enough: A Response to Kathy Rudy, Locavorism, and the Marketing of "Humane" Meat. *The Journal of American Culture, 36*(2), 100–110.

Stone, P. R. (2009). Making Absent Death Present: Consuming Dark Tourism in Contemporary Society. In R. Sharpley & P. R. Stone (Eds.), *The Darker Side of Travel* (pp. 23–38). Bristol; Buffalo; Toronto: Channel View Publications.

Stone, P., & Sharpley, R. (2008). Consuming Dark Tourism: A Thanatological Perspective. *Annals of Tourism Research, 35*(2), 574–595.

Sturken, M., & Cartwright, L. (2009). *Practices of Looking: An Introduction to Visual Culture*. New York; Oxford: Oxford University Press.

Urry, J. (2005). The Place of Emotions within Place. In J. Davidson, L. Bondi, & M. Smith (Eds.), *Emotional Geographies* (pp. 77–83). Aldershot; Burlington, VT: Ashgate Publishing Ltd.

Wadiwel, D. (2008). Three Fragments from a Biopolitical History of Animals: Questions of Body, Soul, and the Body Politics in Homer, Plato, and Aristotle. *Journal for Critical Animal Studies, 6*(1), 17–31.

Wiper, A. P. (2014, January 6). Danish Crown Slaughterhouse, Denmark. *alastairphilipwiper.com*. Online. April 2014.

Wolfe, C. (2012). *Before the Law: Humans and Other Animals in a Biopolitical Frame*. Chicago; London: University of Chicago Press.

Yancy, G. (2008). Colonial Gazing: The Production of the Body as 'Other'. *Western Journal of Black Studies, 32*(1), 1–15.

Zoonen, L. V. (1994). *Feminist Media Studies*. London; Thousand Oaks; New Delhi: SAGE Publications Ltd.

Part V

Conclusion

10

Undoing Cartographies of Meat

Food animals did not choose their form, and yet they have no meaning in the human world that is independent of it. They are afflicted with a space and time prescribed for them by the anthroparchic maps of their bodies that delineate a topography of edibility, usefulness, texture, and taste. It can be argued that the consumption of meat exemplifies Foucault's regime of power/knowledge/pleasure even more than human sexuality, around which he constructed his idea of "a system of legitimate knowledge and of an economy of manifold pleasures" (Foucault 1978: 72). For, as Palmer (2001: 351) observed, there can be no more extreme expression of power than when it "drops off" the spectrum into violence and death. To then go even further and consume the body of the 'other' is the ultimate literal and symbolic assertion of that power and privilege (hooks 1998). In Jamieson and Nadzam's (2015) anthroparchic playground of nature, food animals represent the pinnacle of our narcissistic capacity to extend ourselves and our desires. These carnal cartographies are not easy constructs to overturn. They get to the heart of animals' persistent meatification and meat's persistent, and increasing, edibility (Godfray et al. 2018; Von Massow et al. 2019).

© The Author(s) 2020
P. Arcari, *Making Sense of 'Food' Animals*,
https://doi.org/10.1007/978-981-13-9585-7_10

The emergence of ethical and sustainable meat, and various analogues of 'better' meat, provides a valuable opportunity to illuminate key structural features of the power grid that keeps animals firmly situated on the meat map. These alternatives can be seen as an almost direct response to efforts to highlight the unethical, unsustainable, and unhealthy aspects of mainstream meaty practices. Consequently, as well as being a tangible illustration that these latter efforts missed the point—if that point was to fundamentally challenge meat consumption and/or the use of animals it involves, practices involving ethical and sustainable meat can help identify what it would take to make that point more clearly. For in these re-articulations of 'better' meat production and consumption, associated practices are understood to be more ethical and sustainable not only in comparison with factory farming, but with not eating meat at all. There are clearly associations with meat and/or 'food' animals that transcend ethical, environmental, and health-related concerns—including those based on the most incontrovertible evidence. In undertaking this research, it was my contention that identifying and exploring these associations specifically with producers and consumers of ethical and sustainable meat would reveal the mechanisms of power by which the substantial part of animals' domination is kept hidden even when almost everything about associated practices seems increasingly laid bare, both discursively and visually.

With my participants, I have bored down through their ethical, environmental, social, and health-related concerns associated with the production and consumption of meat to uncover those rhizomatic, embodied mechanisms that insist on the fundamental 'rightness' and 'goodness' of meat. The productive mechanisms of animal edibility that have emerged include: validating discourses, which tend towards the more discursive aspects of Foucault's regime; emotions of distinction, which are more reflective of the regime's emotional and also sensory dimensions; and ethico-aesthetics, which constitute a kind of "aesthetics of existence" that take into account "the pleasure of the other" as well as the self (Foucault 1984: 343–346). In comparison to the validating discourses and distinctive emotions, which also have ethico-aesthetic aspects, the ethico-aesthetic mechanisms constitute parts of a more all-encompassing, life-enhancing existential framework that guides and legitimates my

Table 10.1 Mechanisms of power that maintain the persistent edibility of 'meat' and 'food' animals

Validating discourses	Emotions of distinction	Ethico-aesthetics
• The value of contingent life • Invoking 'nature' • The benevolent 'natural contract'	• Requisite bravery • Cultural omnivorousness	• Moral approval • Respect

participants' relationships to self and 'others' on a more notably visceral level—those 'others' being 'food' animals. These mechanisms, summarised in Table 10.1, constitute the major contribution of this book to understanding how consumers 'make (ethical) sense' of animals and meat, and more effectively challenging practices relating to their persistent edibility.

These mechanisms offer a more potentially destabilising starting point from which to trouble habitual ways of thinking and acting that involve food animals, with the long-term aim of denormalising their domination. Above all, they demonstrate that it is not that these producers and consumers are not informed, do not believe the evidence, or do not care about animals—assumptions that continue to fuel the efforts of animal advocacy organisations. There are bigger nuts to crack relating to cartographies of meat that are profoundly systemic and mostly invisible. This task is the focus of the first section of my conclusion. Based on the mechanisms of power I have identified, I suggest how these might be destabilised to instigate a more radical challenge to the edibility of animals, and, more fundamentally, to the anthropocentric orders that constitute them as such.

My broader conclusions launch from the implications of this challenge, and of Foucault's opening quote. With reference to my openly emancipatory agenda, I consider what a critically posthuman approach might mean for how food animals are then 'made sense of'. Acknowledging the diverse ways in which posthumanism is now conceived and deployed, I re-emphasise here Cudworth and Hobden's reading of the posthuman as denoting neither a world 'after humanity' nor one populated by modified and/or augmented transhumans, but rather a more-than-human world "where interspecies relationships are other than simple control, domination and exploitation" (2018: 7). This is an emancipatory posthumanism that is a direct response to the exceptionalist tendencies of humanism (ibid.: 8). I orient my discussion around my notion of the entitled or human gaze, arguing less for an expansion of this gaze in a sort of benevolent 'becoming with' animals, and rather for its retraction. This respects the possibility that, from a genuinely nonhuman-centric posthuman perspective, animals might rather make no human sense at all. Drawing on Rosi Braidotti, I suggest an alternative way of framing posthuman intentions that foregrounds life instead of the unrelentingly insistent human.

1   Beyond the Grid

What tools, then, might be directed at undoing cartographies of meat, and to what end? With these questions, I turn to my seventh and final objective to consider **whether food animals could be made sense of in other ways, or permitted to make no (human) sense, and ask what sort of dis-ordering of Foucault's nexus of power/knowledge/pleasure this would require.** I tackle these questions in reverse, starting with how to destabilise cartographies of meat.

1.1   Destabilising Cartographies of Meat

From the vantage point of a vegan heterotopia, I have traversed cartographies of meat using a purposively deterritorialising and critically posthuman Foucauldian lens with the aims of challenging the existing and persistent 'order of [animal] things' designated as food. This approach has allowed me to identify some key mechanisms of power, which serve as vital landmarks in the meatified territory, anchoring food animals to their orders of edibility. Destabilising these mechanisms is therefore essential for the production and consumption of meat, and the use of food animals, to be challenged in any radical and lasting way. The first step is to generate awareness of these mechanisms by highlighting when, where, and how they are being exercised. The second is to have rigorous, comprehensive strategies ready to challenge and refute them, not simply piecemeal responses as part of an apparently balanced discussion or debate. Such strategies might involve:

1. Asserting the value of *existing* lives above contingent lives and roundly critiquing the philosophical basis of that argument, drawing on the work of Visak (2013), Visak and Garner (2015), Salt (1914) and others.
2. Breaking associations of meat, food animals, and associated practices with conceptions of nature and 'naturalness'.
3. Demystifying and discrediting the notions of a benevolent, respectful natural contract and benign, mutually beneficial human/animal relations.
4. Problematising the social distinction of masculinised bravery and its association with practices involving food animals, especially killing.
5. Unpacking and challenging the social constitution of practices involving meat and food animals such that 'appropriate' emotional associations become those that prioritise the life that is taken above the life at risk of being offensive, offended, or excluded.

These strategies also highlight potential areas for further research, such as the role of socialised emotions in the constitution and reproduction of practices, and how these emotions are themselves constituted more broadly. This could include building on Wright's (2015) articulation of the 'threatening' vegan body, along with Ahmed's 'affect aliens' and Twine's 'vegan killjoys', to explore the abstract figures in which a bundle of (negative) emotions seem to become invested and almost solidified. Common conceptions and understandings of such 'inappropriate' figures were very evident across my participants' accounts, which suggests these (and other) shadowy spectres exert some agency in the constitution of practices that is yet to be fully recognised. In addition, the enduring association of certain masculinised emotions with meat and other forms of mastery over nature, and how that intersects with feminist priorities, need to be examined more closely. Indeed, in light of Parry's (2010) analysis of the New Carnivore movement, and particularly the feminisation of killing and butchery that he documents, the specific role that gender plays in the (re)constitution of meat and animals as universally persistent elements of everyday practices involving food is a significant topic for further empirical investigation. Stepping back further, and as noted in the introduction, interrogating the way social locations besides gender, such as socioeconomic status, religion, and 'race', intersect with practices involving 'better' meat—similarly conceived through the regime of power/knowledge/pleasure—would also offer valuable and much-needed critical insight into the ethical foodsphere in all its varied expressions and permutations.

In the case of the ethico-aesthetic mechanisms, it is admittedly much harder to devise a strategy that could disentangle constitutions of moral approval and respect. As I have illustrated, they are tied to something much deeper, more dispersed, and yet pivotal to all mechanisms of power by which animals continue to be dominated. I therefore propose, as the final strategy, first highlighting (because it is largely invisible) and then problematising the anthropocentric entitlement that permeates and shapes the ordering, territorialisation, meatification, mapping, control, modification, killing, knowing, feeling, sensing, and eating of animals' bodies. That this entitlement is socially constituted needs to be somehow foregrounded and then extracted from naturalised philosophies and cosmologies of existence that are inherently anthroparchal. Only then will there be the possibility of thinking what is currently unthinkable (Sorenson 2014), and seeing what remains persistently sequestered from the entitled gaze.

With the aim of mobilising material transformations in practices, subjectivities, and institutions—and moving from a politics of edibility to a politics of killability—the conceptual challenges outlined above could be deployed in a number of ways that would help mitigate the risk of interventions achieving merely piecemeal or individualised responses. For example, animal sanctuaries (providing shelter to 'wild' animals and animals ejected from the food, entertainment, research, and sport industries) could make clear and direct links to the multitude of practices that contribute to the existence of such sanctuaries. While a focus on individual animals and their stories is important, linking them to the broader mechanisms of power and systems of nonhuman commodification that shape them would magnify their disruptive potential. Thus, the rescued cow or pig is not simply one of the lucky ones who now enjoys a 'cushy' life within a normalised space that serves to absolve the deaths of the millions of others less 'lucky'. Rather, the naturalised orders and implicit human entitlement that create the conditions of life and death for all such animals can be more directly implicated and called into question. Similarly, pet rescue groups could highlight how the co-option of dogs, cats, and a vast range of domesticated and wild animals into globalised systems of demand, trade, and profit—systems that are underpinned by human entitlement—contributes to the ongoing production of 'refugees' from these systems, whose lives are, for whatever reason, no longer valued.

From these examples, a pattern should be emerging whereby a transient encounter with an individual, dislocated animal is used as an opportunity to create a 'moment of hesitation' or ambiguity in the normalised order by linking him/her to the everyday practices that brought them to their current situation. The same strategy could be adopted by animal advocacy organisations. As well as exposing conditions of mistreatment, cruelty, and abuse linked to specific practices (e.g. live export, animal testing), locations (e.g. factory farms, slaughterhouses), and animals (e.g. greyhounds), they could repeatedly put these conditions in the context of the mechanisms of power that constitute them. In this way, the 'problem' is not limited to just one practice, location, or animal, but is squarely, and repeatedly, located in the entitled human gaze that insists on commodifying these living 'natural' resources.

Only by chipping away at the epistemic construction of all nonhuman animals, and not just their individual/species circumstances, can full advantage be taken of those 'moments' where the opportunity arises to think about and relate to them differently. For this reason, whether promoting meat reduction or veganism, campaigns and interventions that focus on diet, health, and/or the environment may achieve some apparently measurable level of success but are built on unstable foundations because the systemic exploitation and domination of animals across interconnected nexuses of practices remain largely unacknowledged and unchallenged.[1] If the mechanisms that normalise the domination of food animals were simultaneously highlighted and disrupted—the validating discourses, the distinctive emotions, and the ethico-aesthetics—then the possibility of veganism persisting as a philosophical practice, instead of fading as a dietary trend, would make a vegan pantopia at least thinkable.

Until such time, and under the existing terms of an enduringly enti-tled, anthroparchic, or simply human gaze, it is still considered wholly natural, appropriate, and ethical to eat animals. It is also considered vari-ously natural, appropriate, and/or ethical to kill, hunt, watch, race, wear, breed, trap, track, cage, ride, touch, cull, poison, exterminate, train, break, punish, and abuse them, as well as use them for human health, 'education', experiment on them, and profit from them. It is not, how-ever, considered natural, appropriate, or ethical to do these things to humans[2] and that is the definition of speciesism. This illustrates the fun-damental, systemic dualism that a posthumanism that is true to its aspi-rations needs to recognise and address by being critical of just how 'post' it is claiming to be—what the critique of humanism extends to and includes, and what it excludes. In other words, referencing Cudworth (2011: 13), how wide is the circle of the social in social science?

Launching from this question, the rest of my conclusion takes its lead from Pick, who invites:

> the possibility that animals may not want to be looked at, or that in turn we may not have the right to look at them, or we may wish to look differ-ently at them. (2015a: Keynote)

This orients my attention to the first part of my final objective, which is to consider if/how food animals might be 'made sense' of in other ways, or be permitted to make no (human) sense?

---

[1] Recent EU, US, and Australian data showing overall increases in per capita meat consumption, primarily from chickens, supports my contention that the overall edibility of animals has not been significantly challenged by such campaigns and interventions that are more prevalent in Westernised countries (Ritchie and Roser 2017; Ritchie 2019; Taylor and Butt 2017).
[2] Although most of these things have been or are still done to human 'others' they are mostly, except for the last three, considered illegal.

1.2   Stop Making Sense

Is it possible to dis-order the existing order, un-map animals' bodies, and allow them to occupy a broader territory of meaning? The imaginings that these sorts of questions lead to are not entirely alien even to some of my participants. In Part II, I showed how Michael (C) imagines a future where food animals are not seen as something to kill and eat, and not even as pets, but "something that should be viewed, in nature … or something to paint, maybe to pat, maybe to ride on". Similarly, in Part III, Alison (P) demonstrates her preference for a future where "you can just have animals for pets, you wouldn't have to eat them". Comparing the gaze of a meat consumer with her own vegetarian gaze, Florence (P) describes the difference as follows, asking of a hypothetical meat eater:

> what are you thinking about that animal, 'oh, you're looking good for the dinner plate', versus you know, like I think about our cattle, how gorgeous are you and you might outlive me, whatever, that's great. You just enjoy being in each other's company with no guilt or anything, or anything using or abusing of the relationship.

However, even here, the anthroparchic gaze is alive and well in ascribing an albeit non-edible role to these animals that is still predicated on some pleasure-based reward for humans—via looking, painting, patting, riding, or otherwise relating in some way. Though Florence (C) does come closest to not exercising that gaze, even if not removing it altogether. As Pick's comment above indicates, the need for their existence to be humanly justified demonstrates just how deeply rooted the anthroparchic sense of entitlement really is. Would these animals be permitted such a non-edible existence if they were neither watchable, paintable, pat-able, ridable, nor relatable? Based on the following comment from (pescetarian) Sophie (C), I suspect not:

> I like cows, I don't want to eat them, yeah, I think it ended up being, if I could have a relationship with that animal potentially, I don't really want to eat it, so I guess that's why I'm comfortable eating seafood, cos yeah, you're not going to have any relations with a fish or prawn, but a cow or a roo, you know, you can get quite friendly with a cow or a roo.

For Sophie (C), the difference is clear. A food animal is either relatable or edible. Either way, it is making a sense that is decidedly human-centric.

It starts to become apparent just how welded the human gaze is to rhizomatically embodied orders of animal use, whereby different understandings of animals—as food, pet, entertainment, research, fashion—are associated with very different but interconnected regimes of power/knowledge/pleasure. In response, I want to explore the possibility of decoupling the gaze and breaking up these orders. Playing with Carol Adams' (2010 [1990]) notion of the absent referent, which describes the erasure of living animals in everyday discourses and practices involving meat, 'livestock', 'protein', and so on, I propose a sort of counter-mechanism in the form of a 'present deferral'—a caesura in the normalisation of food animals that, rather than absenting them, brings them into sharp focus, stripped of their existing orders and maps.

Such caesuras could be regarded as mechanisms of emancipation. Baa the sheep, from Chap. 4, is one example of a present deferral where the normalised edibility of the collectively understood 'sheep' was suspended as Baa, the individual subject of his owner's love, became more than her word and therefore a 'thing' quite out of order whose body had been divested of its traditional map. Baa and her owner created a little heterotopic counter-site that subverted the normal order. Because of this relinquishing of the worst expressions of power, for me, Baa represents a Foucauldian "glimmer" of the territory beyond the anthroparchic "grid of denominations" (1989: 175), much more so than, for example, Dolly the cloned sheep who sits squarely within this power grid. For Braidotti, Dolly "embodies complexity, this entity which is no longer an animal but yet not yet fully machine", and therefore is, for her, "the icon of the posthuman condition" (2013: 74). However, a posthumanism that celebrates and seeks greater incursions into animals' space and bodies, supposedly making us more than human at the same time as we make animals less than themselves, is not one that I think should be encouraged given the ideologies and practices that have led to the Anthropocene. This constitutes a continuation of the normalised order, not its subversion.

I acknowledge here (again) the connections that could be made in my thesis with Foucault's concept of biopower and also Agamben's notions of 'bare life' and the anthropological machine. However, I restate, and support, my focus on how meat and food animals are made sense of emotionally, sensorially, and viscerally as a way to better understand the persistence of these constitutions and their associated practices, with reference to the following observation by Wadiwel (2002: Online):

> whilst Agamben's analysis of bare life, and Foucault's theory of bio-power, provide a means by which to assess the condition of non-human life with respect to sovereign power, the political project must reach beyond these terms, and embrace an intertwining of the human and the non-human: an intersection which may be found in the animal life shared by both entities.

As discussed, caesuras are breaks, interludes, pauses, or ruptures where something is experienced or seen in a novel rather than a normal way—a suspension of regularity. Steve Baker notes that it is in pursuit of such breaks in normality that artists often focus on the 'abject', for its capacity to disturb "orderings of subject and society alike" (2000: 89). Drawing on philosopher Julia Kristeva, he describes the abject "as 'neither subject nor object', and as that which 'draws me toward the place where meaning collapses' " (Baker, citing Kristeva, 2000: 89). Food animals can be seen to occupy just such an ambiguous and liminal space between subject and object, where their ordered meanings can often collapse. Particularly in Part III, I discussed how animals are sometimes rendered emotionally inedible on account of their becoming relatable individuals. However, the rent caused by these ruptures rarely extends species-wide, and even more rarely to all food animals. If it does, it is eventually repaired and 'normality' returns after a few months or years. In fact, the repair often reinforces normality and decreases the likelihood and impact of further ruptures.

Escaped and individualised food animals become momentary transgressions subsumed within the dominant order. An obvious question is, then, how to extend the effect/affect and the duration of these caesuras? However, given that this question has occupied political and social scholars for decades, attentive to questions and 'places of otherness' (Foucault 1984) especially in relation to 'race', 'gender', gender identification, and sexuality, and given further that progress towards emancipation in these areas is still uneven and unpredictable, perhaps I need to set a more realistic goal for nonhuman animals.

Therefore, acknowledging the value of these "paradoxical spaces" (Hayes-Conroy and Hayes-Conroy 2008), or 'heterotopia'[3] (Foucault 1984) where "an alternative social ordering is performed" (Hetherington 1997: 40), I aim simply to embed an attentiveness to them, and their significance, in a critically posthuman approach to social studies of the Anthropocene and more especially the anthroparchy that supports it. In this way, and to paraphrase Douglas, it might be possible to force thinking into less habitual tracks (2002: 38).

My ultimate, unapologetic, but admittedly unrealistic, agenda is to see the expansion of heterotopic sites of veganism to the point of a vegan pantopia. To this very hopeful/less end, and based on my research and findings regarding mechanisms of power, I propose not only an attentiveness to caesuras, ruptures, paradoxical spaces, and heterotopia, but more especially to the entitlement of the more dominant, normalised gaze that they foreground. This is the critical aspect that is missing from most current conceptions of posthumanism.

The entitled, human gaze, like a laser beam of normalised power/knowledge/pleasure, needs to be powered down before any 'beyond' can be glimpsed through its glare; a beyond in which the 'thing' can be more than its word and more than its map. Pick (2015a: Keynote) provides the most eloquent, almost zen,[4] account of how to achieve this, which had a quite profound impact on my own outlook. It involves simply switching off this gaze entirely. As she explains, the possibility of "letting be" then arises—of purposefully forgoing "the automation and acceleration of the gaze" that denotes animals' unconditional availability. Recognising the structural ties that exist between acts of violence and acts of looking (Pick 2015a: Keynote), this perspective is what allows for "the possibility that animals may not want to be looked at, or that in turn we may not have

---

[3] To restate, Foucault's heterotopia set up "unsettling juxtapositions of incommensurate 'objects' which challenge the way we think, especially the way our thinking is ordered" (Hetherington 1997: 42). Heterotopia are thus "sites of all things displaced, marginal, novel or rejected, or ambivalent", and where "meaning is dislocated through a series of *deferrals* that are established between a signifier and a signified" (ibid.: 46; 43, emphasis added).

[4] In Zen Buddhism, instant enlightenment can be achieved via satori. Suzuki describes satori as "the sudden flashing into consciousness of a new truth hitherto undreamed of. [...] intellectually, it is the acquiring of a new viewpoint. The world now appears as if dressed in a new garment, which seems to cover up all the unsightliness of dualism" (1964: 65).

the right to look at them, or we may wish to look differently at them" (ibid.). In this respect, veganism is rather a rejection of the dominant, anthroparchic orders that decree that these animals' lives have no intrinsic value, that we are entitled to them and their bodies from the moment of their birth (and also before), and that their only purpose in existing is to provide food.

Retracting the entitled gaze altogether is thus the more radical response to issues of meat production and consumption than efforts to produce 'better' meat. It renounces the reliance on, and faith in, market solutions, and asks that humans forgo the economic opportunities that seeing animals as a resource have thus far provided. By rejecting the anthroparchic construction of unlimited access to animals, 'doing nothing' refutes the impulse to own, expose, control, and master, and respects the lives of food animals not through 'good killing' but by considering their lives as their own—a "notion of animal privacy that denies human eyes and their technological proxies unlimited access" (Pick 2015b: 2). This is the opposite of Jamieson and Nadzam's (2015) narcissist's playground.

However, entitlement could be the battlecry of the Anthropocene and I do not hold much hope that the state of domination in which animals exist will change any time soon. As Wadiwel remarks, "when we consider our relationship to animals, sovereignty appears to *precede* ethics" (2015: 22, emphasis in original). Nevertheless, attention to the entitlement of the human gaze will have broader benefits for scholarship, especially in those fields attempting to address dualisms or to move beyond them and advance anti-dualistic and posthuman approaches. Firstly, it highlights the fundamental shift in orientation that is required to truly de-centre the human and adopt an alternate stance that acknowledges the orders of value—the grids—that construct our world and our actions, and the worlds and actions of 'others', and succeeds in seeing beyond them. Secondly, awareness of this gaze leads to a recognition of the depth and scale of human entitlement, and this is a crucial element that needs to be factored into any attempt to understand or generate social change.

Anthroparchy holds human entitlement as sovereign—the right to take what is 'yours' as decreed by religious, natural, secular, and also juridical law. Nature, we have been socialised to believe, belongs to humans, to do with as we please. From food to chimeric organ donors, and from zoos to animal selfies,[5] the situation of animals is the ultimate affirmation of the totalising power of our boundless entitlement to consume both physically and visually. As Kingsolver demonstrates plainly and conclusively in her self-sufficiency memoir, "The farm-liberation fantasy simply reflects a modern cultural confusion about farm animals. They're human property, not just legally but biologically" (2008: 223). Consequently, simply asking from whose perspective a certain approach or action construed as 'co-shaping' or 'becoming with' could be said to be beneficial, respectful, nourishing, life-affirming, and ethical might furnish the corrective necessary to avoid the pitfalls of an uncritical posthumanism that is, as Cudworth says, politically weak, and serves to reinforce systemic systems of oppression and domination. In this regard, meat consumption, which Vint identifies as the "material basis of our culture", is "among the most problematic sites that must be addressed in any transformed vision of posthuman companion species" (2010: 44).[6]

Describing Braidotti's 2013 volume *The Posthuman* as a 'tour-de-force', Herbrechter goes on to define her posthuman figure as "embrac[ing] the risks that becoming-other-than-human brings" (2013: 2). Where, once again, it is a human project of becoming that is centred, I am contending that a necessarily critical posthumanism for the Anthropocene should actually be about pulling back, treading more lightly and considerately, instead of continuing full bore on a path of our own discovery, justified as some kind of enlightened post-dual aspiration. Unless we have been invited into other lives and worlds, a multi-species non-dualism is still a decisively human-directed and human-centred endeavour. Indeed, Braidotti refers to her 'cartography' of vital materialism as "a nomadic *zoe*-centred approach [that] connects human to non-human life so as to develop a comprehensive eco-philosophy of becoming" (2013: 104).

---

[5] This trend, which Bridgeman (2016) refers to as "cultural narcissism", has seen the death of dolphins, sharks, peacocks, snakes, and probably many other animals, and fuels a profitable tourist trade in exotic animals, notably lions and tigers, who are taken from the wild and later killed when they grow too big or aggressive to be used as photo props for tourists (Holloway 2016; Dearden 2014). Christina Best (2015) describes the selfie as a purposeful and, importantly, witnessed, extension of the self into the world, invited or not—one that provides the 'self' new ways "to explore and define his or her own self identify" (61).

[6] Vint is referring specifically to what she perceives as a miss-fire in the posthumanism of Braidotti, Haraway, and some others.

Instead of being attentive to and addressing existing maps, Braidotti is intent on creating another, even though, as Korzybski indicated, they are not the territory and serve to conceal the 'thing'. Perhaps, then, just as there are whiteness studies within critical 'race' studies and masculinity studies within feminist and 'gender' studies, there is a need for further human studies before a posthuman and postspecies world can be properly conceived and approached. Nevertheless, despite being littered with traces of the anthroparchic gaze, Braidotti's work offers much of value in my attempt to articulate how to 'stop making sense' of food animals.

2   (Rhizomatic) Non-conclusion

2.1   To Be Posthuman Is to Be…

How to curb the kleptocratic human impulse to identify and extract every bit of perceived value from the natural world, which we consider has been provided solely for our benefit, is the most salient but also the most intractable question for the Anthropocentric age. Finding ways to draw attention to, undermine, and dismantle the entitlement that supports the anthroparchic gaze in all its systemic and rhizomatic forms, as it is expressed and reproduced through social practices, may provide a new way of approaching it. In the previous section, I suggested how this might be achieved in relation to the mechanisms of power that support meaty practices. But what exactly should we focus on when this entitled gaze is switched off? If the human is de-centred, what is centred?

The posthuman still contains 'the human', and while it provides a useful way to introduce, and acclimatise to, the notion of de-centring the human, I suggest that it is time for a more radical and positive articulation of what it is, instead of what it is not. It is here that Braidotti's suggestion of a 'zoe-centred approach' resonates. Zoe means life, as in the fullness of all life, both biological and spiritual, or what I would regard as the immaterial. It is therefore inherently rhizomatic, alluding to what can only be a non-essentialist account of everything that life on earth involves.

I find it interesting in itself that I have struggled to find a way to represent this stance I am reaching to define—the opposite of narcissism, without falling prey to mawkish sentimentality or vague calls for more compassion, empathy, or an ethics of care. Words do not seem to exist for the territory I am trying to describe, like a room that does not yet have the language to furnish it, and so I risk using the wrong linguistic map and confusing the matter even further. Braidotti (2013) comes close in her discussions of a "zoe-centred egalitarianism" (71), and the process of "becoming-earth" (81), although lingering traces of the exercise, rather than removal, of the anthroparchic gaze in her work slightly mar the intent of these terms. However, a '*zoe*-centred' approach, or '*zoe*-ism', seems well suited to my aspiration for a critically posthuman research approach that regards all life, and all that contributes to the fullness of that life, as interconnected and co-constituted constituents of *zoe*.

Fig. 10.1 Chained 'pet' pig beside a café area in central Paris. Photo credit: Viva Sali (Jillian Gibb), 1979

Whether it is defined as 'zoe-ology', 'zoe-graphy', or zoe-ism', it is not sufficient to think through a zoe-centred approach; it must also be sensed—felt emotionally and sensorially. In removing all mention of the insistent human, 'zoe-ism' performs a subtle energetic shift by not fore-grounding thinking about how to be posthuman, but rather placing that thought within a frame that has already removed the human and put zoe at the centre. Perhaps in this way, orders can start to be dis-ordered, bro-ken down rhizomatically the same way the body breaks down enzymes. For my research has demonstrated that power does not just work affec-tively, as Anderson (2014: 27) suggests, but in ways that are more, less, and not only. Breaking down the embodied enzymes of anthroparchic orders of entitlement might be the catalyst that finally reveals animals' state of domination and eventually permits food animals to make no human sense. To this end, I propose consolidating and expanding studies directed at better understanding what it means to be human—critical human studies—so that posthuman thinking, and social practices, can together proceed on an altogether more emancipatory trajectory, one that recognises that our current orders are "not the only possible ones nor the best ones" (Foucault 1989: xxii).

References

Adams, C. J. (2010). *The Sexual Politics of Meat: A Feminist-vegetarian Critical Theory*. London; New York: Continuum International Publishing Group.

Anderson, B. (2014). *Encountering Affect: Capacities, Apparatuses, Conditions*. Farnham: Ashgate Publishing Limited.

Baker, S. (2000). *Postmodern Animal*. London: Reaktion Books.

Best, C. (2015). Narcissism or Self-Actualization? An Evaluation of 'Selfies' as a Communication Tool. In D. S. Coombs & S. Collister (Eds.), *Debates for the Digital Age: The Good, the Bad, and the Ugly of Our Online World* (pp. 55–76). Santa Barbara; Denver: Praeger.

Braidotti, R. (2013). *The Posthuman*. Cambridge; Malden: Polity Press.

Bridgeman, L. (2016, March 2). Dolphin Selfies and Performing Whales: How Our Cultural Narcissism Is Killing the Planet. *Sonar*. Online. March 2016.

Cudworth, E. (2011). *Social Lives with Other Animals*. Basingstoke; New York: Palgrave Macmillan.

Cudworth, E., & Hobden, S. (2018). *The Emancipatory Project of Posthumanism.* Routledge.

Dearden, L. (2014, July 30). Stop Taking 'Tiger Selfies' that Fund Animal Abuse, Charity Says. *The Independent.* Online. February 2017.

Douglas, M. (2002). *Purity and Danger: An Analysis of Concepts of Pollution and Taboo.* London; New York: Routledge Classics.

Foucault, M. (1978). *The History of Sexuality.* New York: Pantheon Books.

Foucault, M. (1984). *Foucault Reader* (P. Rabinow, Ed.). New York: Pantheon Books.

Foucault, M. (1989). *The Order of Things.* London; New York: Routledge.

Godfray, H. C. J., Aveyard, P., Garnett, T., Hall, J. W., et al. (2018). Meat Consumption, Health, and the Environment. *Science, 361*(6399), eaam5324.

Hayes-Conroy, A., & Hayes-Conroy, J. (2008). Taking Back Taste: Feminism, Food and Visceral Politics. *Gender, Place & Culture, 15*(5), 461–473.

Herbrechter, S. (2013, April). Rosi Braidotti (2013) The Posthuman. *Culture Machine: Reviews,* 1–13.

Hetherington, K. (1997). *The Badlands of Modernity: Heterotopia and Social Ordering.* London; New York: Routledge.

Holloway, K. (2016, February 27). 5 Times People Ruined Animals' Lives for a Selfie. *Alternet.* Online. February 2017.

hooks, b. (1998). Eating the Other: Desire and Resistance. In R. Scapp & B. Seitz (Eds.), *Eating Culture* (pp. 181–200). New York: State University of New York (SUNY) Press.

Jamieson, D., & Nadzam, B. (2015). *Love in the Anthropocene.* New York; London: OR Books.

Kingsolver, B. (2008). *Animal, Vegetable, Miracle: A Year of Food Life.* New York: HarperCollins.

von Massow, M., Weersink, A., & Gallant, M. (2019, March 12). Meat Consumption Is Changing But It's Not Because of Vegans. *The Conversation.* Online. April 2019.

Palmer, C. (2001). 'Taming the Wild Profusion of Existing Things'? A Study of Foucault, Power, and Human/Animal Relationships. *Environmental Communication: A Journal of Nature and Culture, 23,* 339–358.

Parry, J. (2010). *The New Visibility of Slaughter in Popular Gastronomy.* Masters Thesis, Cultural Studies: University of Canterbury.

Pick, A. (2015a). *Vegan Cinema: Looking, Eating and Letting Be.* Conference Keynote Lecture. Sixth Australasian Animal Studies Association Conference:

Animal Publics: Emotions, Empathy, Activism. July 12–15, Melbourne, Australia.

Pick, A. (2015b). Why not Look at Animals? *Necsus European Journal of Media Studies*. Spring, Online. November 2015. 17 pp.

Ritchie, H. (2019, February 4). Which Countries Eat the Most Meat? *BBC News*. Online. April 2019.

Ritchie, H., & Roser, M. (2017). Meat and Seafood Production & Consumption. *OurWorldInData.org*. Online. April 2018.

Salt, H. S. (1914). Logic of the Larder. *Henrysalt.co.uk*. Online. June 2016.

Sorenson, J. (2014). Thinking the Unthinkable. In J. Sorenson (Ed.), *Critical Animal Studies: Thinking the Unthinkable* (pp. xi–xxxiv). Toronto: Canadian Scholars' Press Inc.

Suzuki, D. T. (1964). *An Introduction to Zen Buddhism*. New York: Grove Press.

Taylor, E., & Butt, A. (2017, June 9). Three Charts On: Australia's Declining Taste for Beef and Growing Appetite for Chicken. *The Conversation*. Online. April 2019.

Vint, S. (2010). *Animal Alterity: Science Fiction and the Question of the Animal*. Liverpool: Liverpool University press.

Visak, T. (2013). *Killing Happy Animals: Explorations in Utilitarian Ethics*. Basingstoke; New York: Palgrave Macmillan.

Visak, T., & Garner, R. (2015). *The Ethics of Killing Animals*. New York; Oxford: Oxford University Press.

Wadiwel, D. J. (2002). Cows and Sovereignty: Biopower and Animal Life. *Borderlands e-Journal, 1*(2), 1–8.

Wadiwel, D. J. (2015). *The War Against Animals*. Leiden; Boston: Brill Rodopi.

Wright, L. (2015). *The Vegan Studies Project: Food, Animals, and Gender in the Age of Terror*. Athens, Georgia: University of Georgia Press.

References

Aaltola, E. (2015). Animal Suffering: Representations and the Act of Looking. *Anthrozoos, 27*(1), 19–31.

Acampora, R. (2005). Zoos and Eyes: Contesting Captivity and Seeking Successor Practices. *Society and Animals, 13*(1), 69–88.

Acampora, R. (2016). [Provocations from the Field] Epistemology of Ignorance and Human Privilege. *Animal Studies Journal, 5*(2), 1–20.

ACE. (2017). Length of Adherence to Vegetarianism. *Animal Charity Evaluators.* Online. Retrieved April 2018, from https://animalcharityevaluators.org/research/dietary-impacts/vegetarian-recidivism/.

Adams, C. J. (1991). Ecofeminism and the Eating of Animals. *Hypatia, 6*(1), 125–145.

Adams, C. J. (1994). *Neither Man Nor Beast: Feminism and the Defence of Animals.* New York: The Continuum Publishing Company.

Adams, G. (2006, August 9). Ramsay Reduced to Tears as Pigs Go Under Knife. *The Independent.*

Adams, C. J. (2010). *The Sexual Politics of Meat: A Feminist-vegetarian Critical Theory.* London; New York: Continuum International Publishing Group.

Agamben, G. (1998). *Homo Sacer: Sovereign Power and Bare Life.* Palo Alto: Stanford University Press.

Agamben, G. (2004). *The Open: Man and Animal.* Palo Alto: Stanford University Press.

© The Author(s) 2020
P. Arcari, *Making Sense of 'Food' Animals,*
https://doi.org/10.1007/978-981-13-9585-7

Agger, B. (1991). Critical Theory, Poststructuralism, Postmodernism. *Annual Review of Sociology, 17*, 105–131.

Ahmed, S. (2004a). Affective Economies. *Social Text, 22*(2), 117–139.

Ahmed, S. (2004b). *The Cultural Politics of Emotion*. New York; London: Routledge.

Ahmed, S. (2010a). Feminist Killjoys (And Other Willful Subjects). *The Scholar and Feminist Online, 8*(3), 1–10.

Ahmed, S. (2010b). Killing Joy: Feminism and the History of Happiness. *Signs: Journal of Women in Culture and Society, 35*(3), 571–594.

Ahmed, S. (2010c). *The Promise of Happiness*. Durham; London: Duke University Press Books.

Aird, H. (2017, July 5). Delay in Abattoir Animal Cruelty Investigation Angers Welfare Groups. *ABC News*.

Akbar, J. (2015, February 17). Revealed: Disturbing Footage of Pigs Struggling to Breathe as They're Killed by CO_2 Stunning Method Being Used by Supermarket Abattoirs. *The Daily Mail*. Online. April 2018.

Akerman, P. (2010, November 16). Ethical Beef Eaters Move against the Herd. *The Australian*. Online. January 2014.

Alberts, P. (2011). Responsibility Towards Life in the Early Anthropocene. *Angelaki: Journal of the Theoretical Humanities, 16*(4), 5–17.

Alder, J., Campbell, B., Karpouzi, V., Kaschner, K., et al. (2008). Forage Fish: From Ecosystems to Markets. *Annual Review of Environment and Resources, 33*(1), 153–166.

Alliez, E., & Massumi, B. (2014). Performing the Ethico-aesthetic Paradigm. *Performance Research, 19*(3), 15–26.

Alloun, E. (2015). Ecofeminism and Animal Advocacy in Australia: Productive Encounters for an Integrative Ethics and Politics. *Animal Studies Journal, 4*(1), 148–173.

Anderson, B. (2014). *Encountering Affect: Capacities, Apparatuses, Conditions*. Farnham: Ashgate Publishing Limited.

Anderson, E. C., & Barrett, L. F. (2016). Affective Beliefs Influence the Experience of Eating Meat. *PLoS One, 11*(8), 1–16.

Anon. (2014, February 1). China's Vegetarian Population Touches 50 Million: Report. *Times of India*. Online. February 2017.

Anon. (2016a, August 17). Study: The Lamb Ads Aren't Working as Aussies Turn Vego in Droves. *B&T Magazine*.

Anon. (2016b, August 15). The Slow But Steady Rise of Vegetarianism in Australia. *Roy Morgan Research*. Finding No. 6923. Online. February 2017.

Anon. (2018, April 2). Strong Early Results Claimed in Canada's Certified Sustainable Beef Model. *Beef Central.*

Appadurai, A. (1996). *Modernity at Large: Cultural Dimensions of Globalization.* Minneapolis; London: University of Minnesota Press.

Arcari, P. (2017a). Normalised, Human-Centric Discourses of Meat and Animals in Climate Change, Sustainability and Food Security Literature. *Agriculture and Human Values, 34*(1), 69–86.

Arcari, P. (2017b). Perverse Visibilities? Foregrounding Non-Human Animals in 'Ethical' and 'Sustainable' Meat Consumption. *The Brock Review, 13*(1), 1–30.

Arcari, P. (2018a). 'Dynamic' Nonhuman Animals in Theories of Practice: Views from the Subaltern. In C. Maller & Y. Strengers (Eds.), *Social Practices and Dynamic More-than-humans: Living Things, Unbounded Materials, and Automation.* Basingstoke; New York: Palgrave Macmillan.

Arcari, P. (2018b). The Ethical Masquerade: (Un)Masking Mechanisms of Power Behind 'Ethical' Meat. In M. Phillipov & K. Kirkwood (Eds.), *Alternative Food Politics: From the Margins to the Mainstream* (pp. 169–189). Abingdon, UK: Routledge.

Armstrong, P. (2002). The Postcolonial Animal. *Society and Animals, 10*(4), 413–419.

Asher, K., Green, C., Gutbrod, H., Jewell, M., et al. (2014). *Study of Current and Former Vegetarians and Vegans.*

Atwood, M. (1998). *The Edible Woman.* New York: Anchor Books.

Atwood, M. (2004). *Oryx and Crake.* New York: Anchor Books.

Bai, X., van der Leeuw, A., O'Brien, K., Berkhout, F., et al. (2016). Plausible and Desirable Futures in the Anthropocene: A New Research Agenda. *Global Environmental Change, 39*, 351–362.

Bailey, R. (2014, December 3). Impact of Livestock on Climate Change Cannot Be Ignored. *Chatham House.*

Bailey, R., Froggatt, A., & Wellesley, L. (2014). *Livestock – Climate Change's Forgotten Sector* (Chatham House: Research Paper). London: The Royal Institute of International Affairs.

Baka, J. (2013). The Political Construction of Wasteland: Governmentality, Land Acquisition and Social Inequality in South India. *Development and Change, 44*(2), 409–428.

Baker, S. (2000). *Postmodern Animal.* London: Reaktion Books.

Ball, M., & Friedrich, B. (2009). *The Animal Activists' Handbook: Maximizing Our Positive Impact in Today's World.* New York: Lantern Books.

Banerji, D., & Paranjape, M. R. (2016). *Critical Posthumanism and Planetary Futures*. New Delhi: Springer (India).

Banet-Weiser, S. (2012). *AuthenticTM: The Politics of Ambivalence in a Brand Culture*. New York: NYU Press.

Barad, K. (2003). Posthumanist Performativity: Toward an Understanding of How Matter Comes to Matter. *Signs: Journal of Women in Culture and Society, 28*(3), 801–831.

Barad, K. (2007). *Meeting the Universe Halfway: Quantum Physics and the Entanglement of Matter and Meaning*. Durham; London: Duke University Press.

Barford, V. (2014, February 17). The Rise of the Part-time Vegans. *BBC News*. Online. July 2016.

Barker, T., Bashmakov, I., Bernstein, L., Bogner, J. E., et al. (2007). Technical Summary. In B. Metz, O. R. Davidson, P. R. Bosch, R. Dave, et al. (Eds.), *Climate Change 2007: Mitigation. Contribution of Working Group III to the Fourth Assessment Report of the Intergovernmental Panel on Climate Change*. Cambridge; New York: Cambridge University Press.

Barnett, C., Cafaro, P., & Newholm, T. (2005a). Philosophy and Ethical Consumption. In R. Harrison, T. Newholm, & D. Shaw (Eds.), *The Ethical Consumer* (pp. 11–24). London: SAGE Publications Ltd.

Barnett, C., Cloke, P., Clarke, N., & Malpass, A. (2005b). Consuming Ethics: Articulating the Subjects and Spaces of Ethical Consumption. *Antipode, 37*(1), 23–45.

Barnett, C., Clarke, N., Cloke, P., & Malpass, A. (2005c). The Political Ethics of Consumerism. *Consumer Policy Review, 15*(2), 45–51.

Bar-On, Y. M., Phillips, R., & Milo, R. (2018). The Biomass Distribution on Earth. *Proceedings of the National Academy of Sciences, 115*(25), 6506–6511.

Baroni, L., Cenci, L., Tettamanti, M., & Berati, M. (2006). Evaluating the Environmental Impact of Various Dietary Patterns Combined with Different Food Production Systems. *European Journal of Clinical Nutrition, 61*(2), 279–286.

Bastian, B., & Loughnan, S. (2016). Resolving the Meat-Paradox: A Motivational Account of Morally Troublesome Behavior and Its Maintenance. *Personality and Social Psychology Review, 21*(3), 278–299.

Bastian, B., Loughnan, S., Haslam, N., & Radke, H. R. M. (2012). Don't Mind Meat? The Denial of Mind to Animals Used for Human Consumption. *Personality and Social Psychology Bulletin, 38*(2), 247–256.

Bauman, Z. (2002). *Society under Siege*. Oxford; Malden: Polity Press.

Bauman, Z. (2005). *Living in Utopia.* Presentation: London School of Economics. 27 October.

Bauman, Z., & May, T. (2001). *Thinking Sociologically.* Oxford; Malden: Wiley Blackwell.

Beedie, P., & Hudson, S. (2003). Emergence of Mountain-Based Adventure Tourism. *Annals of Tourism Research, 30*(3), 625–643.

Beef Central. (2018, April 2). *Strong Early Results Claimed in Canada's Certified Sustainable Beef Model.* Beef Central.

Beekman, V. (2006). Feeling Food: The Rationality of Perception. *Journal of Agricultural and Environmental Ethics, 19*(3), 301–312.

Beirne, P. (2004). From Animal Abuse to Interhuman Violence? A Critical Review of the Progression Thesis. *Society and Animals, 12*(1), 39–65.

Bekoff, M. (2007). *Animals Matter: A Biologist Explains Why We Should Treat Animals with Compassion and Respect.* Boston, MA: Shambhala Publications.

Bennett, J. (2010). *Vibrant Matter: A Political Ecology of Things.* Durham; London: Duke University Press Books.

Bennett, C. E., Thomas, R., Williams, M., Zalasiewicz, J., et al. (2018). The Broiler Chicken as a Signal of a Human Reconfigured Biosphere. *Royal Society Open Science, 5*(12), 180325.

Bentley, T. (2015). *Empires of Remorse: Narrative, Postcolonialism and Apologies for Colonial Atrocity.* London; New York: Routledge.

Benton, T. (1993). *Natural Relations: Ecology, Animal Rights and Social Justice.* London: Verso.

Berger, J. (1992). Why Look At Animals? In *About Looking* (pp. 1–14). London: Bloomsbury.

Berger, R. (2014). Now I See It, Now I Don't: Researcher's Position and Reflexivity in Qualitative Research. *Qualitative Research, 15*(2), 219–234.

Bergqvist, J., & Gunnarsson, S. (2013). Finfish Aquaculture: Animal Welfare, the Environment, and Ethical Implications. *Journal of Agricultural and Environmental Ethics, 26*(1), 75–99.

Best, C. (2015). Narcissism or Self-Actualization? An Evaluation of 'Selfies' as a Communication Tool. In D. S. Coombs & S. Collister (Eds.), *Debates for the Digital Age: The Good, the Bad, and the Ugly of Our Online World* (pp. 55–76). Santa Barbara; Denver: Praeger.

Beverland, M. B., & Farrelly, F. J. (2010). The Quest for Authenticity in Consumption: Consumers' Purposive Choice of Authentic Cues to Shape Experienced Outcomes. *Journal of Consumer Research, 36*(5), 838–856.

Beverland, M. B., Lindgreen, A., & Vink, M. W. (2008). Projecting Authenticity Through Advertising: Consumer Judgments of Advertisers' Claims. *Journal of Advertising, 37*(1), 5–15.

Beverley, J. (1999). *Subalternity and Representation: Arguments in Cultural Theory.* Durham; London: Duke University Press.

Biermann, F., Bai, X., Bondre, N., Broadgate, W., et al. (2016). Down to Earth: Contextualizing the Anthropocene. *Global Environmental Change, 39,* 341–350.

Bittman, M. (2013). *VB6: Eat Vegan Before 6:00 to Lose Weight and Restore Your Health … for Good.* New York: Hachette Books.

Black, J. (2009, October 28). The Kinder Side of Veal. *The Washington Post.*

Black, L. L., & Stone, D. (2005). Expanding the Definition of Privilege: The Concept of Social Privilege. *Journal of Multicultural Counseling and Development, 33*(4), 243–255.

Blair, O. (2017, October 18). 6 Former Vegetarians Explain Why They Started Eating Meat Again. *Cosmopolitan.*

Blakemore, S. (2011). Potbellied Pigs: The Right Pet for You? *petful.com.* Online. April 2018.

Bluestone, G. (2014, December 15). Cow Dies Hero's Death After Escaping Slaughterhouse. *Gawker.* Online. October 2016.

Blundell, G. (2016, October 15). Matthew Evans's for the Love of Meat Draws in Viewers. *The Australian.* Online. June 2017.

de Boo, J. (2016, May 19). How Many Vegans? One of the Fastest Growing Lifestyle Movements. *The Huffington Post.* Online. October 2016.

Borges, J. L. (1964). *The Analytical Language of John Wilkins. In Other Inquisitions 1937–1952* (pp. 101–105). Austin: University of Texas Press.

Bost, J. (2012, May 3). The Ethicist Contest Winner: Give Thanks for Meat. *The New York Times.* Online. May 2013.

Braidotti, R. (2013). *The Posthuman.* Cambridge; Malden: Polity Press.

Bramble, B., & Fischer, B. (2015). *The Moral Complexities of Eating Meat.* Oxford: Oxford University Press.

Braschel, N., & Posch, A. (2013). A Review of System Boundaries of GHG Emission Inventories in Waste Management. *Journal of Cleaner Production, 44*(April), 30–38.

Bridgeman, L. (2016, March 2). Dolphin Selfies and Performing Whales: How Our Cultural Narcissism Is Killing the Planet. *Sonar.* Online. March 2016.

Brophy, B. (1965, October 10). The Rights of Animals. *The Sunday Times.* Online. April 2015.

Brown, C. (2015). Fish Intelligence, Sentience and Ethics. *Animal Cognition, 18*(1), 1–17.

Brymer, E., & Schweitzer, R. (2013). Extreme Sports Are Good for Your Health: A Phenomenological Understanding of Fear and Anxiety in Extreme Sport. *Journal of Health Psychology, 18*(4), 477–487.

Buddle, E. A., Bray, H. J., & Ankeny, R. A. (2018). Why Would We Believe Them? Meat Consumers' Reactions to Online Farm Animal Welfare Activism in Australia. *Communication Research and Practice, 19*(2), 1–15.

Bujok, M. (2013). Animals, Women and Social Hierarchies: Reflections on Power Relations. *Deportate, esuli, profughe, 23*, 32–48.

Burgin, V. (2009). *Situational Aesthetics: Selected Writings by Victor Burgin* (A. Streitberger, Ed.). Leuven: Leuven University Press.

Burke, K. (2011, May 31). Shocking Slaughterhouse Abuse Sparks Investigation - Comments. *The Sydney Morning Herald*. Online. January 2014.

Burnett, C. (2001). Whose Game Is It Anyway? Power, Play and Sport. *Agenda, 16*(49), 71–78.

Butler, J. (1990). *Gender Trouble: Feminism and the Subversion of Identity*. London; New York: Routledge.

Butler, J. (1993). *Bodies that Matter: On the Discursive Limits of 'Sex'*. London; New York: Routledge.

Butler, J. (2000). Subjects of Sex/Gender/Desire. In S. Saguaro (Ed.), *Psychoanalysis and Woman: A Reader* (pp. 309–322). New York: NYU Press.

Calarco, M. (2008). Facing the Other Animal: Levinas. In *Zoographies: The Question of the Animal from Heidegger to Derrida* (pp. 55–78). New York; Chichester: Columbia University Press.

Calvino, I. (1983). *Mr Palomar*. San Diego; New York; London: A Harvest Book.

Campbell, S. (1994). Being Dismissed: The Politics of Emotional Expression. *Hypatia, 9*(3), 46–65.

Cannane, S. (2012, February 9). NSW Abattoir Closed over Slaughter Practices. *ABC News Lateline*. Online. January 2014.

Cansdale, D. (2018, June 17). Ethically-Sourced Meat: Kill Your Own Pig and Reap the Tasty Reward. *ABC News*.

Capiola, A., & Raudenbush, B. (2012). The Effects of Food Neophobia and Food Neophilia on Diet and Metabolic Processing. *Food and Nutrition Sciences, 3*, 1397–1403.

Carlet, J., Jarlier, V., Harbarth, S., Voss, A., et al. (2012). Ready for a World without Antibiotics? The Pensières Antibiotic Resistance Call to Action. *Antimicrobial Resistance and Infection Control, 1*(1), 11.

Carlyon, P. (2016, October 14). Tasmanian Abattoir Investigated over Animal Cruelty Claims. *ABC News*. Online. February 2017.

Carnicelli-Filho, S., Schwartz, G. M., & Tahara, A. K. (2010). Fear and Adventure Tourism in Brazil. *Tourism Management, 31*(6), 953–956.

Carpio, C. E., Wohlgenant, M. K., & Boonsaeng, T. (2008). The Demand for Agritourism in the United States. *Journal of Agricultural and Resource Economics, 33*(2), 254–269.

Carrier, J. G. (2008). Think Locally, Act Globally: The Political Economy of Ethical Consumption. *Research in Economic Anthropology, 28*, 31–51.

Carson, R. (1962). *Silent Spring.* Boston; New York: Houghten Mifflin.

Castelfranchi, C., & Miceli, M. (2011). Anticipation and Emotion. In P. Petta, C. Pelachaud, & R. Cowie (Eds.), *Emotion-Oriented Systems. Cognitive Technologies* (pp. 483–500). Heidelberg: Springer Berlin.

Castree, N., & Nash, C. (2004). Mapping Posthumanism: An Exchange. *Environment and Planning A, 36*(8), 1341–1363.

Castricano, J., & Simonsen, R. R. (2016). *Critical Perspectives on Veganism.* Cham, Switzerland: Palgrave Macmillan.

Cavalieri, P. (2008). A Missed Opportunity: Humanism, Anti-humanism and the Animal Question. In J. Castricano (Ed.), *Animal Subjects: An Ethical Reader in a Posthuman World* (pp. 97–123). Waterloo, ON: Wilfrid Laurier University Press.

Caviola, L., Everett, J. A. C., & Faber, N. S. (2018). The Moral Standing of Animals: Towards a Psychology of Speciesism. *Journal of Personality and Social Psychology, 116*(6), 1011–1029.

CCA. (2013). *Beef 2015 & Beyond.* Cattle Council of Australia. Report.

Chambers, P. G., & Grandin, T. (2001). *Guidelines for Humane Handling, Transport and Slaughter of Livestock.* Food and Agricultural Organisation of the United Nations and the Humane Society International. Regional Office for the Asia Pacific.

Chan, E. Y., & Zlatevska, N. (2019). Jerkies, Tacos, and Burgers: Subjective Socioeconomic Status and Meat Preference. *Appetite, 132*(1), 257–266.

Chasan, A. (2015, June 1). Cow Escapes Slaughterhouse, Later Shot and Killed. *New York Daily.* Online. October 2016.

Chau, A. Y. (2008). The Sensorial Production of the Social. *Ethnos, 73*(4), 485–504.

Chemnitz, C., & Becheva, S. (Eds.). (2014). *Meat Atlas: Facts and Figures about the Animals We Eat.* Berlin; Brussels: Heinrich Boll Stiftung and Friends of the Earth Europe.

Chiles, R. M. (2013). Intertwined Ambiguities: Meat, *in Vitro* Meat, and the Ideological Construction of the Marketplace. *Journal of Consumer Behaviour, 12*(6), 472–482.

Chrulew, M. (2012). Animals in Biopolitical Theory: Between Agamben and Negri. *New Formations, 76,* 53–68.

Chrulew, M., & Wadiwel, D. (2016). Editors' Introduction: Foucault and Animals. In *Foucault and Animals* (pp. 1–18). Brill: Leden.

Citrin, L. B., Roberts, T.-A., & Fredrickson, B. L. (2004). Objectification Theory and Emotions: A Feminist Psychological Perspective on Gendered Affect. In L. Z. Tiedens & C. W. Leach (Eds.), *The Social Life of Emotions* (pp. 203–226). Cambridge; New York: Cambridge University Press.

Clarke, N., Cloke, P., Barnett, C., & Malpass, A. (2008). The Spaces and Ethics of Organic Food. *Journal of Rural Studies, 24*(3), 219–230.

Clune, S., Crossin, E., & Verghese, K. (2017). Systematic Review of Greenhouse Gas Emissions for Different Fresh Food Categories. *Journal of Cleaner Production, 140*(Part 2), 766–783.

Clymer, J. (2012). *Family Money: Property, Race, and Literature in the Nineteenth Century.* Oxford; New York: Oxford University Press.

Coghlan, S. (2016). Moral Individualism and Relationalism: A Narrative-Style Philosophical Challenge. *Ethical Theory and Moral Practice, 19*(5), 1241–1257.

Cole, M. (2011). From "Animal Machines" to "Happy Meat"? Foucault's Ideas of Disciplinary and Pastoral Power Applied to 'Animal-Centred' Welfare Discourse. *Animals, 1*(4), 83–101.

Cole, M., & Stewart, K. (2014). *Our Children and Other Animals.* London; New York: Routledge.

Coles. (2011, April). RSPCA Approved Chicken to Hit Shelves in Victoria, South Australia and Tasmania for the First Time. *Coles News.* Online. June 2015.

Columpar, C. (2002). The Gaze as Theoretical Touchstone. *Women's Studies Quarterly, 30*(1/2), 25–44.

Comninou, M. (1995). Speech, Pornography, and Hunting. In C. J. Adams & J. Donovan (Eds.), *Animals and Women: Feminist Theoretical Explorations* (pp. 126–148). Durham; London: Duke University Press Books.

Conaty, D. (1998, January 17). Flying Pigs Saved from the Abattoir. *The Irish Times,* Online. October 2016.

Conway, A. (2015, November 23). Global Egg Consumption to Rise Worldwide through 2024. *WATTAgNet.* Online. October 2017.

Cook, I., & Crang, P. (1996). The World on a Plate: Culinary Culture, Displacement and Geographical Knowledges. *Journal of Material Culture, 1*(2), 131–153.

Craig, W. J., & Mangels, A. R. (2009). Position of the American Dietetic Association: Vegetarian Diets. *Journal of the American Dietetic Association, 109*(7), 1266–1282.

Crang, P. (1996). Displacement, Consumption and Identity. *Environment and Planning A, 28,* 47–67.

Crawford, E. (2015, March 17). Vegan Is Going Mainstream, Trend Data Suggests. *Food navigator.com.* October 2016.

Cruickshank, B. (2014). Review: Disciplining the Poor: Neoliberal Paternalism and the Persistent Power of Race. *Contemporary Political Theory, 13*(1), e1–e3.

Crutzen, P. J. (2002). Geology of Mankind. *Nature, 415*(3 January), 23.

Cudworth, E. (2003). *Environment and Society.* London: Routledge.

Cudworth, E. (2005). *Developing Ecofeminist Theory: The Complexity of Difference.* Basingstoke; New York: Palgrave Macmillan.

Cudworth, E. (2008). Most Farmers Prefer Blondes: "The Dynamics of Anthroparchy in Animals" Becoming Meat. *Journal for Critical Animal Studies, 6*(1), 32–45.

Cudworth, E. (2010). 'The Recipe for Love'? Continuities and Changes in the Sexual Politics of Meat. *Journal for Critical Animal Studies, 8*(4), 78–100.

Cudworth, E. (2011). *Social Lives with Other Animals.* Basingstoke; New York: Palgrave Macmillan.

Cudworth, E. (2014). Beyond Speciesism: Intersectionality, Critical Sociology and the Human Domination of Other Animals. In N. Taylor & R. Twine (Eds.), *The Rise of Critical Animals Studies: From the Margins to the Centre* (pp. 19–35). Taylor & Francis.

Cudworth, E., & Hobden, S. (2014a). Civilisation and the Domination of the Animal. *Millennium - Journal of International Studies, 42*(3), 746–766.

Cudworth, E., & Hobden, S. (2014b). Liberation for Straw Dogs? Old Materialism, New Materialism, and the Challenge of an Emancipatory Posthumanism. *Globalizations, 12*(1), 134–148.

Cudworth, E., & Hobden, S. (2018). *The Emancipatory Project of Posthumanism.* Routledge.

Cutcliffe, J. R. (2003). Reconsidering Reflexivity: Introducing the Case for Intellectual Entrepreneurship. *Qualitative Health Research, 13*(1), 136–148.

Cuthbert, D. (2016, September 21). Mobile Abattoir Improves Animal Welfare and Meat Quality in France. *Food & Drink International.*

Cyranoski, D. (2015). Gene-Edited Pigs to Be Sold as Pets. *Nature, 526*(7571), 18.

Dalziell, J., & Wadiwel, D. J. (2016). Live Exports, Animal Advocacy, Race and 'Animal Nationalism'. In A. Potts (Ed.), *Meat Culture* (pp. 73–89). Leiden; Boston: BRILL.

Dave, N. (2014). Witness: Humans, Animals, and the Politics of Becoming. *Cultural Anthropology, 29*(3), 433–456.

De Leersnyder, J., Boiger, M., & Mesquita, B. (2015). Cultural Differences in Emotions. In *An Interdisciplinary, Searchable, and Linkable Resource* (pp. 1–15). Hoboken, NJ: John Wiley & Sons, Inc.

Dean, T. (2014, January 13). China's Vegan Population Is Largest in the World. *VegNews Daily*. Online. February 2017.

Dearden, L. (2014, July 30). Stop Taking 'Tiger Selfies' that Fund Animal Abuse, Charity Says. *The Independent*. Online. February 2017.

Deckha, M. (2008). Intersectionality and Posthumanist Visions of Equality. *Wisconsin Journal of Law, Gender and Society, 23*(1, Spring), 249–267.

Deckha, M. (2012). Toward a Postcolonial, Posthumanist Feminist Theory: Centralizing Race and Culture in Feminist Work on Nonhuman Animals. *Hypatia, 27*(3), 527–545.

Dehaene, M., & De Cauter, L. (2008). *Heterotopia and the City*. London; New York: Routledge.

Deleuze, G., & Guattari, F. (1987). *A Thousand Plateaus: Capitalism and Schizophrenia*. London; New York: Continuum.

Deller, R. (2016). The Animated Aesthetics of Cultured Steak. In K. Sellberg & L. Wanggren (Eds.), *Corporeality and Culture: Bodies in Movement* (pp. 67–79). London; New York: Routledge.

Demers, C. (2019, January 9). Three Lucky Calves Recovering at Farm Sanctuary. *WENY News*.

Derrida, J. (2008). *The Animal that Therefore I Am*. New York: Fordham University Press.

Derrida, J., & Roudinesco, E. (2004). Violence Against Animals. In *For What Tomorrow … A Dialogue* (pp. 62–76). Palo Alto: Stanford University Press.

Despret, V. (2004). The Body We Care For: Figures of Anthropo-Zoo-Genesis. *Body & Society, 10*(2–3), 111–134.

Dess, N. K., & Chapman, C. D. (1998). 'Humans and Animals'? On Saying What We Mean. *Psychological Science, 9*(2), 156–157.

Devi, M. (1997). *Rudali, from Fiction to Performance*. Kolkata: Seagull Book Pvt. Ltd.

Diamond, J. (1987, May 1). The Worst Mistake in the History of the Human Race. *Discover Magazine*. Online. January 2018.

DoH. (2017). Call to Action on Antimicrobial Resistance (AMR).

Donaldson, S. (2017, December 31). Five Years a Vegetarian - and Now I'm Back Eating Meat. *The Independent.*

Donovan, J., & Adams, C. J. (Eds.). (2007). *The Feminist Care Tradition in Animal Ethics: A Reader.* New York; Chichester: Columbia University Press.

Douglas, M. (2002). *Purity and Danger: An Analysis of Concepts of Pollution and Taboo.* London; New York: Routledge Classics.

Dowsett, E., Semmler, C., Bray, H., Ankeny, R. A., & Chur-Hansen, A. (2018). Neutralising the Meat Paradox: Cognitive Dissonance, Gender, and Eating Animals. *Appetite, 123,* 280–288.

Dunayer, J. (1995). Sexist Words, Speciesist Roots. In C. J. Adams & J. Donovan (Eds.), *Animals and Women: Feminist Theoretical Explorations* (pp. 11–31). Durham; London: Duke University Press Books.

Dunayer, J. (2001). *Animal Equality.* Derwood, MD: Ryce Publishing.

EASAC. (2007). *Tackling Antibacterial Resistance in Europe* (European Academics Scientific Advisory Council). London: The Royal Society.

Eckhardt, G. M., Belk, R., & Devinney, T. M. (2010). Why Don't Consumers Consume Ethically? *Journal of Consumer Behaviour, 9*(6), 426–436.

Eisnitz, G. A. (2006). *Slaughterhouse.* New York: Prometheus Books.

Ellingson, T. (2001). *The Myth of the Noble Savage.* Berkeley; Los Angeles; London: University of California Press.

Ellis, R., & Waterton, C. (2005). Caught between the Cartographic and the Ethnographic Imagination: The Whereabouts of Amateurs, Professionals, and Nature in Knowing Biodiversity. *Environment and Planning D: Society and Space, 23*(5), 673–693.

Embury-Dennis, T. (2018, February 16). Cow Escapes on Way to Slaughterhouse, Smashes through Metal Fence, Breaks Arm of Man Trying to Catch Her Then Swims to Safety on Island in Lake. *The Independent.* Online. February 2018.

Emel, J., & Wolch, J. (1998). Witnessing the Animal Moment. In J. Wolch & J. Emel (Eds.), *Animal Geographies: Place, Politics, and Identity in the Nature-Culture Borderlands* (pp. 507–531). New York: Verso Books.

English, N. (2014, January 17). I Was a Vegetarian for Five Years and Today I Slaughtered a Chicken. *Greatist.* Online. June 2015.

Epstein, R. A. (2002). *Animals as Objects, or Subjects, of Rights.* John M. Olin Law and Economics Working Paper No. 171. The Law School: The University of Chicago.

Esposito, R. (2008). *Bíos: Biopolitics and Philosophy.* Minneapolis; London: University of Minnesota Press.

Evans, M. (2016). *For the Love of Meat*. Narrator and Co-author: Matthew Evans. Producer, Director and Co-author: Stephen Oliver. Distributor/ Broadcaster: SBS One.

Evans, A. B., & Miele, M. (2012). Between Food and Flesh: How Animals Are Made to Matter (and Not Matter) within Food Consumption Practices. *Environment and Planning D: Society and Space, 30*(2), 298–314.

Evershed, N., & Wahlquist, C. (2018, April 10). Live Exports: Mass Animal Deaths Going Unpunished as Holes in System Revealed. *The Guardian*.

Fairlie, S. (2010). *Meat: A Benign Extravagance*. Hampshire: Chelsea Green Publishing.

FAO. (2006, November). *Livestock Impacts on the Environment*. Food and Agricultural Organisation of the United Nations. Spotlight. Online. September 2014.

FAO. (2014, November). *Meat & Meat Products*. Food and Agricultural Organisation of the United Nations: Animal Production and Health. Online. June 2015.

FAO. (2016). *The State of World Fisheries and Aquaculture: Contributing to Food Security and Nutrition for All*. Food and Agricultural Organisation of the United Nations, Rome, 200 pp.

Faruqi, S. (2015). *Project Animal Farm: An Accidental Journey into the Secret World of Farming and the Truth About Our Food*. Pegasus Books.

Fassin, D. (2011). Morals and Moralities: A Critical Perspective from the Social Sciences. *Institute for Advanced Studies*. Fall issue. Online. May 2015.

Fearnley-Whittingstall, H. (2008, January 20). Poultry Is Not a Class Issue. *The Guardian*.

Ferguson, S. (2011, 30 May). A Bloody Business. *abc.net.au*. Online. January 2014.

Feskens, E. J. M., Sluik, D., & van Woudenbergh, G. J. (2013). Meat Consumption, Diabetes, and Its Complications. *Current Diabetes Reports, 13*(2), 298–306.

Fiala, N. (2008). Meeting the Demand: An Estimation of Potential Future Greenhouse Gas Emissions from Meat Production. *Ecological Economics, 67*(3), 412–419.

Fiddes, N. (1992). *Meat: A Natural Symbol*. London; New York: Routledge.

Fiorucci, M. (2018, September 28). Veal: The Greener (and Rosier) Side. *The Healthy Butcher*. Online.

Fischler, C. (1988). Food, Self and Identity. *Social Science Information, 27*(2), 275–292.

Fitzgerald, A. J., Kalof, L., & Dietz, T. (2009). Slaughterhouses and Increased Crime Rates: An Empirical Analysis of the Spillover from 'The Jungle' into the Surrounding Community. *Organization and Environment, 22*(2), 158–184.

Flower, S. (2018). *The Part-time Vegan*. Hachette Books.

Foer, J. S. (2009). *Eating Animals*. New York; Boston; London: Little, Brown and Company.

Foucault, M. (1967). Of Other Spaces: Utopias and Heterotopias. *Architecture/Mouvement/Continuite*, (October), 1–9.

Foucault, M. (1973). *The Birth of the Clinic: An Archaeology of Medical Perception* (R. D. Laing, Ed.). London and New York: Routledge.

Foucault, M. (1977). *Discipline and Punish: The Birth of the Prison*. New York: Vintage Books.

Foucault, M. (1978). *The History of Sexuality*. New York: Pantheon Books.

Foucault, M. (1979a, October 10 and 16). Omnes et Singulatim: Towards a Criticism of 'Political Reason'. *The Tanner Lectures on Human Values*. Stanford University, Palo Alto.

Foucault, M. (1979b) *Power, Truth, Strategy* (M. Morris & P. Patton, Eds.). Feral Publications Sydney.

Foucault, M. (1980). *Power/Knowledge: Selected Interviews and Other Writings, 1972–1977*. New York: Pantheon Books.

Foucault, M. (1981). *The Order of Discourse* (R. Young, Ed.). London and New York: Routledge.

Foucault, M. (1982a). The Subject and Power. *Critical Inquiry, 8*(4), 777–795.

Foucault, M. (1982b). *The Archaeology of Knowledge and the Discourse on Language*. New York: Pantheon Books.

Foucault, M. (1984). *Foucault Reader* (P. Rabinow, Ed.). New York: Pantheon Books.

Foucault, M. (1985). *The Use of Pleasure*. New York: Random House.

Foucault, M. (1989). *The Order of Things*. London; New York: Routledge.

Foucault, M. (1994a). *Ethics, Subjectivity and Truth* (P. Rabinow, Ed.). New York: The New Press.

Foucault, M. (1994b). *Aesthetics, Method, and Epistemology* (J. D. Faubion, Ed.). New York: The New Press.

Foucault, M. (1997). *Society Must Be Defended* (M. Bertani & A. Fontana, Eds.). New York: Picador.

Freeman, C. P. (2010). Meat's Place on the Campaign Menu: How US Environmental Discourse Negotiates Vegetarianism. *Environmental Communication: A Journal of Nature and Culture, 4*(3), 255–276.

Freeman, C. P., & Tulloch, S. (2013). Was Blind but Now I See: Animal Liberation Documentaries' Deconstruction of Barriers to Witnessing Injustice. In A. Pick & G. Narraway (Eds.), *Screening Nature: Cinema Beyond the Human* (pp. 110–126). New York; Oxford: Berghahn Books.

Freud, S. (1955). *The Standard Edition of the Complete Psychological Works of Sigmund Freud.* London: The Hogarth Press.

Friend, C. (2009). *The Compassionate Carnivore.* Philadelphia: Da Capo Lifelong Books.

Frye, M. (1983). In and Out of Harm's Way: Arrogance and Love. In *The Politics of Reality: Essays in Feminist Theory* (pp. 52–83). Santa Cruz: Crossing Press.

Fudge, E. (2002). *Animal.* New York: Reaktion Books.

Fudge, E. (2013a). Milking Other Men's Beasts. *History and Theory, 52*(December), 13–28.

Fudge, E. (2013b). The Animal Face of Early Modern England. *Theory, Culture and Society, 30*(7–8), 177–198.

Gaard, G. (2013). Toward a Feminist Postcolonial Milk Studies. *American Quarterly, 65*(3), 595–618.

Gabardi, W. (2017). *The Next Social Contract: Animals, the Anthropocene, and Biopolitics.* Philadelphia: Temple University Press.

Gallagher, W. (2011). *New: Understanding Our Need for Novelty and Change.* New York: Penguin Books.

Geen, E. (2016). *The Many Selves of Katherine North.* London: Bloomsbury Publishing.

Genel, K. (2006). The Question of Biopower: Foucault and Agamben. *Rethinking Marxism, 18*(1), 43–62.

Genovese, E. D., & Fox-Genovese, E. (2011). *Fatal Self-Deception: Slaveholding Paternalism in the Old South.* New York: Cambridge University Press.

Gherardi, S. (1994). The Gender We Think, the Gender We Do in Our everyday Organizational Lives. *Human Ecology Review, 47*(6), 591–610.

Gillespie, K. (2016). Witnessing Animal Others: Bearing Witness, Grief, and the Political Function of Emotion. *Hypatia, 31*(3), 572–588.

Gilligan, C. (1982). *In a Different Voice.* Cambridge, MA; London: Harvard University Press.

Giovannucci, D., Scherr, S., Nierenberg, D., Hebebrand, C., et al. (2012). *Food and Agriculture: The Future of Sustainability.* A Strategic Input to the Sustainable Development in the 21st Century (SD21) Project. New York: United Nations Department of Economic and Social Affairs, Division for Sustainable Development. 80 pp.

Giraud, E. (2013). 'Beasts of Burden': Productive Tensions between Haraway and Radical Animal Rights Activism. *Culture, Theory and Critique, 54*(1), 102–120.

Glenn, C. (2017). Complicating the Theory of the Male Gaze: Hitchcock's Leading Men. *New Review of Film and Television Studies, 15*(4), 496–510.

Godfray, H. C. J., Aveyard, P., Garnett, T., Hall, J. W., et al. (2018). Meat Consumption, Health, and the Environment. *Science, 361*(6399), eaam5324.

Gold, J. (2011, September 10). 99 Essential Restaurants. *LA Times*. Online. April 2018.

Goodland, R., & Anhang, J. (2009, November/December). Livestock and Climate Change. *Worldwatch*.

Goodman, M. K. (2004). Reading Fair Trade: Political Ecological Imaginary and the Moral Economy of Fair Trade Foods. *Political Geography, 23*(7), 891–915.

Goodman, M. K., Maye, D., & Holloway, L. (2010). Ethical Foodscapes?: Premises, Promises, and Possibilities. *Environment and Planning A, 42*(8), 1782–1796.

Goodman, D., DuPuis, E. M., & Goodman, M. K. (2012). *Alternative Food Networks: Knowledge, Practice, and Politics*. Routledge.

Gordon, L., & Baroke, S. (2014, September 20). Veal: Evolving from 'Cruel Meat' to Ethical Choice. *Euromonitor International*.

Goud, N. H. (2005). Courage: Its Nature and Development. *The Journal of Humanistic Counseling, 44*(1), 102–116.

Gray, M. (2013). *Labor and the Locavore - The Making of a Comprehensive Food Ethics*. Berkeley; Los Angeles; London: University of California Press.

Greger, M. (2007). The Long Haul: Risks Associated With Livestock Transport. *Biosecurity and Bioterrorism: Biodefense Strategy, Practice, and Science, 5*(4), 301–312.

Grigg, K., & Halford, J. (2013, May 31). Clearing More Land: We All Lose. *The Conversation*. Online. December 2016.

Gruen, L. (2014). *Entangled Empathy: An Alternative Ethic for Our Relationships with Animals*. New York: Lantern Books.

Guattari, F. (1995). *Chaosmosis: An Ethico-aesthetic Paradigm*. Bloomington; Indianapolis: Indiana University Press.

Gunderson, R. (2013). Problems with the Defetishization Thesis: Ethical Consumerism, Alternative Food Systems, and Commodity Fetishism. *Agriculture and Human Values, 31*(1), 109–117.

Gunderson, R. (2014). The First-generation Frankfurt School on the Animal Question. *Sociological Perspectives, 57*(3), 285–300.

Guthman, J. (2003). Fast Food/Organic Food: Reflexive Tastes and the Making of 'Yuppie Chow'. *Social & Cultural Geography, 4*(1), 45–58.

Guthman, J. (2008). Bringing Good Food to Others: Investigating the Subjects of Alternative Food Practice. *Cultural Geographies, 15*(4), 431–447.

Gutjahr, J. (2013). The Reintegration of Animals and Slaughter into Discourses of Meat Eating. In H. Röcklinsberg & P. Sandin (Eds.), *The Ethics of Consumption: The Citizen, the Market and the Law* (pp. 379–385). Wageningen, Netherlands: Wageningen Academic Publishers.

Halkier, B., & Jensen, I. (2011). Methodological Challenges in Using Practice Theory in Consumption Research. Examples from a Study on Handling Nutritional Contestations of food Consumption. *Journal of Consumer Culture, 11*(1), 101–123.

Halkier, B., Katz-Gerro, T., & Martens, L. (2011). Applying Practice Theory to the Study of Consumption: Theoretical and Methodological Considerations. *Journal of Consumer Culture, 11*(1), 3–13.

Hall, S. M. (2011). Exploring the 'Ethical Everyday': An Ethnography of the Ethics of Family Consumption. *Geoforum, 42*(6), 627–637.

Hankivsky, O. (2014). *Intersectionality 101*. Institute for Intersectionality Research and Policy, Canada, 34 pp.

Harari, Y. N. (2011). *Sapiens: A Brief History of Humankind*. New York: HarperCollins.

Haraway, D. (1990). *Simians, Cyborgs and Women: The Reinvention of Nature*. New York; London: Routledge.

Haraway, D. (2008). *When Species Meet*. Minneapolis; London: University of Minnesota Press.

Haraway, D. (2010). When Species Meet: Staying with the Trouble. *Environment and Planning D: Society and Space, 28*(1), 53–55.

Haraway, D. (2016). *Staying with the Trouble: Making Kin in the Chthulucene*. Combined Academic Publishers.

Harper, A. B. (2011). Vegans of Color, Racialized Embodiment, and Problematics of the 'Exotic'. In A. H. Alkon & J. Agyeman (Eds.), *Cultivating Food Justice: Race, Class, and Sustainability* (pp. 221–238). Cambridge, MA; London: The MIT Press.

Harwatt, H., & Hayek, M. (2019). *Eating Away at Climate Change with Negative Emissions: Repurposing UK Agricultural Land to Meet Climate Goals*. Cambridge, MA: Harvard Law School, Animal Law and Policy Program.

zur Hausen, H. (2012). Red Meat Consumption And Cancer: Reasons to Suspect Involvement of Bovine Infectious Factors in Colorectal Cancer. *International Journal of Cancer (Journal international du cancer), 130*(11), 2475–2483.

Hayes-Conroy, A., & Hayes-Conroy, J. (2008). Taking Back Taste: Feminism, Food and Visceral Politics. *Gender, Place & Culture, 15*(5), 461–473.

Hayes-Conroy, A., & Hayes-Conroy, J. (2010). Visceral Difference: Variations in Feeling (Slow) Food. *Environment and Planning A, 42*(12), 2956–2971.

Healy, S. (2014). Animal Farming in Australia: Consumer Awareness, Concern and Action. In G. L. Burns & M. Paterson (Eds.), *Engaging with Animals: Interpretations of a Shared Existence* (pp. 185–204). Sydney: Sydney University Press.

Heaton, T. (2017, April 14). Should New Zealanders Eat More Veal? *Stuff.co.nz.*

Held, V. (2006). *The Ethics of Care: Personal, Political, and Global.* Oxford: Oxford University Press.

Herbrechter, S. (2013a). *Posthumanism – A Critical Analysis.* London; New Delhi; New York; Sydney: Bloomsbury Academic.

Herbrechter, S. (2013b, April). Rosi Braidotti (2013) The Posthuman. *Culture Machine: Reviews,* 1–13.

Herzog, H. (2014, December 2). 84% of Vegetarians and Vegans Return to Meat. Why? *Psychology Today.* Online. February 2017.

Hetherington, K. (1997). *The Badlands of Modernity: Heterotopia and Social Ordering.* London; New York: Routledge.

Hickmann, T., Partzsch, L., Pattberg, P., & Weiland, A. (2018). *The Anthropocene Debate and Political Science.* Abingdon: Routledge.

Hideg, I., & Ferris, D. L. (2016). The Compassionate Sexist? How Benevolent Sexism Promotes and Undermines Gender Equality in the Workplace. *Journal of Personality and Social Psychology, 111*(5), 706–727.

Hinrichs, C. C., & Allen, P. (2008). Selective Patronage and Social Justice: Local Food Consumer Campaigns in Historical Context. *Journal of Agricultural and Environmental Ethics, 21*(4), 329–352.

Hirschman, E. C., & Stern, B. B. (2001). Do Consumers' Genes Influence Their Behavior? Findings on Novelty Seeking and Compulsive Consumption. *Advances in Consumer Research, 28,* 403–410.

Hochschild, A. R. (1979). Emotion Work, Feeling Rules, and Social Structure. *American Journal of Sociology, 85*(3), 551–575.

Holbrook, M. B., & Hirschman, E. C. (1982). The Experiential Aspects of Consumption: Consumer Fantasies, Feeling, and Fun. *Journal of Consumer Research, 9*(September), 132–140.

Holbrook, M. B., & Hirschman, E. C. (2015). Experiential Consumption. In D. T. Cook & J. M. Ryan (Eds.), *The Wiley Blackwell Encyclopedia of Consumption and Consumer Studies* (pp. 1–3). Oxford: John Wiley & Sons, Ltd.

Holloway, K. (2016, February 27). 5 Times People Ruined Animals' Lives for a Selfie. *Alternet*. Online. February 2017.

Hollows, J. (2003). Oliver's Twist. *International Journal of Cultural Studies, 6*(2), 229–248.

Hook, D. (2007). *Foucault, Psychology and the Analytics of Power*. Basingstoke; New York: Palgrave Macmillan.

hooks, b. (1998). Eating the Other: Desire and Resistance. In R. Scapp & B. Seitz (Eds.), *Eating Culture* (pp. 181–200). New York: State University of New York (SUNY) Press.

hooks, b. (2000). *Feminism Is for Everybody: Passionate Politics*. London: Pluto Press.

Horkheimer, M. (1978). *Der Wolkenkratzer*. Dawn and Decline: Notes 1926–1931 and 1950–1969, New York 1978, S. 66.

Hornborg, A. (2017). Dithering While the Planet Burns: Anthropologists' Approaches to the Anthropocene. *Reviews in Anthropology, 46*(2–3), 61–77.

Hsu, E. (2008). The Senses and the Social: An Introduction. *Ethnos, 73*(4), 433–443.

Hui, A., Schatzki, T., & Shove, E. (2016). *The Nexus of Practices: Connections, Constellations, Practitioners*. London; New York: Routledge.

Humphery, K. (2011). The Simple and the Good: Ethical Consumption as Anti-consumerism. In T. Lewis & E. Potter (Eds.), *Ethical Consumption: A Critical Introduction* (pp. 40–53). London; New York: Routledge.

Hunt, B. (2018, February 1). Stories from People Who Have Given Up Vegan Diets. *benhunt.com*.

Hurlstone, L. D. (2011). *Performing Marginal Identities: Understanding the Cultural Significance of Tawa'if and Rudali Rough the Language of the Body in South Asian Cinema*. Dissertation and Thesis. Paper 154. Master of Science in Communication. Portland State University.

Hynes, M. (2013). The Ethico-aesthetics of Life: Guattari and the Problem of Bioethics. *Environment and Planning A, 45*(8), 1929–1943.

IAASTD. (2009). Agriculture at a Crossroads: Global Report. In B. D. McIntyre, H. R. Herren, & J. Wakhungu (Eds.), *International Assessment of Agricultural Knowledge, Science and Technology for Development*. Washington: Island Press. 590 pp.

Illouz, E. (2009). Emotions, Imagination and Consumption: A New Research Agenda. *Journal of Consumer Culture, 9*(3), 377–413.

Ingram, J., Shove, E., & Watson, M. (2007). Products and Practices: Selected Concepts from Science and Technology Studies and from Social Theories of Consumption and Practice. *Design Issues, 23*(2), 3–16.

IPCC. (2007). *Climate Change 2007: Synthesis Report* (Core Writing Team, Bernstein et al., Eds.). Geneva: Intergovernmental Panel on Climate Change. 73 pp.

IPCC. (2014). Summary for Policymakers. In O. Edenhofer et al. (Eds.), *Climate Change 2014, Mitigation of Climate Change. Contribution of Working Group III to the Fifth Assessment Report of the Intergovernmental Panel on Climate Change.* Cambridge, UK; New York: Cambridge University Press.

Iveson, R. (2013). Deeply Ecological Deleuze and Guattari: Humanism's Becoming-Animal. *Humanimalia, 4*(2), 34–53.

Iveson, R. (2014). *Zoogenesis: Thinking Encounter with Animals.* London: Pavement Books.

Jacob, C. (2006). *The Sovereign Map: Theoretical Approaches in Cartography Throughout History.* Chicago; London: University of Chicago Press.

Jacobson, M. H. (2013). Solid Modernity, Liquid Utopia – Liquid Modernity, Solid Utopia. In A. Elliott (Ed.), *The Contemporary Bauman* (pp. 217–240). London; New York: Taylor & Francis.

Jamieson, D., & Nadzam, B. (2015). *Love in the Anthropocene.* New York; London: OR Books.

Jamison, S. W. (1998). Rhinoceros Toes, Manu V.17–18, and the Development of the Dharma System. *Journal of the American Oriental Society, 118*(2), 249–256.

Ji, M., Wong, I. A., Eves, A., & Scarles, C. (2016). Food-Related Personality Traits and the Moderating Role of Novelty-Seeking in Food Satisfaction and Travel Outcomes. *Tourism Management, 57,* 387–396.

Johnson, L. (2012). *Power, Knowledge, Animals.* Basingstoke; New York: Palgrave Macmillan.

Jorgensen, M., & Phillips, L. J. (2002). *Discourse Analysis as Theory and Method.* SAGE Publications Ltd.

Joshi, A. D., Corral, R., Siegmund, K. D., Haile, R. W., et al. (2009). Red Meat and Poultry Intake, Polymorphisms in the Nucleotide Excision Repair and Mismatch Repair Pathways and Colorectal Cancer Risk. *Carcinogenesis, 30*(3), 472–479.

Joy, M. (2009). *Why We Love Dogs, Eat Pigs, and Wear Cows: An Introduction to Carnism.* San Francisco: Red Wheel, Weiser.

Kagan, S. (2015). Singer on Killing Animals. In T. Visak & R. Garner (Eds.), *The Ethics of Killing Animals* (pp. 136–153). Oxford; New York: Oxford University Press.

Kanji, A. (2017). Colonial Animality: Constituting Canadian Settler Colonialism through the Human-Animal Relationship. In M. Woons & S. Weier (Eds.), *Critical Epistemologies of Global Politics* (pp. 63–78). Bristol: E-International Relations Publishing.

Katsnelson, A. (2015, January 13). Will McDonald's "Sustainable Beef" Burgers Really Be any Better? *The Guardian*. Online. June 2015.

Keith, L. (2009). *Vegetarian Myth: Food, Justice, and Sustainability*. Oakland: PM Press.

Kemper, T. D. (1978). *A Social Interactional Theory of Emotions*. Hoboken, NJ: John Wiley and Sons.

Kim, C. J. (2015). *Dangerous Crossings: Race, Species and Nature in a Multicultural Age*. Cambridge, UK; New York: Cambridge University Press.

King, B. J. (2017). *Personalities on the Plate: The Lives and Minds of Animals We Eat*. Chicago; London: University of Chicago Press.

Kingsolver, B. (2008). *Animal, Vegetable, Miracle: A Year of Food Life*. New York: HarperCollins.

Kinsella, E. L., Ritchie, T. D., & Igou, E. R. (2017). On the Bravery and Courage of Heroes: Considering Gender. *Heroism Science, 2*(1), 1–14.

Kirkwood, K. (2016, May 26). Dude Food vs Superfood: We're Cultural Omnivores. *The Conversation*. Online. October 2016.

Knight, D. (1999). Why We Enjoy Condemning Sentimentality: A Meta-aesthetic Perspective. *The Journal of Aesthetics and Art Criticism, 57*(4), 411.

Kolcaba, K. (2003). *Comfort Theory and Practice: A Vision for Holistic Health Care and Research*. New York: Springer Publishing Company.

Koneswaran, G., & Nierenberg, D. (2008). Global Farm Animal Production and Global Warming: Impacting and Mitigating Climate Change. *Environmental Health Perspectives, 116*(5), 578–582.

Korsmeyer, C. (2002). *Making Sense of Taste*. New York: Cornell University Press.

Korzybski, A. (1958). *Science and Sanity: An Introduction to Non-Aristotelian Systems and General Semantics*. Englewood, NJ: Institute of General Semantics.

Kowitt, B. (2017, February 27). McDonald's Is Exploring a New Menu Item: Sustainable Beef. *Fortune*. Online. April 2018.

Krause, J. (2015, January 6). Runaway Chicken Escapes Deadly Yom Kippur Ceremony to Live Out Her Dream Life. *Animal Scoop*. Online. October 2016.

Kwan, M. (2004). Beyond Difference: From Canonical Geography to Hybrid Geographies. *Annals of the Association of American Geographers, 94*(4), 756–763.

Larsen, C. S. (2006). The Agricultural Revolution as Environmental Catastrophe: Implications for Health and Lifestyle in the Holocene. *Quaternary International, 150*(1), 12–20.

Larsen, J., & Roney, J. M. (2013, June 12). Farmed Fish Production Overtakes Beef. *Earth Policy Institutei.* Online. February 2017.

Latimer, J. (2013). Being Alongside: Rethinking Relations amongst Different Kinds. *Theory, Culture and Society, 30*(7–8), 77–104.

Latimer, J., & Miele, M. (2013). Naturecultures? Science, Affect and the Non-human. *Theory, Culture and Society, 30*(7–8), 5–31.

Latour, B. (1993). *We Have Never Been Modern.* Cambridge, MA; London: Harvard University Press.

LaVeck, J. (2006, September 8–11). Compassion for Sale? Doublethink Meets Doublefeel as 'Happy Meat' Comes of Age. *Satya.*

Laxminarayan, R. (2002). How Broad Should the Scope of Antibiotics Patents Be? *American Journal of Agricultural Economics, 84*(5), 1287–1292.

Laxminarayan, R., Duse, A., Wattal, C., Zaidi, A. K. M., et al. (2013). The Lancet Infectious Diseases Commission: Antibiotic Resistance—the Need for Global Solutions. *The Lancet, 13*(12), 1057–1098.

Layton, L. (2010, June 20). As Demand Grows for Locally Raised Meat, Farmers Turn to Mobile Slaughterhouses. *The Washington Post.*

Leigh, T. W. (2006). The Consumer Quest for Authenticity: The Multiplicity of Meanings within the MG Subculture of Consumption. *Journal of the Academy of Marketing Science, 34*(4), 481–493.

Lein, A. (2017). After Being Vegan for 3 Years, I Went Back to Meat. And This Happened… Healthline. Online, December 19.

Leip, A., Billen, G., Garnier, J., Grizzetti, B., et al. (2015). Impacts of European Livestock Production: Nitrogen, Sulphur, Phosphorus and Greenhouse Gas Emissions, Land-Use, Water Eutrophication and Biodiversity. *Environmental Research Letters, 10*(11), 115004.

Lennon, C. (2007, August). Why Vegetarians Are Eating Meat. *Food and Wine.* Online. July 2015.

Levi-Strauss, C. (1991). *Totemism.* London: Merlin Press.

Lewis, T. (2008). Transforming Citizens? Green Politics and Ethical Consumption on Lifestyle Television. *Continuum, 22*(2), 227–240.

Lewis, T. (2011). The Ethical Turn in Commodity Culture: Consumption, Care and the Other. *sic: Journal of Literature, Culture and Literary Translation,* Vol. 2. Online. March 2013.

Lewis, T. (2016). Ethical Consumers and Sustainability Citizenship. In R. Horne, J. Fien, B. Beza, & A. Nelson (Eds.), *Sustainability Citizenship in Cities: Theory and Practices* (pp. 199–208). London: Routledge.

Lewis, T., & Huber, A. (2015). A Revolution in an Eggcup? Supermarket Wars, Celebrity Chefs, and Ethical Consumption. *Food, Culture and Society: An International Journal of Multidisciplinary Research, 18*(2), 289–307.

Lewis, T., & Potter, E. (Eds.). (2011). *Ethical Consumption: A Critical Introduction*. London; New York: Routledge.

Lewis, P., & Simpson, R. (2007). *Gendering Emotions in Organizations*. Basingstoke; New York: Palgrave.

Lim, N. (2016). Cultural Differences in Emotion: Differences in Emotional Arousal Level between the East and the West. *Integrative Medicine Research, 5*(2), 105–109.

Linne, T. (2014). Grazing the Green Fields of Social Media. In E. A. Cederholm, A. Bjorck, K. Jennbert, & A.-S. Lonngren (Eds.), *Exploring the Animal Turn: Human-Animal Relations in Science, Society and Culture* (pp. 19–32). Lund: The Pufendorf Institute for Advanced Studies.

Lisle, D. (2004). Gazing at Ground Zero: Tourism, Voyeurism and Spectacle. *Journal for Cultural Research, 8*(1), 3–21.

Littler, J. (2008). *Radical Consumption: Shopping for Change in Contemporary Culture*. Maidenhead, UK: Open University Press.

Locher, J. L., Yoels, W. C., Maurer, D., & van Ells, J. (2005). Comfort Foods: An Exploratory Journey into the Social and Emotional Significance of Food. *Food and Foodways, 13*(4), 273–297.

Lockwood, A. (2019, January 7). Do 84% of Vegans and Vegetarians Really Go Back to Eating Meat? *Plant Based News*. Online. April 2019.

Lopez, P. J., & Gillespie, K. (2015). *Economies of Death: Economic Logics of Killable Life and Grievable Death*. London: Routledge.

Loughnan, S., Bastian, B., & Haslam, N. (2014). The Psychology of Eating Animals. *Current Directions in Psychological Science, 23*(2), 104–108.

Lövbrand, E., Beck, S., Chilvers, J., Forsyth, T., et al. (2015). Who Speaks for the Future of Earth? How Critical Social Science Can Extend the Conversation on the Anthropocene. *Global Environmental Change, 32*, 211–218.

Lupton, D. (1996). *Food, the Body and the Self*. London; Thousand Oaks; New Delhi: SAGE Publications Ltd.

Lury, C. (2011). *Consumer Culture*. Cambridge; Malden: Policy Press.

Lyon, D. (2006). 9/11. Synopticon, and Scopophilia: Watching and Being Watched. In R. V. Ericson & K. D. Haggerty (Eds.), *The New Politics of Surveillance and Visibility* (pp. 35–54). Toronto: University of Toronto Press Incorporated.

Machin, D., & Mayr, A. (2012). *How to Do Critical Discourse Analysis*. London; Thousand Oaks; New Delhi: SAGE Publications Ltd.

Mackenzie, M. (2015, January 17). Mobile Butchers in Demand as Consumers Seek Fresh, Ethically Slaughtered Meat. *ABC Radio National.* Online. June 2015.

Major, T. (2017). Mini Pigs with Big Appeal: Popularity Soars Despite Ownership Restrictions. *ABC Rural.*

Malins, P. (2004). Machinic Assemblages: Deleuze, Guattari and an Ethico-aesthetics of Drug Use. *Janus Head, 7*(1), 84–104.

Malins, P. (2011). An Ethico-aesthetics of Heroin Chic: Art, Cliché and Capitalism. In L. Guillaume & J. Hughes (Eds.), *Deleuze and the Body* (pp. 165–187). Edinburgh: Edinburgh University Press.

Maller, C. J. (2015). Understanding Health Through Social Practices: Performance and Materiality in Everyday Life. *Sociology of Health & Illness, 37*(1), 52–66.

March, S. (2011, November 30). Abattoir Closed Over Animal Cruelty Concerns. *ABC News.* Online. February 2017.

Mark, J. (2017, February 24). Toward a Moral Case for Meat Eating. *The National Magazine of the Sierra Club.* Online. April 2018.

Martin, P. Y. (2003). "Said and Done" Versus "Saying and Doing" Gendering Practices, Practicing Gender at Work. *Gender & Society, 17*(3), 342–366.

Martin, M. J., Thottathil, S. E., & Newman, T. B. (2015). Antibiotics Overuse in Animal Agriculture: A Call to Action for Health Care Providers. *American Journal of Public Health, 105*(12), 2409–2410.

von Massow, M., Weersink, A., & Gallant, M. (2019, March 12). Meat Consumption Is Changing But It's Not Because of Vegans. *The Conversation.* Online. April 2019.

Major, T. (2017). *Mini Pigs with Big Appeal: Popularity Soars Despite Ownership Restrictions.* ABC Rural.

Matchar, E. (2015). *Homeward Bound: Why Women Are Embracing the New Domesticity.* New York: Simon & Schuster.

Matsumoto, N. (2016, June 14). The Grassfed Burger Gap. *Civil Eats.* Online. August 2016.

Maxwell, J. A. (2012). *A Realist Approach for Social Research.* London; Thousand Oaks; New Delhi: SAGE Publications.

May, T. (2014). Moral Individualism, Moral Relationalism, and Obligations to Non-human Animals. *Journal of Applied Philosophy, 31*(2), 155–168.

Mayes, C. (2015). *The Biopolitics of Lifestyle: Foucault, Ethics and Healthy Choices.* London; New York: Routledge.

McCosker, A. (2018, May 8). Beef Australia 2018: What Lies Ahead for the Industry as 'Locavores' and Digital Disruption Loom Closer. *ABC News: Rural.*

McGehee, N. G., & Kim, K. (2004). Motivation for Agri-tourism Entrepreneurship. *Journal of Travel Research, 43*(2), 161–170.

McKibben, P. (2019). *Love Notes: For a Politics of Love.* New York: Lantern Books.

McKie, R. (2014, December 14). Earth Faces Sixth 'Great Extinction' with 41% of Amphibians Set to Go the Way of the Dodo. *The Guardian.* Online. December 2014.

McLisky, C. (2015). "And They'll Know We Are Christians by Our Love": Exploring the Role of Christian Love on Maloga Mission, 1874–1888. *Journal of Religious History, 39*(3), 333–351.

McMichael, A. J., Powles, J. W., Butler, C. D., & Uauy, R. (2007). Food, Livestock Production, Energy, Climate Change, and Health. *The Lancet, 370*(9594), 1253–1263.

Meininger, H. P. (2013). Inclusion as Heterotopia: Spaces of Encounter between People with and without Intellectual Disability. *Journal of Social Inclusion, 4*(1), 24–44.

Mekonnen, M. M., & Hoekstra, A. Y. (2012). A Global Assessment of the Water Footprint of Farm Animal Products. *Ecosystems, 15*(3), 401–415.

Memery, J., Megicks, P., Angell, R., & Williams, J. (2012). Understanding Ethical Grocery Shoppers. *Journal of Business Research, 65*(9), 1283–1289.

Menely, T. (2007). Zoophilpsychosis: Why Animals Are What's Wrong with Sentimentality. *Symploke, 15*(1/2), 244–267.

Merchant, C. (1990). *The Death of Nature: Women, Ecology, and the Scientific Revolution.* New York: HarperCollins.

Merchant, C. (2005). *Radical Ecology: The Search for a Liveable World.* London; New York: Routledge.

Merleau-Ponty, M. (2002). *Phenomenology of Perception.* London; New York: Routledge.

Meryment, E. (2011, June 5). Sickened Meat-Lovers Turn to 'Ethical' Beef. *The Sunday Telegraph.* Online. April 2013.

Mesirow, B. (2011, January 6). Kill it, Cook It, Eat It, Slaughtering in Your Living Room in HD. *LA Weekly.* Online. April 2018.

Mesquita, B., & Frijda, N. H. (1992). Cultural Variations in Emotions: A Review. *Psychol Bulletin, 112*(2), 179–204.

Mesquita, B., & Walker, R. (2003). Cultural Differences in Emotions: A Context for Interpreting Emotional Experiences. *Behaviour Research and Therapy, 41*(7), 777–793.

Meyers, D. T. (2017). Commentary on *Entangled Empathy* by Lori Gruen. *Hypatia, 32*(2), 415–427.

Micheletti, M., & Stolle, D. (2010). Vegetarianism - A Lifestyle Politics? In M. Micheletti & A. S. McFarland (Eds.), *Creative Participation: Responsibility-Taking in the Political World* (pp. 125–145). Boulder, CO; London: Paradigm Publishers.

Micheletti, M., Cheng, S.-L., Stolle, D., Olsen, W., et al. (2012). Habits of Sustainable Citizenship: The Example of Political Consumerism. In A. Warde & D. Southerton (Eds.), *The Habits of Consumption* (Collegium - Studies across Disciplines in the Humanities and Social Sciences) (pp. 141–163). Helsinki: Helsinki Collegium for Advanced Studies.

Miller, J. (2012). In Vitro Meat: Power, Authenticity and Vegetarianism. *Journal for Critical Animal Studies, 10*(4), 41–63.

Mitchell, R., & Hall, M. C. (2003). Consuming Tourists: Food Tourism Consumer Behaviour. In M. C. Hall, L. Sharples, R. Mitchell, N. Macionis, et al. (Eds.), *Food Tourism Around the World* (pp. 60–80). London; New York: Routledge.

MLA. (2010). Strategic Plan 2010–2015. *Meat and Livestock Australia*. 48 pp.

Molz, J. G. (2007). Eating Difference: The Cosmopolitan Mobilities of Culinary Tourism. *Space and Culture, 10*(1), 77–93.

Monaco, E. (2017, April 28). Why It's More Important to Be an Ethical Omnivore than a Vegetarian. *Good Housekeeping*.

Monastersky, R. (2014, December 10). Biodiversity: Life – a Status Report. *Nature: News Feature*. Online. December 2014.

Monastersky, R. (2015). The Human Age. *Nature, 519*(7542), 144–147.

Moody, L. (2018, May). 8 Real People Share Why They Stopped Being Vegan. *Mindbodygreen*.

Morgan, K., & Cole, M. (2011). The Discursive Representation of Nonhuman Animals in a Culture of Denial. In B. Carter & N. Charles (Eds.), *Human and Other Animals: Critical Perspectives* (pp. 112–132). Basingstoke; New York: Palgrave Macmillan.

Mulvey, L. (1989). *Afterthoughts on 'Visual Pleasure and Narrative Cinema' Inspired by King Vidor's Duel in the Sun (1946)* (pp. 29–38). London: Palgrave Macmillan.

Mulvey, L. (1999). Visual Pleasure and Narrative Cinema. In L. Braudy & M. Cohen (Eds.), *Film Theory and Criticism: Introductory Readings* (pp. 833–844). New York; Oxford: Oxford University Press.

Mulvey, L. (2001). Unmasking the Gaze: Some Thoughts on New Feminist Film Theory and History.

Murdoch, L. (2016, September 26). Duterte's Drug Crackdown Hits Home as Man Shot Dead in Front of His Pregnant Wife, Family. *The Age*. Online. September 2016.

Mykletun, R. J., & GyimOthy, S. (2010). Beyond the Renaissance of the Traditional Voss Sheep's-Head Meal: Tradition, Culinary Art, Scariness and Entrepreneurship. *Tourism Management, 31*(3), 434–446.

Narula, S.K. (2014, April 9). What's Wrong with Sentimentality? *The Atlantic.* Online. January 2018.

NHMRC. (2013). *Australian Dietary Guidelines.* Australian Government, National Health and Medical Research Council. Department of Health and Aging. Canberra: Commonwealth of Australia. 226 pp.

Nibert, D. A. (2013). *Animal Oppression and Human Violence: Domesecration, Capitalism, and Global Conflict.* New York; Chichester: Columbia University Press.

Nicholson, J. (2012). *The Meat Fix: How a Lifetime of Healthy Eating Nearly Killed Me!* Biteback Publishing.

Noddings, N. (1984). *Caring, a Feminine Approach to Ethics & Moral Education.* Berkeley; Los Angeles; London: University of California Press.

Noone, R. (2014, April 2). Video Shows Abattoir Staff Abusing Pigs. *The Daily Telegraph.* Online. February 2017.

Noske, B. (1989). *Humans and Other Animals: Beyond the Boundaries of Anthropology.* London: Pluto Press.

Noske, B. (1997). *Beyond Boundaries: Humans and Animals.* Montreal: Black Rose Books.

O'Brien, S. J. (2012). *Unnerving Images: Cinematic Representations of Animal Slaughter and the Ethics of Shock.* Doctoral Thesis, The Centre for Comparative Literature, University of Toronto. 274 pp.

O'Sullivan, S. D. (2008). Academy: The Production of Subjectivity. In I. Rogoff (Ed.), *Academy* (pp. 238–244). Revolver: Frankfurt.

O'Sullivan, S. (2011). *Animals, Equality and Democracy.* New York: Palgrave Macmillan.

Ocejo, R. E. (2014). Show the Animal: Constructing and Communicating New Elite Food Tastes at Upscale Butcher Shops. *Poetics, 47,* 106–121.

OECD-FAO. (2017). *OECD-FAO Agricultural Outlook 2017–2026.* Paris: OECD Publishing. 142 pp. https://doi.org/10.1787/agr_outlook-2017-en.

Oliver, K. (2001). *Witnessing: Beyond Recognition.* Minneapolis, MN: University of Minnesota Press.

Oliver, K. (2004). Witnessing and Testimony. *Parallax, 10*(1), 78–87.

Oliver, K. (2017). The Male Gaze Is More Relevant, and More Dangerous, than Ever. *New Review of Film and Television Studies, 15*(4), 451–455.

Olsen, N. (2017, December 19). After Being Vegan for 3 Years, I Went Back to Meat. And This Happened…. *Healthline.*

Pachirat, T. (2011). *Every Twelve Seconds: Industrialized Slaughter and the Politics of Sight*. New Haven; London: Yale Agrarian Studies Series.

Pachirat, T. (2015). The Glass Walls Fallacy: Reflections from an Industrialized Kill Floor on the Promises and Pitfalls of Transparency. Keynote: Animal Publics: Emotions, Empathy, Activism. Australasian Animal Studies Association Conference, July 12–15, Melbourne, Australia.

Padva, G., & Buchweitz, N. (Eds.). (2014). *Sensational Pleasures in Cinema, Literature and Visual Culture: The Phallic Eye*. New York: Palgrave Macmillan.

Palmer, C. (2001). 'Taming the Wild Profusion of Existing Things'? A Study of Foucault, Power, and Human/Animal Relationships. *Environmental Communication: A Journal of Nature and Culture, 23*, 339–358.

Pan, A., Sun, Q., Bernstein, A., Schultze, M. B., et al. (2012). Red Meat Consumption and Mortality. *Archives of Internal Medicine, 172*(7), 555–563.

Parry, J. (2010a). *The New Visibility of Slaughter in Popular Gastronomy*. Masters Thesis, Cultural Studies: University of Canterbury.

Parry, J. (2010b). Gender and Slaughter in Popular Gastronomy. *Feminism & Psychology, 20*(3), 381–396.

Passafaro, P., Rimano, A., Piccini, M. P., Metastasio, R., et al. (2014). The Bicycle and the City: Desires and Emotions Versus Attitudes, Habits and Norms. *Journal of Environmental Psychology, 38*, 76–83.

Patterson, M., & Elliott, R. (2010). Negotiating Masculinities: Advertising and the Inversion of the Male Gaze. *Consumption, Markets & Culture, 5*(3), 231–249.

Pearsall, J. L. S. (2015). *Praise the Pig: Loin to Belly, Shoulder to Ham—Pork-Inspired Recipes for Every Meal*. New York: Skyhorse Publishing.

Pederson, H. (2011). Release the Moths: Critical Animal Studies and the Posthumanist Impulse. *Culture, Theory and Critique, 52*(1), 65–81.

Peggs, K. (2012). *Animals and Sociology*. Basingstoke; New York: Palgrave Macmillan.

Pellandini-Simanyi, L. (2014). *Consumption Norms and Everyday Ethics*. Basingstoke; New York: Palgrave Macmillan.

Pepper, M., Jackson, T., & Uzzell, D. (2009). An Examination of the Values that Motivate Socially Conscious and Frugal Consumer Behaviours. *International Journal of Cultural Studies, 33*(1), 126–136.

Petenko, E. (2018, December 31). The Cow that Escaped the Slaughterhouse Gave Birth, and Her New Baby Is Udderly Adorable. *NJ Advance Media*.

Peterson, R. A., & Kern, R. M. (1996). Changing Highbrow Taste: From Snob to Omnivore. *American Sociological Review, 61*(5), 900–907.

Phillip, S., Hunter, C., & Blackstock, K. (2010). A Typology for Defining Agritourism. *Tourism Management, 31*(6), 754–758.

Phillipov, M., & Goodman, M. K. (2017). The Celebrification of Farmers: Celebrity and the New Politics of Farming. *Celebrity Studies, 8*(2), 346–350.

Pick, A. (2012). Turning to Animals Between Love and Law. *New Formations, 76*, 68–86.

Pick, A. (2015a). *Vegan Cinema: Looking, Eating and Letting Be.* Conference Keynote Lecture. Sixth Australasian Animal Studies Association Conference: Animal Publics: Emotions, Empathy, Activism. July 12–15, Melbourne, Australia.

Pick, A. (2015b). Why not Look at Animals? *Necsus European Journal of Media Studies.* Spring, Online. November 2015. 17 pp.

Pickles, J. (2004). *A History of Spaces: Cartographic Reason, Mapping and the Geo-Coded World.* London; New York: Routledge.

Pimentel, D., & Pimentel, M. H. (2007). *Food, Energy, and Society.* London; New York: Taylor & Francis.

Pimentel, D., Pimentel, D., Pimentel, M., & Pimentel, M. (2003). Sustainability of Meat-Based and Plant-Based Diets and the Environment. *The American Journal of Clinical Nutrition, 78*(3), 660S–663S.

Piqueras-Fiszman, B., & Spence, C. (2015). Sensory Expectations Based on Product-Extrinsic Food Cues: An Interdisciplinary Review of the Empirical Evidence and Theoretical Accounts. *Food Quality and Preference, 40*, 165–179.

Pleasance, C. (2014, June 3). This Little Piggy's Not Going to Market: "Babe" the Porker Escapes Slaughterhouse Van by Leaping 16ft to Freedom … and Avoids the Chop after Being Adopted. *The Daily Mail Australia.* Online. October 2016.

Pluhar, E. B. (2009). Meat and Morality: Alternatives to Factory Farming. *Journal of Agricultural and Environmental Ethics, 23*(5), 455–468.

Plumwood, V. (1994). *Feminism and the Mastery of Nature.* London; New York: Routledge.

Plumwood, V. (2001). *Environmental Culture: The Ecological Crisis of Reason.* London; New York: Routledge.

Plumwood, V. (2002). Decolonising Relationships with Nature. *PAN: Philosophy Activism Nature, 2*, 7–30.

Plumwood, V. (2004). Gender, Eco-Feminism and the Environment. In R. White (Ed.), *Controversies in Environmental Sociology* (pp. 43–60). Cambridge, UK; New York: Cambridge University Press.

Plumwood, V. (2007). Human Exceptionalism and the Limitations of Animals: A Review of Raimond Gaita's the Philosopher's Dog. *Australian Humanities Review, 42*(August), 1–7.

Plumwood, V. (2013a). *The Eye of the Crocodile*. Canberra: ANU E Press.

Plumwood, V. (2013b). Animals and Ecology: Towards a Better Integration. In L. Shannon (Ed.), *The Eye of the Crocodile* (pp. 77–90). ANU E Press.

Postill, J. (2010). Introduction: Theorising Media and Practice. In B. Brauchler & J. Postill (Eds.), *Theorising Media and Practice* (pp. 1–33). New York; Oxford: Berghahn Books.

Pottinger, L. (2013). Ethical Food Consumption and the City. *Geography Compass, 7*(9), 659–668.

Potts, A. (Ed.). (2016). *Meat Culture*. Leiden; Boston: Brill.

Potts, A., & Parry, J. (2010). Vegan Sexuality: Challenging Heteronormative Masculinity through Meat-free Sex. *Feminism & Psychology, 20*(1), 53–72.

Pratt, J. (2008). Food Values: The Local and the Authentic. In G. De Neve, P. Luetchford, J. Pratt, & D. C. Wood (Eds.), *Research in Economic Anthropology* (pp. 53–70). Bingley: Emerald Group Publishing Limited.

Probyn, E. (2000). *Carnal Appetites: FoodSexIdentities*. London; New York: Routledge.

Probyn-Rapsey, F. (2017). *The Cultural Politics of Eradication*. Conference Keynote Lecture. Seventh Australian Animal Studies Association Conference: Animal Intersection. 3–5 July, Adelaide, Australia.

Quinn, S. (2016, April 18). Number of Vegans in Britain Rises by 360% in 10 Years. *The Telegraph*. Online. February 2017.

Quinn, E., & Westwood, B. (Eds.). (2018). *Thinking Veganism in Literature and Culture: Towards a Vegan Theory*. Cham: Palgrave Macmillan.

Rabinow, P., & Rose, N. (2006). Biopower Today. *BioSocieties, 1*(2), 195–217.

Reckwitz, A. (2002). Toward a Theory of Social Practices: A Development in Culturalist Theorizing. *European Journal of Social Theory, 5*(2), 243–263.

Reckwitz, A. (2016). Practices and Their Affects. In A. Hui, T. Schatzki, & E. Shove (Eds.), *The Nexus of Practices: Connections, Constellations, Practitioners* (pp. 114–125). London; New York: Taylor & Francis.

Redman, C. L. (1999). *Human Impact on Ancient Environments*. Tucson: The University of Arizona Press.

Reed, B. (2013). *The Ethical Butcher: How Thoughtful Eating Can Change Your World*. Berkeley, CA: Soft Skull Press.

Regan, T. (2004). *The Case for Animal Rights*. Berkeley; Los Angeles: University of California Press.

Richards, E., Signal, T., & Taylor, N. (2013). A Different Cut? Comparing Attitudes toward Animals and Propensity for Aggression within Two Primary Industry Cohorts—Farmers and Meatworkers. *Society and Animals, 21*(4), 395–413.

Rickards, L. (2015). Metaphor and the Anthropocene: Presenting Humans as a Geological Force. *Geographical Research, 53*(3), 280–287.

Ripple, W. J., Smith, P. T., Haberl, H., Montzka, S. A., et al. (2014). Ruminants, Climate Change and Climate Policy. *Nature Climate Change, 4,* 2–5.

Ritchie, H. (2019, February 4). Which Countries Eat the Most Meat? *BBC News.* Online. April 2019.

Ritchie, H., & Roser, M. (2017). Meat and Seafood Production & Consumption. *OurWorldInData.org.* Online. April 2018.

Roberts, A. M. (2016, May 26). The True Confessions of an Ex-Vegan. *POPSUGAR.* Online. February 2017.

Robinson, A. (2016, October 22). 'Ethical' Meat Production Concept Launched in France. *Farmers Guardian.*

Robinson, T. P., Bu, D. P., Carrique-Mas, J., Fèvre, E. M., et al. (2016). Antibiotic Resistance Is the Quintessential One Health Issue. *Transactions of The Royal Society of Tropical Medicine and Hygiene, 110*(7), 377–380.

Rodosthenous, G. (2015). *Theatre as Voyeurism: The Pleasures of Watching.* New York: Springer.

Rose, N. (2009). *The Politics of Life Itself: Biomedicine, Power, and Subjectivity in the Twenty-First Century.* Princeton; Oxford: Princeton University Press.

Rousseau, O. (2016, April 13). Food Trends: Meat Consumption up, Beef Declines. *GlobalMeatnews.com.* Online. February 2017.

Ruby, M. B., & Heine, S. J. (2011). Meat, Morals, and Masculinity. *Appetite, 56*(2), 447–450.

Russell, B. (1986). The Forms of Power. In S. Lukes (Ed.), *Power* (pp. 19–27). New York: New York University Press.

Ryan, S. (1996). *The Cartographic Eye: How Explorers Saw Australia.* Cambridge, UK; New York: Cambridge University Press.

Salt, H. S. (1894). *Animals' Rights: Considered in Relation to Social Progress.* New York: Macmillan & Co.

Salt, H. S. (1914). Logic of the Larder. *Henrysalt.co.uk.* Online. June 2016.

Sanbonmatsu, J. (2014a). Interview with John Sanbonmatsu (S. Rodriguez, Ed., pp. 1–12). Online. Direct Action Everywhere.

Sanbonmatsu, J. (2014b). The Animal of Bad Faith: Speciesism as an Existential Project. In J. Sorenson (Ed.), *Critical Animal Studies: Thinking the Unthinkable* (pp. 29–45). Brown Bear Press.

Sanford, V. (2006). Introduction. In V. Sanford & A. Angel-Ajani (Eds.), *Engaged Observer: Anthropology, Advocacy, and Activism* (p. 272). Rutgers University Press.

Sassatelli, R. (2011). Interview with Laura Mulvey. *Theory, Culture and Society, 28*(5), 123–143.

Satya. (2006, October). The Satya Interview with Peter Singer. *Satya*. Online. July 2016.

Scarborough, P., Appleby, P. N., Mizdrak, A., Briggs, A. D. M., et al. (2014). Dietary Greenhouse Gas Emissions of Meat-Eaters, Fish-Eaters, Vegetarians and Vegans in the UK. *Climatic Change, 125*(2), 179–192.

Schatzki, T. R. (1996). *Social Practices: A Wittgensteinian Approach to Human Activity and the Social*. Cambridge, UK; New York: Cambridge University Press.

Schatzki, T. R. (1997). Practices and Actions: A Wittgensteinian Critique of Bourdieu and Giddens. *Philosophy of the Social Sciences, 27*(3), 283–308.

Schatzki, T. R. (2001). *Practice Theory*. London; New York: Routledge.

Schatzki, T. R. (2005). Peripheral Vision: The Sites of Organizations. *Organization Studies, 26*(3), 465–484.

Schatzki, T. R. (2006). On Organizations as They Happen. *Organization Studies, 27*(12), 1863–1873.

Schatzki, T. (2013). The Edge of Change: On the Emergence, Persistence, and Dissolution of Practices. In E. Shove & N. Spurling (Eds.), *Sustainable Practices: Social Theory and Climate Change* (pp. 79–110). London; New York: Routledge.

Schatzki, T. (2016a). Practice Theory as Flat Ontology. In G. Spaargaren, D. Weenink, & M. Lamers (Eds.), *Practice Theory and Research: Exploring the Dynamics of Social Life* (pp. 28–42). London: Routledge.

Schatzki, T. R. (2016b). Practices, Governance, and Sustainability. In Y. Strengers & C. Maller (Eds.), *Social Practices, Intervention and Sustainability: Beyond Behaviour Change* (pp. 15–30). London; New York: Routledge.

Schelling, A. (2015, July 1). Desperate Cow Does the Unthinkable to Escape Slaughter. *The Dodo*. Online. October 2016.

Schor, J. B., & Fitzmaurice, C. J. (2015). Collaborating and Connecting: The Emergence of the Sharing Economy. In L. A. Reisch & J. Thogersen (Eds.), *Handbook of Research on Sustainable Consumption* (pp. 410–425). Cheltenham; Northampton, MA: Edward Elgar Publishing.

Schwarzer, S., Witt, R., & Zommers, Z. (2012, October 1–10). Growing Greenhouse Gas Emissions Due to Meat Production. *UNEP Global Environmental Alert Services (GEAS)*. Online. June 2013.

Scott, J. (2017, September 5). Why I Gave Up Being Vegan. *BBC News*.

Sedgwick, E. K. (2003). *Touching Feeling: Affect, Pedagogy, Performativity.* Durham; London: Duke University Press.

Seidman, S. (1998). *Contested Knowledge: Social Theory in the Postmodern Era.* New York: John Wiley and Sons.

Seltzer, L. F. (2015, October 21). The Complex Emotion of Courage: Do You Really Understand It? *Psychology Today.* Online. October 2016.

Sergie, M. A. (2018, April 8). Trump Calls Assad 'Animal', Blames Putin after Alleged Chemical Attack. *The Age.* Online. April 2018.

Sexton, A. E. (2018). Eating for the Post-Anthropocene: Alternative Proteins and the Biopolitics of Edibility. *Transactions of the Institute of British Geographers, 43*(4), 586–600.

Shove, E. (2003). Converging Conventions of Comfort, Cleanliness and Convenience. *Journal of Consumer Policy, 26*(4), 395–418.

Shove, E. (2010). Beyond the ABC: Climate Change Policy and Theories of Social Change. *Environment and Planning A, 42*(6), 1273–1285.

Shove, E. (2012). Habits and Their Creatures. In A. Warde & D. Southerton (Eds.), *The Habits of Consumption* (Collegium - Studies across Disciplines in the Humanities and Social Sciences) (pp. 100–112). University of Helsinki.

Shove, E., & Warde, A. (2002). Inconspicuous Consumption: The Sociology of Consumption, Lifestyles, and the Environment. In R. E. Dunlap (Ed.), *Sociological Theory and the Environment: Classical Foundations, Contemporary* (pp. 230–251). Lanham, MD: Rowman and Littlefield Publishers.

Shove, E., & Spurling, N. (Eds.). (2013). *Sustainable Practice: Social Theory and Climate Change.* London; New York: Routledge.

Shove, E., Pantzar, M., & Watson, M. (2012). *The Dynamics of Social Practice: Everyday Life and How It Changes.* London; Thousand Oaks; New Delhi: SAGE Publications Ltd.

Silbergeld, E. K., Graham, J., & Price, L. B. (2008). Industrial Food Animal Production, Antimicrobial Resistance, and Human Health. *Annual Review of Public Health, 29*(1), 151–169.

Silver, J. J., & Hawkins, R. (2014). I'm Not Trying to Save Fish, I'm Trying to Save Dinner': Media, Celebrity and Sustainable Seafood as a Solution to Environmental Limits. *Geoforum, 84*(August), 218–227.

Simunaniemi, A.-M., Sandberg, H., Andersson, A., & Nydahl, M. (2013). Normative, Authentic, and Altruistic Fruit and Vegetable Consumption as Weblog Discourses. *International Journal of Consumer Studies, 37*(1), 66–72.

Singer, P. (1975). *Animal Liberation.* New York: Random House.

Singer, H. (2016). Writing the Fleischgeist. *Animal Studies Journal, 5*(2), 183–201.

Singh, J., Khanna, A., & Khanna, P. K. (2014). Rudali' as an Epitome of Caste, Class and Gender Subalternity: An Analysis of Mahasweta Devi's Rudali. *Indian Journal of Applied Research, 4*(7), 282–283.

Slim, H., Thompson, P., & Cross, N. (1993). Ways of Listening: The Art of Collecting Oral Testimony. In O. Bennett & N. Cross (Eds.), *Listening for a Change: Oral Testimony and Development*. London: Panos Publications.

Slocum, R. (2007). Whiteness, Space and Alternative Food Practice. *Geoforum, 38*(3), 520–533.

Smaill, B. (2014). New Food Documentary: Animals, Identification, and the Citizen Consumer. *Film Criticism, 39*(2), 79–102.

Smith, P., Martino, D., Cai, Z., Gwary, D., et al. (2007). Agriculture. In *Climate Change 2007: Mitigation. Contribution of Working Group III to the Fourth Assessment Report of the IPCC*. Cambridge, UK; New York: Cambridge University Press.

Smith, F. A., Smith, R. E. E., Lyons, K., & Payne, J. L. (2018). Body Size Downgrading of Mammals over the Late Quaternary. *Science, 360*, 310–313.

Sobal, J. (2005). Men, Meat, and Marriage: Models of Masculinity. *Food and Foodways, 13*(102), 135–158.

Solomon, R. C. (2004). *In Defence of Sentimentality*. New York; Oxford: Oxford University Press.

Sorenson, J. (2014). Thinking the Unthinkable. In J. Sorenson (Ed.), *Critical Animal Studies: Thinking the Unthinkable* (pp. xi–xxxiv). Toronto: Canadian Scholars' Press Inc.

Spaargaren, G., & Mol, A. (2008). Greening Global Consumption: Redefining Politics and Authority. *Global Environmental Change, 18*(3), 350–359.

Spellberg, B., Guidos, R., Gilbert, D., Bradley, J., et al. (2008). The Epidemic of Antibiotic-Resistant Infections: A Call to Action for the Medical Community from the Infectious Diseases Society of America. *Clinical Infectious Diseases, 46*(2), 155–164.

Spivak, G. C. (1988). Can the Subaltern Speak? In Marxism and the Interpretation of Culture. In C. Nelson & L. Grossberg (Eds.), *Marxism and the Interpretation of Culture* (pp. 271–313). Basingstoke: Macmillan Education.

Springgay, S. (2011). The Ethico-aesthetics of Affect and a Sensational Pedagogy. *Journal of the Canadian Association for Curriculum Studies, 9*(1), 66–82.

Springmann, M., Clark, M., Mason-D'Croz, D., Wiebe, K., et al. (2018). Options for Keeping the Food System within Environmental Limits. *Nature, 562*(7728), 519–525.

Stallen, M., & Sanfey, A. G. (2015). The Neuroscience of Social Conformity: Implications for Fundamental and Applied Research. *Frontiers in Neuroscience, 9*, 337.

Stănescu, V. (2010). 'Green' Eggs and Ham? The Myth of Sustainable Meat and the Danger of the Local. *Journal for Critical Animal Studies*. VIII, (1/2), 8–32.

Stănescu, V. (2013). Why "Loving" Animals Is Not Enough: A Response to Kathy Rudy, Locavorism, and the Marketing of "Humane" Meat. *The Journal of American Culture, 36*(2), 100–110.

Steffen, W., Crutzen, P. J., & McNeill, J. R. (2007). The Anthropocene: Are Humans Now Overwhelming the Great Forces of Nature? *Ambio, 36*(8), 614–621.

Steinbuch, Y. (2016, January 22). Cow that Ran from Slaughterhouse Gets Cushy New Life. *New York Post*. Online. October 2016.

Steinfeld, H., Gerber, P., Wassenaar, T., Castel, V., et al. (2006). *Livestock's Long Shadow: Environmental Issues and Options*. Rome: Food and Agricultural Organisation of the United Nations. 416 pp.

Stephens, N. (2010). In Vitro Meat: Zombies on the Menu? *SCRIPTed, 7*(2), 394–401.

Stephenson, A. (2016, June 1). McDonald's Unveils Results of "Sustainable Beef" Pilot Project in Canada. *Calgary Herald*. Online. July 2016.

Stibbe, A. (2001). Language, Power and the Social Construction of Animals. *Society and Animals, 9*(2), 145–161.

Stone, P. R. (2009). Making Absent Death Present: Consuming Dark Tourism in Contemporary Society. In R. Sharpley & P. R. Stone (Eds.), *The Darker Side of Travel* (pp. 23–38). Bristol; Buffalo; Toronto: Channel View Publications.

Stone, P., & Sharpley, R. (2008). Consuming Dark Tourism: A Thanatological Perspective. *Annals of Tourism Research, 35*(2), 574–595.

Strengers, Y., & Maller, C. (Eds.). (2014). *Social Practices, Intervention and Sustainability: Beyond Behaviour Change*. New York; London: Routledge.

Stull, D. D., & Broadway, M. J. (2012). *Slaughterhouse Blues: The Meat and Poultry Industry in North America*. Belmont, CA: Wadsworth/Thompson.

Sturken, M., & Cartwright, L. (2009). *Practices of Looking: An Introduction to Visual Culture*. New York; Oxford: Oxford University Press.

Sullivan, K. (2018, March 7). Riverside Meats Closes Echuca Abattoir as Probe Continues. *The Weekly Times*.

Sutton, P. W. (2004). *Nature, Environment and Society*. Palgrave Macmillan.

Sutton, D. E. (2010). Food and the Senses. *Annual Review of Anthropology, 39*, 209–223.

Suzuki, D. T. (1964). *An Introduction to Zen Buddhism*. New York: Grove Press.

Swyngedouw, E. (2013). The Non-political Politics of Climate Change. *ACME: An International E-journal for Critical Geographies, 12*(1), 1–8.

Swyngedouw, E. (2015). Depoliticized Environments and the Promises of the Anthropocene. In R. L. Bryant (Ed.), *The International Handbook of Political Ecology*. Cheltenham, UK: Edward Elgar Publishing.

Tanentzap, A. J., Lamb, A., Walker, S., & Farmer, A. (2015). Resolving Conflicts between Agriculture and the Natural Environment. *PLoS Biology, 13*(9), e1002242.

Tarlow, P. E. (2007). Dark Tourism: The Appealing 'Dark' Side of Tourism and More. In M. Novelli (Ed.), *Niche Tourism: Contemporary Issues, Trends and Cases* (pp. 47–58). London; New York: Routledge.

Taylor, C. (2010). Foucault and the Ethics of Eating. *Foucault Studies*, (9), 71–88.

Taylor, C. (2013a). Foucault and Critical Animal Studies: Genealogies of Agricultural Power. *Philosophy Compass, 8*(6), 539–551.

Taylor, N. (2013b). *Humans, Animals, and Society: An Introduction to Human-Animal Studies*. Columbia University Press.

Taylor, N., & Twine, R. (Eds.). (2014). *The Rise of Critical Animal Studies: From the Margins to the Centre*. London; Thousand Oaks; New Delhi: SAGE Publications Ltd.

Taylor, E., & Butt, A. (2017, June 9). Three Charts On: Australia's Declining Taste for Beef and Growing Appetite for Chicken. *The Conversation*. Online. April 2019.

Thomas, M. D. (2016, November 24). Secret Video Inside Victoria's Riverside Meats Abattoir Reveals Shocking Abuse. *The Age*. Online. February 2017.

Thompson, C. J. (2011). Understanding Consumption as Political and Moral Practice: Introduction to the Special Issue. *Journal of Consumer Culture, 11*(2), 139–144.

Todd, A. M. (2009). Happy Cows and Passionate Beefscapes: Nature as Landscape and Lifestyle in Food Advertisments. In J. A. Sandlin & P. McLaren (Eds.), *Critical Pedagogies of Consumption: Living and Learning in the Shadow of the 'Shopocalypse'* (pp. 169–179). New York; London: Routledge.

Tone, E. (2014, August 26). Why I Stopped Being a Vegetarian: The Harm-Free Way to Eat Meat. *Elite Daily*.

Townsend, M. (2014, August 18). Baa, the Springvale Sheep Whose Owner Fought a High Court Battle to Keep Her in His Backyard, Dies after a Lengthy Illness. *The Herald Sun*. Online. May 2016.

Tronto, J. (1993). *Moral Boundaries: A Political Argument for an Ethic of Care*. New York: Routledge.

Turner, J. H., & Stets, J. E. (2005). *The Sociology of Emotions*. Cambridge University Press.

Turner, J. H., & Stets, J. E. (2006). Sociological Theories of Human Emotions. *Annual Review of Sociology, 32*(1), 25–52.

Twine, R. (2010a). *Animals as Biotechnology: Ethics, Sustainability and Critical Animal Studies.* London; Washington, DC: Earthscan.

Twine, R. (2010b). Intersectional Disgust? Animals and (Eco)Feminism. *Feminism & Psychology, 20*(3), 397–406.

Twine, R. (2012). Revealing the 'Animal-Industrial Complex' - A Concept & Method for Critical Animal Studies. *Journal for Critical Animal Studies, 10*(1), 12–39.

Twine, R. (2013). *Exploring Veganism as a Social Innovation in Eating Practices.* 11th Nordic Environmental Social Science Conference, pp. 1–14.

Twine, R. (2014). Vegan Killjoys at the Table—Contesting Happiness and Negotiating Relationships with Food Practices. *Societies, 4*, 623–639.

UK Veterinary Medicines Directorate. (2019). *UK One Health Report – Joint Report on Antibiotic Use and Antibiotic Resistance, 2013–2017.* New Haw, Addlestone: Veterinary Medicines Directorate.

UNEP. (2012). Avoiding Future Famines: Strengthening the Ecological Foundation of Food Security through Sustainable Food Systems. *United Nations Environment Programme (UNEP).* Nairobi, Kenya: UNEP.

UNEP. (2018, November 8). What's in Your Burger? More than You Think. *UN Environment.*

Urry, J. (2005). The Place of Emotions within Place. In J. Davidson, L. Bondi, & M. Smith (Eds.), *Emotional Geographies* (pp. 77–83). Aldershot; Burlington, VT: Ashgate Publishing Ltd.

Vannini, P., Waskul, D., & Gottschalk, S. (2013). *The Senses in Self, Society, and Culture: A Sociology of the Senses.* New York; London: Routledge.

Vergnaud, A. C., Norat, T., Romaguera, D., Mouw, T., et al. (2010). Meat Consumption and Prospective Weight Change in Participants of the EPIC-PANACEA Study. *American Journal of Clinical Nutrition, 92*(2), 398–407.

Vidot, A., & Conifer, D. (2016, May 20). Investigation Launched into Claims Australian Cattle Were Slaughtered with Sledgehammer in Vietnam. *ABC News.* Online. July 2016.

Vigors, B. (2018). Reducing the Consumer Attitude–Behaviour Gap in Animal Welfare: The Potential Role of 'Nudges'. *Animal, 8*(2), 232.

Villas, J. (2011). *Pig: King of the Southern Table.* Hoboken, NJ: John Wiley & Sons Inc.

Vint, S. (2010). *Animal Alterity: Science Fiction and the Question of the Animal.* Liverpool: Liverpool University press.

Visak, T. (2013). *Killing Happy Animals: Explorations in Utilitarian Ethics.* Basingstoke; New York: Palgrave Macmillan.

Visak, T., & Garner, R. (2015). *The Ethics of Killing Animals*. New York; Oxford: Oxford University Press.

Voiceless. (2015). *The Life of the Dairy Cow: A Report on the Australian Dairy Industry*. Paddington, NSW: Voiceless. 92 pp.

Wadiwel, D. J. (2002). Cows and Sovereignty: Biopower and Animal Life. *Borderlands e-Journal, 1*(2), 1–8.

Wadiwel, D. (2008). Three Fragments from a Biopolitical History of Animals: Questions of Body, Soul, and the Body Politics in Homer, Plato, and Aristotle. *Journal for Critical Animal Studies, 6*(1), 17–31.

Wadiwel, D. J. (2015). *The War Against Animals*. Leiden; Boston: Brill Rodopi.

Wadiwel, D. (2017). *The Werewolf in the Room: Animals and Capitalism*. Conference Keynote Lecture. Seventh Australian Animal Studies Association Conference: Animal Intersection. 3–5 July, Adelaide, Australia.

Wahlquist, C. (2017, August 10). WA Seeks Powers to Prosecute Live Exporters after 3,000 Sheep Die on Ship. *The Guardian*. Online. April 2018.

Waite, R., & Vennard, D. (2018, October 17). Without Changing Diets, Agriculture Alone Could Produce Enough Emissions to Surpass 1.5°C of Global Warming. *World Resources Institute*.

Walker, G. (2013). Inequality, Sustainability and Capability: Locating Justice in Social Practice. In E. Shove & N. Spurling (Eds.), *Sustainable Practices: Social Theory and Climate Change* (pp. 181–196). London; New York: Routledge.

Walker, K. (2016, June 17). Become a Part-Time Vegan and Get Healthy. *Body and Soul*.

Walker, P., Rhubart-Berg, P., McKenzie, S., Kelling, K., et al. (2007). Public Health Implications of Meat Production and Consumption. *Public Health Nutrition, 8*(4), 348–356.

Walls, J. (2016, November 24). Echuca Abattoir Should Be 'Shut Down Immediately'. *Bendigo Advertiser*. Online. February 2017.

Walsh, B. (2011, July 7). The End of the Line. *Time Magazine - Science and Space*. Online. March 2013.

Wang, Y., & Beydoun, M. A. (2009). Meat Consumption Is Associated with Obesity and Central Obesity among US Adults. *International Journal of Obesity, 33*(6), 621–628.

Wansink, B., Cheney, M. M., & Chan, N. (2003). Exploring Comfort Food Preferences Across Age and Gender. *Physiology & Behavior, 79*(4–5), 739–747.

Warde, A. (2005). Consumption and Theories of Practice. *Journal of Consumer Culture, 5*(2), 131–153.

Warde, A., Wright, D., & Gayo-Cal, M. (2007). Understanding Cultural Omnivorousness: Or, the Myth of the Cultural Omnivore. *Cultural Sociology, 1*(2), 143–164.

Warren, K. (1997). *Ecofeminism: Women, Culture, Nature.* Bloomington: Indiana University Press.

Warren, K. (2000). *Ecofeminist Philosophy: A Western Perspective on What It Is and Why It Matters.* Lanham; Boulder; New York; Oxford: Rowman & Littlefield.

Watson, M. (2016). Placing Power in Practice Theory. In A. Hui, E. Shove, & T. Schatzki (Eds.), *The Nexus of Practices: Connections, Constellations and Practitioners* (pp. 169–182). London; New York: Routledge.

Watters, N. (2015). 16 Million People in the US Are Now Vegan or Vegetarian! *The Raw Food World.* Online. February 2017.

Weisberg, Z. (2009). The Broken Promises of Monsters: Haraway, Animals and the Humanist Legacy. *Journal for Critical Animal Studies, 7*(2), 22–62.

Weiss, B. (2012). Configuring the Authentic Value of Real Food: Farm-to-fork, Snout-to-tail, and Local Food Movements. *American Ethnologist, 39*(3), 614–626.

Wellesley, L., Happer, C., & Froggatt, A. (2015). *Changing Climate, Changing Diets.* p. 64.

Westwood, S. (2002). *Power and the Social.* London; New York: Routledge.

Wetherell, M. (2012). *Affect and Emotion: A New Social Science Understanding.* London; Thousand Oaks; New Delhi: SAGE Publications Ltd.

Whatmore, S. (2002). *Hybrid Geographies: Natures Cultures Spaces.* London; Thousand Oaks; New Delhi: SAGE Publications Ltd.

Whatmore, S. (2013). Dissecting the Autonomous Self: Hybrid Cartographies for a Relational Ethics. *Environment and Planning D: Society and Space, 15,* 37–53.

Wheal, C. (2013, May 24). Animal Farm. *Chriswheal.com.* Online. February 2017.

Wilk, R. (2001). Consuming Morality. *Journal of Consumer Culture, 1*(2), 245–260.

Wilk, R. (2002). Consumption, Human Needs, and Global Environmental Change. *Global Environmental Change, 12,* 5–13.

Wilk, R. (2009). The Edge of Agency: Routines, Habits and Volition. In E. Shove, F. Trentmann, & R. Wilk (Eds.), *Time, Consumption and Everyday Life: Practice, Materiality and Culture* (pp. 143–155). London; New York; Delhi; Sydney: Bloomsbury.

Williams, A. M., & Irurita, V. F. (2006). Emotional Comfort: The Patient's Perspective of a Therapeutic Context. *International Journal of Nursing Studies, 43*(4), 405–415.

Williams, A. M., Lester, L., Bulsara, C., Petterson, A., et al. (2017). Patient Evaluation of Emotional Comfort Experienced (PEECE): Developing and Testing a Measurement Instrument. *BMJ Open, 7*, e012999.

Wilson, M. W. (2011). 'Training the Eye': Formation of the Geocoding Subject. *Social & Cultural Geography, 12*(4), 357–376.

Wilson, B. (2019, January 4). Protein Mania: The Rich World's New Diet Obsession. *The Guardian.*

Wiper, A. P. (2014, January 6). Danish Crown Slaughterhouse, Denmark. *alastairphilipwiper.com.* Online. April 2014.

Woginrich, J. (2011, January 20). My Beef Isn't with Beef: Why I Stopped Being a Vegetarian. *The Guardian.* Online. August 2016.

Wolfe, C. (2003). *Animal Rites: American Culture, the Discourse of Species, and Posthumanist Theory.* Chicago and London: University of Chicago Press.

Wolfe, C. (2010). *What Is Posthumanism?* Minneapolis; London: University of Minnesota Press.

Wolfe, C. (2012). *Before the Law: Humans and Other Animals in a Biopolitical Frame.* Chicago; London: University of Chicago Press.

Wooffitt, R. (2005). *Conversation Analysis & Discourse Analysis.* London; Thousand Oaks; New Delhi: SAGE Publications Ltd.

Worldwatch Institute. (2013). *Grain Harvest Sets Record, But Supplies Still Tight.* Washington, DC: Worldwatch Institute. Product No. VST 101. Online. December 2014.

Worldwatch Institute. (2015). *Global Meat Production and Consumption Continue to Rise.* Washington, DC: worldwatch.org. Online. June 2015.

Wright, L. (2015). *The Vegan Studies Project: Food, Animals, and Gender in the Age of Terror.* Athens, Georgia: University of Georgia Press.

WWF. (2017). Appetite for Destruction.

Yancy, G. (2008). Colonial Gazing: The Production of the Body as 'Other'. *Western Journal of Black Studies, 32*(1), 1–15.

Yates, R. (2010). Language, Power and Speciesism. *Critical Society, 3*(Summer), 11–19.

Young, I. (1990). *Justice and the Politics of Difference.* Princeton; Oxford: Princeton University Press.

Younger, J. (2015). *Breaking Vegan.* Fair Winds Press.

Zembylas, M. (2005). *Teaching with Emotion: A Postmodern Enactment.* Connecticut: Information Age Publishing.

Zoonen, L. V. (1994). *Feminist Media Studies.* London; Thousand Oaks; New Delhi: SAGE Publications Ltd.

Index[1]

---

[1] Note: Page numbers followed by 'n' refer to notes.

© The Author(s) 2020
P. Arcari, *Making Sense of 'Food' Animals*,
https://doi.org/10.1007/978-981-13-9585-7

The manufacturer's authorised representative in the EU is Springer
Nature Customer Service Centre GmbH, Europaplatz 3, 69115 Heidelberg,
Germany. If you have any concerns regarding our products, please
contact ProductSafety@springernature.com

Printed and bound by CPI Group (UK) Ltd, Croydon, CR0 4YY
05/05/2026
02102981-0002